ela *annals*

environment &
landscape architecture

环境与景观

ela annals 1

Publisher Whee-Young Oh · Chan-Wook Park
Editor ela Korea Editorial team
Project manager Jeong-Hee Baek
Editorial Design Sun-Ah Park

Address 529-5, Munbal-ri, Gyoha-eup, Paju-si, Gyeonggi-do, South Korea(#413-756)
TEL 82-31-955-4966~8 **FAX** 82-31-955-4969
e-mail klam@chol.com http://www.lafent.com

Published by ela Korea
First Published August 1, 2009

Fixed price $70

Printed in Korea

Environment & Landscape Architecture of Korea(ELA, Korea), which is the leading magazine in Korean landscape architecture profession, have the 27th anniversary on July, 2009. With the anniversary as a momentum, we publish the first <ela annals> and it is very meaningful for us.

It is certain that Korea was in a major period of nation-wide land development and industrialization from late 1960s to early 1990s. During that time, Korean landscape architecture profession played a major role to reduce environmental degradation as an professional practice.

However, the circumstances are entering into the information-oriented society from the industrial society. Also the status and the vision of Korean landscape architecture profession is going to a new phase.

They would be our assignment that overhauling today's Korean landscape architecture field through the last 27 years of actual results and preparing tomorrow's Korean landscape architecture. By all means, we hope this <ela annals> could play a important role in offering a new vision for the future of this field, at the same time, could contribute as the motive for the internationalization of Korean landscape architecture.

Thank you very much.

2009. 8. 1.
Publisher Whee-Young Oh
Chan-Wook Park

Preface

<环境与景观>作为韩国景观界的代表杂志，已经在韩国热销了近27年，
值此创刊27年之际，特别企划了新刊<ela annals>.
在这近27年韩国社会发展进程中，韩国景观界一直在一定的社会领域中担负着很重要的社会角色.
随着信息时代的飞速发展，韩国景观界迎来了一个新的契机，期待着新的飞跃.
27年的发展使韩国景观界有着很深厚的历史积淀，目前如何开创未来是景观界面临的新的课题.
希望本期刊为从事景观的设计师与学生带来较高的参考和使用价值.
谢谢！

2009. 8. 1.
发行了 吳輝永，朴贊旭

대한민국 조경분야의 대표지 〈환경과조경〉이 2009년 7월로서 창간 27주년을 맞이하게 되며, 이를 계기로 <ela annals>의 첫 출간을 하게 되어 매우 뜻깊게 생각합니다. 한국 근대 산업화 30년 과정에서, 한국조경은 나름대로 전문분야로서의 역할과 책임을 충실하게 하였다고 자부합니다. 그러나 현대 산업화 과정을 뛰어 넘어 새로운 정보화 사회로 급속히 진입하고 있는 상황에서, 한국조경의 위상과 중장기 발전의 비전은 새로운 국면을 맞이하고 있습니다.
지난 27년을 통하여 오늘의 한국조경을 점검하고, 내일을 위하여 어떻게 준비하여야 할 것인가 하는 것은 오늘의 한국조경분야를 이끌고 있는 우리들의 과제입니다. 모쪼록 본 책자가 미래를 보는 창으로서 새로운 희망과 꿈을 제시하는 동시에 한국조경의 세계화를 지향하는 모티브로써 활용될 수 있기를 기대합니다.
감사합니다.

2009. 8. 1.
발행인 오휘영 · 박찬욱

Foreword

首先，祝贺代表韩国景观界最高水准的杂志 <环境与景观> 中文版面世.
在日新月异的知识经济社会中，传统行业间泾渭分明的界限已逐渐模糊，行业间的壁垒逐渐被打破，
同时行业之间的交叉也更加密集，各行各业通过信息交流与合作促进共同发展已是一个必然趋势.
这就需要行业之间建立起能够自由沟通和交流的氛围，以通过良性竞争达到共同进步，共同发展的目标.
因此，景观界需要齐心协力，通过积极地交流与合作，实现景观行业的多元化和国际化发展.
在这样的背景下，已有27年历史的代表韩国景观界最高水准的杂志《环境与景观》国际版面世了，
相信这将对亚洲景观界产生较为广泛的影响. 愿《环境与景观》编辑部编制的 <ela annals>.
成为景观界高收藏价值的，专业人士首选的景观刊物，愿杂志对亚洲景观发展产生积极的影响，
愿杂志在世界景观界的交流中发挥重要的作用，并带领亚洲景观走向世界.

2009. 8. 1.
《风景园林》 主编
北京林业大学园林学院教授，副院长
北京多义景观规划设计事务所主持设计师 王向荣

한국 최고의 조경매체인 〈환경과조경〉의 해외판 출간을 진심으로 축하드립니다.
급변하는 현대 지식기반사회 속에서 모든 전문분야들은 분야간 정보교류 및 협력을 통해 조화로운 발전을 이루어 나가야하는 필연적인 상황에 처해있습니다. 전통적인 업역의 경계가 점차 모호해지는 가운데 연관분야와의 중첩은 피해갈 수 없는 현실이 되었으며, 오히려 과감히 그리고 자유롭게 서로의 경계를 넘나들며 선의의 경쟁을 통해 각자의 분야를 발전시키고 있습니다. 이에 적극적인 국제적 교류와 협력 속에 원림, 조경분야 역시 다각적인 발전을 이뤄나갈 수 있도록 힘을 모아야 할 것입니다. 이러한 시기에 27년이란 긴 시간동안 한국 조경분야의 역사와 함께해 온 대표 매체 〈환경과조경〉의 해외판 <ela annals>가 출간되어 아시아 지역의 원림, 조경정보의 교류에 앞장서는 것은 무척 의미있는 일이 아닐 수 없습니다. 전문 편집진이 제작한 <ela annals>가 누구나 소장할만한 가치가 있는 원림, 조경분야의 필독 참고서로써 아시아 원림, 조경분야의 발전에 일조하며, 더 나아가서는 전세계 원림, 조경분야를 선도하는 매체로서 앞으로 더욱 활발한 국제적 정보교류의 교두보가 되기를 기대합니다.

2009. 8. 1.
풍경원림학회 부회장, 북경임업대학 부학장 王向荣

ela annals 1

발행인 오휘영 · 박찬욱
엮은이 환경과조경 편집부
편집기획 백정희
편집디자인 박선아

주소 경기도 파주시 교하읍 문발리 파주출판도시 529-5 환경과조경사옥
TEL 82-31-955-4966~8 **FAX** 82-31-955-4969
e-mail klam@chol.com **홈페이지** www.lafent.com

발행일 2009년 8월 1일
발행처 환경과조경
출력 한결그래픽스
인쇄 백산하이테크

금액 70,000원

Printed in Korea

Contents

ea
annals

WORKS OF LANDSCAPE ARCHITECTURE

PENTAGON MEMORIAL | SEATTLE OLYMPIC SCULPTURE PARK | SYDNEY AND WALDA BESTHOFF SCULPTURE GARDEN | SANTA MONICA AIRPORT PARK | JINJI LAKE PARK | BALSLEY PARK | CAPITOL PLAZA | TIDE POINT | WEST HARLEM PIERS PARK | COPENHAGEN HARBOUR BATH | MUSEO DEL ACERO HORNO^3 MONTERREY | SASAKWA HILL LODGE | LOST DOG WASH TRAILHEAD | NICOLAI SCHOOL CENTRE | GUDBRANDSJUVET VIEWING PLATFORMS & BRIDGES | ERMAN RESIDENCE GARDEN | BARODA GARDEN | WEST VILLAGE TOWNHOUSE GARDEN

Pentagon Memorial

In response to the September 11, 2001 terrorist attacks on the Pentagon, a core group of family members related to the 184 individuals whose lives were lost conceived of and embarked upon a mission to realize a memorial that would honor their loved ones. Eight months later, during the summer of 2002, an international design competition for the Pentagon Memorial was announced. Led by a team from the United States Army Corps of Engineers, the competition gathered together an esteemed group of jury members charged with judging the entries. Upon evaluating more than 1,100 entries from 50 different countries, the jury selected six finalists to advance to the final competition round.

Site Plan

Location _ Arlington, Virginia, U.S.A
Size _ approximately 2 acres
Landscape Architects / Architect of Record _ Lee + Papa and Associates
Client _ The Pentagon Memorial Fund, Inc.
Owner _ US Department of Defense / Washington Headquarters Services
Concept Design _ KBAS(Keith Kaseman & Julie Beckman)
Completion _ 2008. 9. 11
Courtesy of KBAS

On March 3, 2003, The Army Corp of Engineers, along with the United States Department of Defense, announced the unanimous selection of KBAS' design as the winning scheme.

Adjacent to the point of impact of American Airlines Flight 77, the Pentagon Memorial is a place like no other. Inviting personal interpretation on the part of the visitor, the Memorial invites thought and interpretation while not prescribing what to think or how to feel. Both individual and collective in nature, the Memorial intends to record the sheer magnitude of the tragic events of September 11, 2001 by embedding layers of specificity that begin to tell the story of those whose lives were taken.

Organized by a timeline based on the ages of these individuals, 184 Memorial Units are uniquely placed along Age Lines parallel with the trajectory of Flight 77-each marking a birth-year, ranging from 1998 to 1930. Highly articulate in its form and placement, the Memorial Unit is the heart of the project, as each Unit demarcates a special place dedicated to each individual. As such, directional orientation inherent to the cantilevered Unit specifies whether an individual was aboard Flight 77 or in the Pentagon at the time of impact. Each individual's name is engraved at the end of the cantilever, hovering above a flowing pool of water that glows with light at night.

While thousands of people have contributed to the project at large, almost two hundred have been involved specifically with the development of the Memorial Unit. Fully designed in a 3-dimensional computer modeling environment and manifested through extensive research and development, this zone of the collaborative team consists of twenty-eight companies, fabrication shops, testing/research labs, organizations and consultants – all working to refine the interwoven techniques and processes necessary to produce the eight primary components that, when assembled, comprise a Memorial Unit. Precision detailing throughout the Memorial park articulates the three primary building materials – finely finished stainless steel, granite and concrete – emphasizing the sublime design and overall simplicity.

The Memorial Units are dispersed throughout a tactile, sensory driven environment with all materials contributing to an emphasis on life. A porous stabilized gravel system not only enables visitors to hear their own footsteps and those of others, but also allows the grove of Paperbark Maple trees to thrive and grow directly through the gravel without protective grates. Consistently shading the Memorial Units, brilliant foliage will create a dynamic canopy of light and color throughout the day and seasons, while elegantly exfoliating bark will register change long into the future. Finally, the Memorial Park is surrounded by a continuous perimeter bench which is backed by a soft border of ornamental grasses - together with the Memorial Units there is over 2000 linear feet of seating surface creating a comfortable environment for several people or several thousand.

The Memorial was dedicated to the public on the seventh anniversary of that tragic day and is now open twenty-four hours a day.

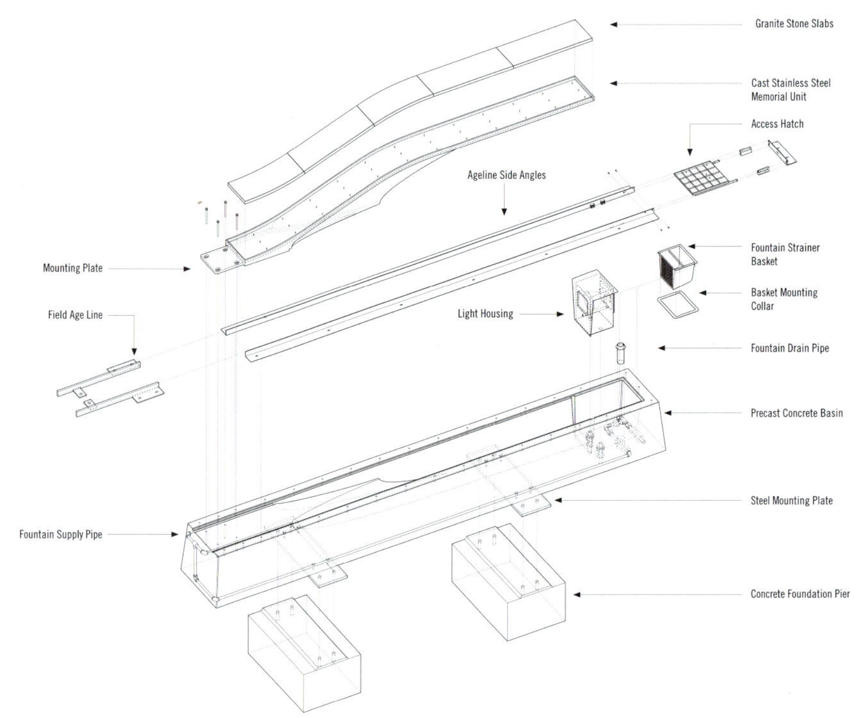

Memorial Unit Exploded Axon

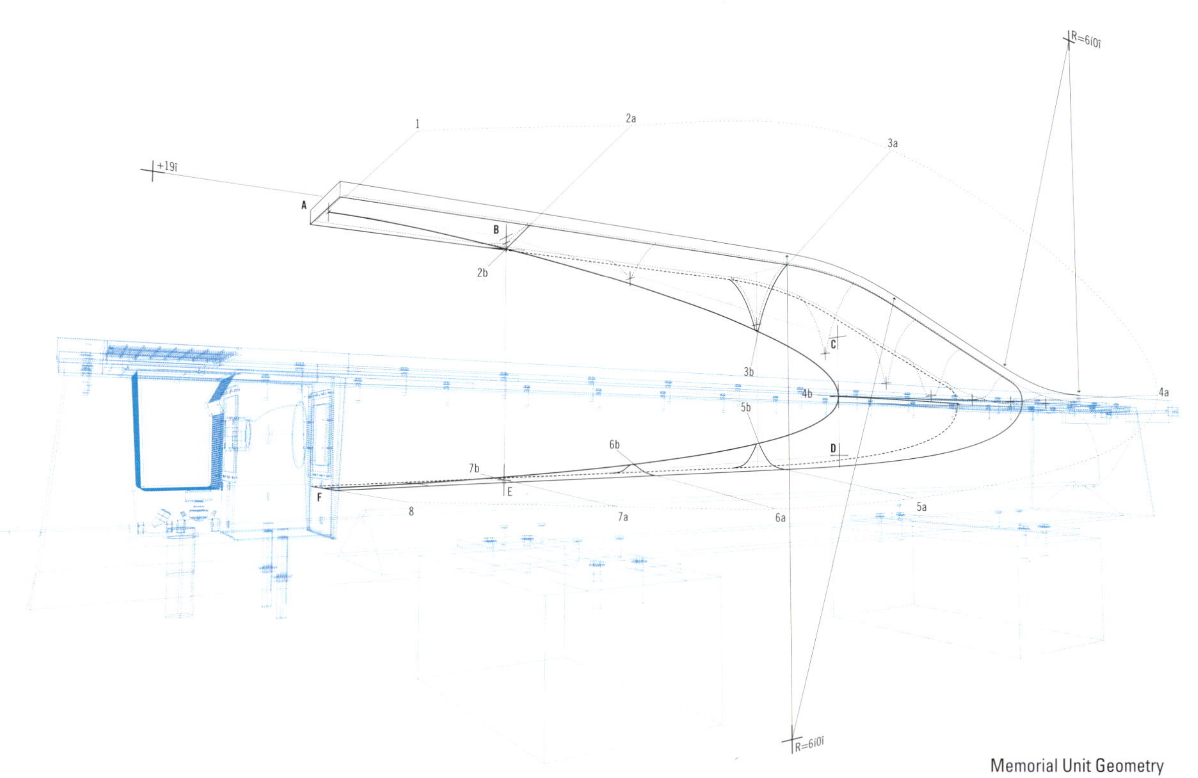

Memorial Unit Geometry

지난 2001년 9월 11일 미국 국방부를 향했던 테러리스트의 공격으로 희생당한 184명의 유족대표단은 그들이 사랑했던 이들을 명예롭게 할 메모리얼을 건립하기 위한 계획을 구상하고 착수하기에 이르렀다. 그로부터 8개월 후인 2002년 여름, 펜타곤 메모리얼 국제설계공모가 발표되었고, 미 육군공병단에 의해 공모에 참가한 작품들을 평가할 심사위원들이 구성되었다. 심사위원단은 세계 50개국으로부터 접수된 1천 1백개 이상의 참가작들을 평가하여, 최종 결선에 진출할 6개의 작품을 선별하였다. 이후 2003년 3월 3일 미국 국방부와 미 육군공병단은 만장일치로 KBAS의 계획안을 당선작으로 발표하였다.

아메리칸항공 77편의 충돌 지점에 인접한 펜타곤 메모리얼은 다른 어떤 장소와도 같을 수 없는 특별한 곳이다. 개인적인 해석은 방문객의 몫으로 남기면서, 무엇을 어떻게 느껴야 하는지 미리 정하지 않은 설계를 통해 숙고와 판단을 요구한다. 사실상 개인과 집단 모두에게, 2001년 9월 11일 일어난 비극적인 사건의 숭고한 중요성을 증언하기를 바라고 있으며, 이는 생명을 빼앗긴 이들의 이야기를 시작하는 특수한 요소들이 설치됨으로써 이루어졌다.

희생자들의 나이에 근거한 시간표가 작성되었고, 184개의 추모의자들Memorial Units이 나이의 선Age line을 따라 77편의 궤적에 평행하게 배치되었으며, 이는 1930년에서 1988년까지 분포하는 각 희생자들의 태어난 해를 표시한다. 형태와 배치가 매우 유기적이며, 각 추모의자는 각자에게 헌납된 특별한 장소로 분리됨으로써 이번 프로젝트의 가장 중요한 부분이 되고 있다. 이처럼 캔틸레버(외팔보) 형식의 추모의자의 방향은 그 사람이 충돌의 순간에 미 국방부 건물에 있었거나 77편에 탑승했었다는 것을 알려준다. 각 개인의 이름이 추모의자의 끝에 새겨졌는데, 밤에는 조명을 밝힌 풀pool로 인해 물 위에 떠있는 것처럼 보인다.

추모의자는 3D 컴퓨터 모델링 환경에서 충분히 디자인된 후 광범위한 연구 개발을 통하여 구체화되었고, 제작 공장들, 시험/실험실, 기구와 컨설턴트 등 28개의 회사가 참여했으며, 이들 모두는 추모의자 조립 및 구성시 필요한 8개의 주요한 구성요소들을 생산하는데 필요한 기술과 과정들을 정밀하게 만들기 위해 노력했다. 메모리얼 공원 전체 디테일의 명확성은 아름답게 마감된 스테인리스 스틸, 화강암, 콘크리트 등 3개의 주요 재료를 통해 뚜렷하게 표현되는데, 각 재료들은 장엄한 디자인과 전체적인 단순함을 강조하고 있다.

도처에 흩어져 있는 추모의자들은 모든 재료와 함께 촉각과 감각이 이끄는 주변환경과 더불어 삶을 강조한다. 다공성의 안정화된 자갈 시스템은 방문객들이 그들 자신과 다른 사람의 발소리를 들을 수 있도록 할 뿐만 아니라, 적피단풍Paperbark Maple 숲이 번성하고, 보호 그레이팅 없이도 자갈 사이로 나무가 성장할 수 있도록 한다. 지속적으로 추모의자에 그늘을 드리우는 잎들은 나무가 성장해가면서, 날들과 계절을 따라 다이내믹한 빛과 색깔의 캐노피를 연출할 것이다. 마지막으로, 메모리얼 공원은 잔디의 부드러운 경계에 의해 뒤로 물러나 있는, 연결된 벤치에 의해 둘러싸여 있는데, 몇몇 사람들 또는 수천 명의 사람들을 위한 편안한 환경을 만들어주는 좌석 표면을 가진 2천피트 길이가 넘는 추모의자들과 함께 한다.

이 메모리얼은 충돌 사건이 있은 지 7년째 되는 날에 공공에게 개방되었고, 24시간 늘 열려 있다.

Seattle Olympic Sculpture Park

Envisioned as a new urban model for sculpture parks, this project is located on Seattle's last undeveloped waterfront property—an industrial brownfield site sliced by train tracks and an arterial road. The design connects three separate sites with an uninterrupted Z-shaped "green" platform, descending forty feet from the city to the water, capitalizing on views of the skyline and Elliott Bay, and rising over existing infrastructure to reconnect the urban core to the revitalized waterfront.

Location _ Seattle, U.S.A
Size _ 9 acres
Lead Designer _ Weiss/Manfredi Architecture/Landscape/Urbanism
Client _ Seattle Art Museum
Structural and Civil Engineering _ Magnusson Klemencic Associates
Lighting Design Consultant _ Brandston Partnership Inc.
Geotechnical Engineering _ Hart Crowser
Environmental _ Aspect Consulting
Aquatic Engineering _ Anchor Environmental
Project Management _ Barrientos LLC
Architectural Site Representation _ Owens Richards Architects
Courtesy of Weiss/Manfredi

©Benjamin Benscheider

© Benjamin Benscheider

© Benjamin Benscheider

Formerly owned by Union Oil of California(Unocal), the area was used as an oil transfer facility. Before construction of the park, over 120,000 tons of contaminated soil were removed. The remaining petroleum contaminated soil is capped by a new landform with over 200,000 cubic yards of clean fill, much of it excavated from the Seattle Art Museum's downtown expansion project.

Winner of an international design competition, the design for the Olympic Sculpture Park capitalizes on the forty-foot grade change from the top of the site to the water's edge. Planned as a continuous landscape that wanders from the city to the shoreline, this Z-shaped hybrid landform provides a new pedestrian infrastructure. Built with a system of mechanically stabilized earth, the enhanced landform re-establishes the original topography of the site, as it crosses the highway and train tracks and descends to meet the city. Layered over the existing site and infrastructure, the scheme creates a dynamic link that makes the waterfront accessible. The main pedestrian route is initiated at an 18,000-square-foot exhibition pavilion and descends as each leg of the path opens to radically different views. The first stretch crosses a highway, offering views of the Olympic Mountains; the second crosses the train tracks, offering views of the city and port; and the last descends to the water, opening views of the newly created beach. This pedestrian landform now allows free movement between the city's urban center and the restored beaches at the waterfront.

As the route descends from the pavilion to the water, it links three distinct settings: a dense and temperate evergreen forest; a deciduous forest of seasonally changing characteristics; and a shoreline garden including a series of new tidal terraces for salmon habitat and saltwater vegetation. Throughout the park, landforms and plantings collaborate to direct, collect, and cleanse storm water as it moves through the site before being discharged into Elliott Bay.

©Benjamin Benscheider

©Paul Warchol

©Benjamin Benscheider

As a "landscape for art", the Olympic Sculpture Park defines a new experience for modern and contemporary art outside the museum walls. The topographically varied park provides diverse settings for sculpture of multiple scales. Deliberately open-ended, the design invites new interpretations of art and environmental engagement, reconnecting the fractured relationships of art, landscape, and urban life.

이 지역은 Unocal(Union Oil of California)의 소유지로 원래는 원유를 이송하는 업체가 사용하던 공간이었다. 따라서 프로젝트를 실행하기 전 12만톤에 이르는 오염된 지역을 제거하고 작업을 본격적으로 시작하였다고 한다. 약 20만 평방야드의 제거된 오염지역에 새로운 흙이 포설되었는데 이렇게 가져온 흙의 대부분은 시애틀 예술박물관을 확장하는 과정에서 얻은 것이라고 한다.

Olympic Sculpture Park 부지는 국제 디자인 공모전을 통해 디자인 컨셉을 잡아 변화를 꾀하였으며, 도시에서 해안까지 전망을 볼 수 있는 구조와 Z형상의 지형 등 보행자를 위한 기초적인 시설을 갖추게 되었다. Olympic Sculpture Park는 기계적으로 안정화된 지반에 형성된 구조의 증진된 지질학적인 특성과 함께 고속도로와 철로가 도심에서 만날 수 있는 역동적인 시스템이 특징이다.

보행자 전용도로는 전시용 퍼골라에서 시작되어 주변의 다양한 볼거리를 감상하면서 이동할 수 있다.

첫번째 경로는 고속도로를 지나Olympic Mountain을 감상할 수 있고, 두번째 경로는 철로를 지나며 도시와 항만을 구경할 수 있도록 하였으며, 마지막 경로는 최근에 조성된 해변가를 느낄 수 있다. 즉 보행자도로는 도심지와 새롭게 복원된 해변가를 자유롭게 이동할 수 있는 길인 셈이다. 전시용 퍼골라에서부터 해안가로 이르는 경로는 계절에 의해 특색을 보여주는 상록수림, 해안선을 따라 조성된 연어 서식지와 바닷속 생물을 위한 정원, 엘리엇만까지 지형을 활용해 구조적으로 잘 연계된 공원을 느낄 수 있어 크게 각광받고 있다. 예술성을 두루 갖춘 조경이 만들어낸 작품, Olympic Sculpture Park는 현대예술에 있어서 새로운 경험을 줄 수 있는 환경을 제시한다. 다양한 공원들이 제공하는 다채로운 환경, 디자인과 문화적 부분이 만나는 연관성 등 이 프로젝트의 특징들은 예술, 조경, 그리고 도시적 라이프에 있어서 큰 토대를 이어줄 수 있는 관계를 성립해 준 좋은 사례로 꼽히고 있다.

Sydney and Walda Besthoff Sculpture Garden

The Sydney and Walda Besthoff Sculpture Garden at the New Orleans Museum of Art occupies approximately five acres in City Park adjacent to the Museum. Atypical of most sculpture gardens, this Garden is located within a mature existing landscape of Pines, Magnolias and Live Oaks. Designed in association with Ledbetter Fullerton Architects of New Orleans, the garden design creates outdoor viewing spaces within this picturesque landscape. A re-configured lagoon bisects the site and creates two distinct halves: a mature Pine and Magnolia grove adjacent to the Museum, and a more open area of two-hundred year old Live Oaks across the lagoon near the New Orleans Botanical Garden. The Sculpture Garden is designed for approximately sixty sculptures, most of them donated to the New Orleans Museum of Art by Sydney and Walda Besthoff.

LOVE

A pair of cast stone and bronze pavilions with clerestory windows marks the entry to the Sydney and Walda Besthoff Sculpture Garden at the New Orleans Museum of Art. The pavilions, which face a large meadow adjacent to the Museum's entry drive, flank a large plaza containing Henry Moore's 'Reclining Mother and Child'. The pavilions and entry plaza also establish a centerline through the depth of the garden that becomes an organizing element of the design.

The garden is bisected by a lagoon which has been re-shaped into two basins on either side of the central pedestrian bridge. These lagoon basins each contain a sculpture, Kenneth Snelson's 'Virlane Tower' and Arman's 'Pablo Casal's Obelisk'. By locating these vertical sculptures in the water, their reflections become part of the compositions. A total of three pedestrian bridges cross the lagoon and connect the two sides of the garden. The entries to the bridges are marked by simple, cast stone pylons that support bronze lanterns. Side rails constructed of bronze and stainless steel mesh reflect the light and allow views

through the bridges. This combination of cast stone lantern pylons and bronze with stainless mesh is repeated in the form of monumental gates at the main entry and two secondary entries.

A picturesque composition of curvilinear paths compliments the mature trees while creating smaller viewing areas, or 'galleries', within the garden. These paths also connect a series of garden features including the Overlook Terrace, the Exedra at the end of one of the pedestrian bridges, the monumental Water Steps at one of the secondary entrances, the elliptical Sculpture Theater containing smaller pieces of sculpture and the Cascade Garden Pool containing Robert Graham's 'Source Figure'. The pathways also frame the elliptical Pine Grove immediately off of the main entry plaza and the large Oak Lawn at the opposite end of the garden. With the exception of the lagoon, the Oak Lawn is the most open space in the garden and contains some of the larger sculptures including works by Louise Bourgeois, Deborah Butterfield, Elizabeth Frink, Joel Shapiro and Claes Oldenburg.

Location _ 1 Collins Diboll Circle, City Park, New Orleans, Louisiana 70179

Size _ 4.76 acres, including 127,000 square feet of planted area, 50,000 square feet of sod area, and 23,230 square feet of paved areas

Landscape Architect _ Sawyer/Berson Architecture & Landscape Architecture, LLP

Client _ New Orleans Museum of Art

Primary Donor _ The Sydney and Walda Besthoff Foundation

Architecture _ LF Architects, New Orleans, Louisiana

Lighting Design_ Cline Bettridge Bernstein Lighting Design, Inc., New York, New York

Structural & Civil Engineering _ Burk-Kleinpeter, Inc., New Orleans, Louisiana

Electrical Engineer _ Huseman Wang Consulting Engineers, New Orleans, Louisiana

GPS CAD Consultant _ Vollmer Associates LLP, New York, New York

Consulting Arborist _ Bayou Tree Service, New Orleans, Louisiana

Photographer _ Richard Sexton

Courtesy of Sawyer/Berson

뉴올리언즈의 건축가 Ledbetter Fullerton과 Sawyer/Berson사가 공동으로 디자인하여 그림 같은 풍경 속에서 새로운 공간을 보여주는 정원 디자인을 창조했다. 기존의 연못형태를 변화시켜 부지전체를 크게 양분하여 미술관 인접지역에는 소나무와 태산목을 심었고, 뉴올리언즈 식물원에 인접한 연못을 가로지는 지역에는 200년 이상된 떡갈나무를 심어 초점이 되도록 했다. 조각공원에는 대략 60개정도의 조각상을 두었는데, 대부분 Sydney and Walda Besthoff가 뉴올리언즈 미술관에 기부한 것들이라고 한다.

한 쌍의 캐스트 스톤cast stone과 고딕형식의 채광창으로 밖을 볼 수 있는 청동색의 전시관은 뉴올리언즈 미술관에서 시드니 & 월다 베스트호프 조각공원으로 가는 입구를 나타낸다. 전시관은 미술관의 진입로에 근접한 풀밭을 마주보고 있으며, 옆면에는 서양미술가, Henry Moore의 작품인 'Reclining Mother and Chind' 가 있는 대형 플라자가 자리잡고 있다. 전시관과 플라자는 또한 디자인에서의 한 요소로 구성되어 정원 전체의 깊이를 보여주는 중심축이 되고 있다.

정원은 연못을 중심으로 양분되었다고 하였는데 그 중심선에는 보행자용 다리를 두어 양옆으로 두 개의 분지으로 재구성하였다. 이 지역에는 조각상 같은 것을 둘 수 있도록 디자인하였으며, 조각가인 Kenneth Snelson의 'Virlane Tower' 와 Arman의 'Pablo Casal' s Obelisk' 등과 같은 유명한 조각상을 두어 방문객들의 시선을 끈다. 연못주변 조각상들의 위치 역시, 부지의 전체적인 균형에 있어서 중요한 요소가 되고 있다.

연못을 가로지르는 보행자용 다리는 모두 세 개가 있는데, 그 중에서 두 개의 다리가 정원과 연결된다. 다리의 진입부는 심플한 캐스트 스톤으로 제작되어 있으며 청동으로 마무리된 정탑들로 구성되어 있다. 다리 난간은 청동과 그물형태의 스테인리스 철강으로 제작되어 빛을 반사할 수 있도록 하였다. 캐스트 스톤으로 제작된 정탑과 이 난간과의 조합은 첫번째의 입구와 두번째 입구에서 기념비 형태로 반복되도록 조성하였다.

곡선을 이루는 오솔길의 아름다운 경관은 정원 내에서 다양한 나무와 함께 즐길 수 있는 새로운 소규모의 갤러리로서, 보는 이로 하여금 감탄사가 나오도록 디자인되었다. 모든 오솔길은 전망이 좋은 테라스와 다리 끝에 자리잡은 반원형태의 실외벤치, 물이 흐르는 계단과 미술가 Robert Graham의 작품인 'Source Figure' 를 볼 수 있는 폭포가 있는 연못, 자그마한 조각상들이 있는 조각극장 등을 포함해 특징적인 공간에 연속적인 경관의 매개체로서 중요한 역할을 하고 있다. 또한, 정원 끝 맞은편에 위치한 떡갈나무 잔디밭과 플라자의 주진입부 오솔길은 하나의 틀을 만들어주는 역할도 하고 있다. 연못을 제외한, 떡갈나무 잔디밭은 공원에서 가장 개방적인 공간으로 Louise Bourgeois, Deborah Butterfield, Elizabeth Frink, Joel Shapiro, Claes Oldenburg 등과 같은 유명한 조각가들의 작품들이 전시되어 있다.

SANTA MONICA AIRPORT PARK

Bordered on one side by world-famous beaches, the City of Santa Monica, California seems an unlikely candidate for the designation "park-poor."

The beachside city, however, has historically experienced a shortage of open and green spaces. In the past 10 years since adoption of the City's first parks and recreation master plan, Santa Monica has been on a drive to spruce up its existing parks, having completed major improvements to Virginia Avenue Park, Palisades Park and Douglas Park, as well as several new playgrounds and a new skate park.

General
Facility

On April 29, 2007, the City dramatically increased its park area with the opening of Airport Park, the first city-built "ground-up" park to open in Santa Monica in 24 years.

Designed by Ah′be landscape architects of Culver city, California, the park occupies one of the last remaining large open spaces for park development in Santa Monica. Sited on the south side of the Santa Monica Airport, the 8.3-acre site on Airport Avenue and Centinela/Bundy was most recently used as a shuttle parking lot for Santa Monica Community College and as a storage area for a local car dealership.

The park project is a large-scale conversion of an area that has great historical significance in the aviation industry. Santa Monica Airport was home to many landmark events, including the first circumnavigations of the world by airplane in 1924, as well as the founding of Douglas Aircraft Corporation, which manufactured the DC-3. "The site's history as well as the surrounding industrial architecture informed and shaped the design of the park as well as our selection of site furniture and fixtures," says Calvin Abe, FASLA, principal, Ah′be landscape architects.

"For example, in choosing benches and tables for the picnic areas and lighting for the site and sports fields, we selected galvanized steel for its industrial look instead of the traditional colorful, rustic-looking park furniture and finishes."

The site's location also impacted the park's design at the conceptual level. "An airport connects the city to other parts of the world", says Ah′be, whose firm was involved with all phases of the project, from planning through construction. "In designing the park, we expressed this notion through a series of pathways that abstractly represent people making connections. These pathways also suggest taxiways and runways."

In addition to providing a new green space for the residents of Santa Monica, the park functions as a visual buffer between the airport and the adjacent commercial and residential neighborhoods. The new $7-million Airport Park includes a state-of-the-art soccer field with 80′-tall lighting standards with shielded lamps, an off-leash dog park planted with fragrant ground cover and shrubbery that features separate areas for large and small dogs, a children's playground, a picnic area with six barbecues as well as informal open space and restroom and storage facilities.

In keeping with the City's commitment to sustainable development and quality design, ah'be landscape architects incorporated sustainable techniques and features that address storm water management, landscape planting and irrigation. Synthetic turf was installed on the sports field using a $1.5-million state grant obtained by the City. Infiltration beds under the sports field will detain storm water from portions of the airport and the park, improving the City's storm water management. The parking lots were paved in permeable asphalt pavement for storm-water retention.

The dog park was designed as a series of three terraces, each of which drains back into the slope, thereby capturing 100 percent of the water. A French drain system includes a network of filters that prevent any overflow or runoff from entering the city storm drain.

While there are existing dog parks in Santa Monica and the surrounding communities such as Venice, the park at Airport Park is a far cry from the vacant parcels of land covered with gravel or mulch and surrounded by a chain-link fence, where the only "amenity" might be a dog's drinking fountain.

"It makes sense that people are clamoring for a dog park like the one at Airport Park.," says Abe, who toured new dog parks with members of the design team in order to come up with a plan that was cutting-edge and environmentally sound. "At Airport Park, we designed the dog park for people as well as their dogs. There are areas for both large and small dogs - similar to playground 'tot lots' - as well as seating, shade trees and shade structures. Instead of being an afterthought, the dog park was just as important as the soccer field. In fact, because of the added issue of dog waste disposal, it had to be even more sensitively planned."

지역적인 면에서 바닷가 근처는 잔디를 심을 수 있는 지역이 부족하다던가 드넓은 공간이 부족한 것은 사실이나 불명예스럽게도 세계에서 유명한 해변가 중 하나인 캘리포니아 산타모니카Santa Monica는 '공원이 드문 해변가'로 불리울 정도로 공원이 흔하지 않다. 그러나 최근 버지니아 에비뉴 공원Virginia Avenue Park, 팔리세이드 공원Palisades Park 그리고 더글래스 공원Douglas Park 등이 새로운 형태의 놀이시설과 스케이트 공원 등 복합적으로 조성되며, 이들 공원과 함께 산타모니카 에어포트 파크가 말끔한 모습으로 새단장을 하였다.
이 공원은 캘리포니아주 컬버시Culver city, California에 위치한 Ah'be landscape architects에서 계획하였으며, 공원이 자리잡은 곳은 산타모니카 공항 남쪽에 위치한 8.3에이커의 면적으로 산타모니카 대학을 왕복하기 위한 셔틀버스 정류장과 자동차 딜러샵들의 주차공간이 밀집되어 있었다.

공원계획은 항공산업의 주요성을 지닌 지역을 대규모 프로젝트로 전환하는데 있었다. 산타모니카 공항은 역사적인 사건을 지닌 명소로서 그 중에는 1924년 최초로 더글라스 항공사에서 제작한 비행기를 타고 세계일주순회를 하는 것을 포함하고 있다.
따라서 이러한 역사적인 면모를 유지할 수 있도록 조형물과 시설물들을 공원디자인에 적용할 수 있도록 하였다. 예를 들면, 피크닉을 즐기기 위한 장소에는 벤치와 테이블을 선택하는 것과 스포츠를 즐길 수 있는 코트에 들어가는 외부 조명을 설치하는 것을 고려하여, 설계자는 전통적인 스타일, 즉 시골풍의 외형스타일과 색상을 벗어나 산업단지에 적합하면서도 알맞은 아연철판을 이용한 디자인과 제품을 고려했다고 한다.
본 프로젝트에 있어서 중점을 둔 사항은 필요하면서도 획기적인 디자인과 공항에서 다른 지역으로의 이동이 용이하도록 하는데 목적을 두었다.

공원을 디자인하는데 있어서 다양한 스타일의 오솔길을 두어 사람이 다니기 쉽게 하였으며, 오솔길 뿐만 아니라 유도로誘導路와 주로走路도 많은 고려를 하였다. 산타모니카 인근에 거주하는 사람들에게 새로운 잔디공간을 제공함으로써 이웃주민들이 모여 있는 주택지역과 상업단지, 그리고 공항간의 상관관계에 있어서도 이 공원은 일종의 완충장치로 작용하고 있다.

7백만불의 추가예산으로는 축구장과 80인치 높이의 외부 가로등을 설치하였고, 애완견들을 위한 지역과 어린이 놀이터, 바비큐 파티 및 피크닉을 즐길 수 있는 공간을 따로 분리하여 잔디의 보호와 청결함을 유지할 수 있도록 배려했다.

아베에서 근무하는 디자이너들과 설계자들은 축적된 기술력과 지형을 고려한 높은 퀄리티의 디자인으로 특히 도시내 수처리 관리, 공원내 식재 그리고 수로시설에 많은 신경을 썼다고 한다.

시에서 양도받은 150만 달러의 기금을 바탕으로 야외운동경기장에는 인조잔디를 설치하였는데, 경기장 아래 설치된 빗물흡수 패드는 우천시에 물을 저장해두는 역할을 하였다가 후에 공원내 수처리에 긴밀히 사용이 되기도 한다. 주차장 또한 수분흡수가 가능한 투수콘 포장을 함으로써 이 또한 수처리 관리에 도움이 많이 될 수 있다고 전한다.

애완견 공원은 총 3개의 테라스로 설계되었는데, 경사면의 테라스 아래 부분에는 물탱크가 있어 100% 저장이 완벽하다. 이러한 수처리 관련시스템은 프랑스의 수도관리시설에서 많은 도움을 받았다고 하며, 공원내 물공급 부족시기가 다가오면 이렇게 저장해둔 물을 유용하게 사용할 수 있도록 했다.

아베사의 대표는 "이 공원을 계획함에 있어서 우리가 중시했던 부분이 바로 애완견을 위한 공원을 만드는 것" 이라고 말하면서, 환경적인 면에서의 안정성과 최신 스타일의 기술을 접목시키기 위해 디자인 팀원들에게 직접 개를 데리고 산책을 할 것을 권장하였다고 한다. 특이한 점은 애완견들이 크기에 따라 즐길 수 있는 놀이터 개념의 장소를 조성하여 개를 사랑하는 사람들의 마음을 공원디자인에 적용하였다는 것과 단지 배설물을 처리할 수 있는 장소 이상의 의미를 두어, 축구장만큼 중요한 기능을 담당하도록 한 것이다.

Location _ Culver city, California, U.S.A
Size _ 8.3 acres
Landscape Design _ Ah′ be Landscape Architects
Completion _ April 29, 2007
Courtesy of Ah′ be Landscape Architects

STOP
STOP

Jinji Lake Park

Location _ Suzhou, China
Size _ 513 ha
Landscape Design _ EDAW
Courtesy of EDAW

Jinji Lake is located in the heart of a new Suzhou urban development. It is about 4km outside of the famous historic center of Suzhou. The master plan covers 513 ha and can easily be one of the largest waterfront development project. The Jinji lakefront district is one of the five special economic development zones that were designated by Chinese government to attract U.S. and Europe′ s Fortune 500 companies. One of the primary purposes of Jinji Lake master plan was to act as a catalyst for the forthcoming economic developments. To this end, EDAW came up with a master plan with 8 distinctively unique waterfront neighborhoods that are continuously connected through lakefront open spaces.

Suzhou is an ancient historic city best known for its traditional gardens, but at the same time it is one of the fastest growing industrial cities in China. The master plan of Jinji Lake tries to speak to both sides of the city : its continuity of historic culture as an ancient city and its contemporary aesthetics as an up and coming new city. To this end, the lake area was divided into two different areas. The northwest side of the lake was designed as more urban, young and energetic space with active programs such as urban plaza, city harbor, promenade and marina.

The southeast side of the lake was designed as more natural, quiet and peaceful space with more passive green space. Similar strategy was applied to the urban design program for the surrounding development. The northwest side of the lake was developed as high-rise residentialwith cultural, commercial and entertainment programs, and the southeast of the lake was developed as low-rise residential with passive natural programs such as wetlands. Water's edge was manipulated to accommodate programs such as lawn, marina and islands. However, all of these interventions to the nature were conducted with a common goal of representing Suzhou's contemporary and historic characteristics as well as its urban and natural characteristics.

진지호수가 위치한 수변개발 지구는 중국 정부가 미주와 유럽의 포춘지 선정 500대 기업을 유치하기 위해 지정한 5개 경제특구 중의 하나로서, 호수공원 개발의 주요한 목적은 곧이어 시작될 주변지역 경제개발의 촉진제로서의 선도적 역할을 하는 것이었다. 이에 대한 해답으로 이도EDAW는 오픈스페이스로 연결된 8개의 각기 다른 특색을 가진 수변지구의 마스터플랜을 마련하였으며, 이 계획안은 향후 신시가지 전체 도시설계의 밑바탕이 되었다.
소주에서의 이도의 첫 인연은 진지호수공원 설계에 개입되기 전인 1996년 이도가 독자적으로 소주의 환경문제, 특히 운하의 수질개선에 관심을 갖고 중국의 건축전공 학생들과 도시계획가들 그리고 이도 내의 환경, 도시 및 조경 전문가들을 초청해 현지 워크샵을 개최하면서 시작되었다. 워크샵이 끝난 후 소주시 관계자들은 이도에 당시의 진지호수 개발계획을 검토해 줄 것을 요청하였고 궁극적으로 이도에게 프로젝트를 맡기게 되었다. 그러나 이도가 발주처인 소주산업단지 개발위원회Suzhou Industrial Park Administration Committee로부터 1997년 처음 설계의뢰를 받았을 때 과업의 규모는 단지 30㎢에 해당하는 호수주변 100여 미터의 수변공간을 설계하는 3개월짜리 용역이었다.

이도의 설계팀은 대상지의 환경적 · 도시적 중요성을 역설하여 토지이용, 개발밀도, 교통 및 전체 8개 지구의 녹지체계와 건축지침의 재평가를 과업에 포함하도록 소주시를 설득하였고 이러한 노력은 결과적으로 전혀 새로운 복합적인 시각의 마스터플랜을 가능하게 하였다.

소주는 유서깊은 정원도시로서의 역사성과 중국에서 가장 신속히 성장하는 경제도시로서의 현대성을 동시에 지니고 있다. 진지호수공원의 마스터플랜은 이와 같은 소주의 문화 · 역사적 연속성을 지키는 동시에 현대적 설계미학을 반영하는 것을 목적으로 하였다. 이를 위해 수변지구는 두 개의 각기 다른 성격을 지닌 공간으로 구분되었다. 호수의 북서측은 도회적이고 젊음과 활기가 넘치는 공간으로 도심광장, 항구, 프롬나드, 마리나 등과 같은 적극적인 프로그램이 제안되었고, 호수의 남동측은 이에 반해 보다 자연적이고 조용함과 휴식을 즐길 수 있는 녹지공간으로 제안되었다. 주변 도시개발 프로그램 또한 동일한 원칙이 적용되어 호수의 북서측 지구에는 고밀 주거단지를 바탕으로 문화, 상업, 위락시설 등이, 남동측으로는 저밀 주거지와 함께 습지정원 등과 같은 자연적인 공원 프로그램이 제안되었다. 호수의 수변경계부는 잔디광장, 마리나, 인공섬 등과 같은 다양한 프로그램을 수용하기 위해 절토를 통해 깎여지기도 하고 성토를 통해 새로 메워지기도 하였으나 경관에서의 이러한 모든 인위적인 개입은 궁극적으로는 소주의 현대성과 역사성 또는 도시성과 자연성을 표현하는 것을 전제로 하였다.

Balsley Park

Location _ New York, NY, U.S.A
Landscape Design _ Thomas Balsley Associates
Client _ Rose Associates
Courtesy of Thomas Balsley Associates

The plaza had become a receptor of social ills, partially because the original design never recognized the dynamics of this corner and failed to reach out to a broad constituency. Given the plaza's unpleasant history, Thomas Balsley Associates had to build a level of trust and responsiveness with the neighborhood's design task force. Careful listening during a series of discussions quickly led to consensus on the new program, the central theme of which became "a green yet accessible park serving a broad constituency". Over time, the designer was rewarded with the freedom and encoura-gement to translate the programmatic elements into artistic gestures of form and color that would strike a distinctive pose for this prominent corner.

The design itself is infused with elements that suggest an escape from the order of the city, from the gird, and from all of the confinements that the urban order foists upon the human imagination. The path ignores the grid, the slope is park-like, and attention is diverted from the walls of the buildings that frame the edge by the creation of a living backdrop of trees. This dense evergreen grove runs the entire length of the park and is fronted by a colorful ribbon-like curving wall that breaks down into a sculptural series of pipes with spaces of five inches or so in between. A counterpoint to the edge of evergreens, this permeable wall clearly shows park users that no dangerous activity is going on behind it. Seatwalls and walls that edge the terraces are curvilinear and colorful, the curve again destroying the sense of the grid. The kiosk itself is cylindrical in shape and its associated tables and chairs add another level of comfort while encouraging another user group. The cafe kiosk and overlook terrace remains open all year and generates a steady flow of foot traffic.

This park stands as testimony to the dramatic transformations that design creativity can bring to our urban lives when it is guided by sound public open space design principles and a meaningful public review process. The park serves as a prototype for the transformation of barren useless plazas, giving hope to the hundreds of spaces in Manhattan and elsewhere that, if given a second chance, could bring civic life and pride to the neighborhoods they have been disappointed and scarred by failed spaces. In an unprecedented acknowledgement of his contribution to public space in the City of New York, the client renamed the park in Mr. Balsley's honor.

Balsa Scale Model

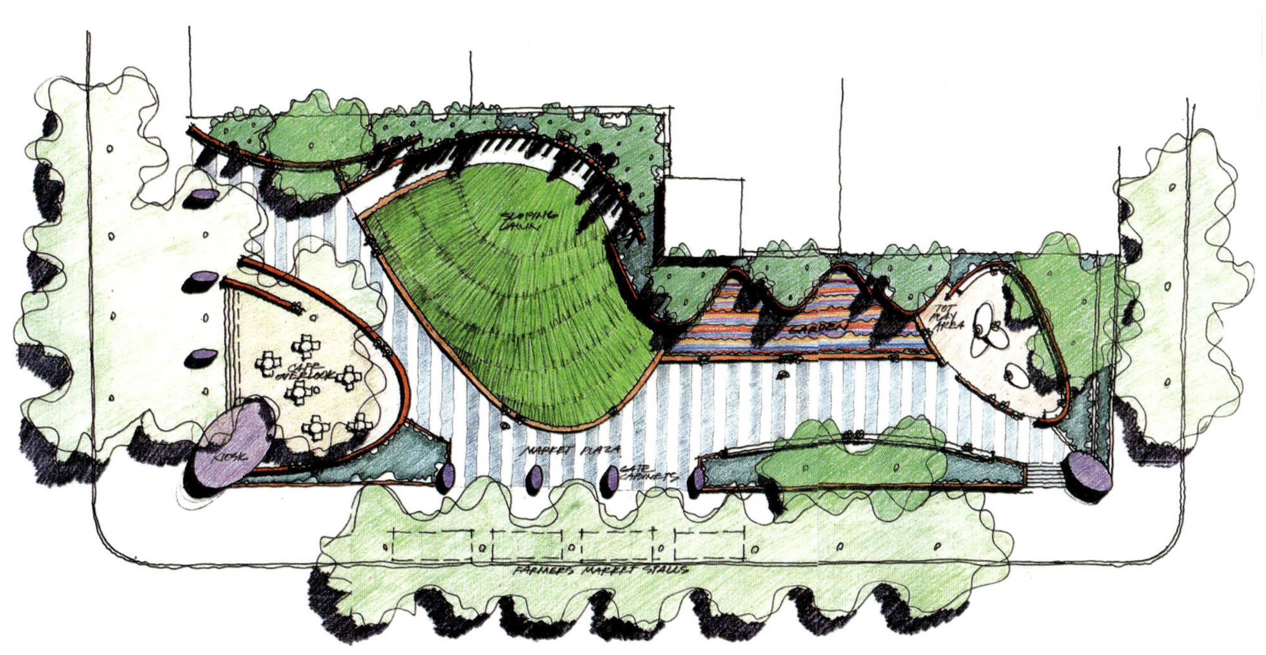

Schematic Design Plan

그동안 이 공간은 사회악의 수용처가 되어 왔다. 그 이유 중의 하나는 기존 디자인이 이 모퉁이의 역동성을 제대로 인지하지 못했고, 지역 주민들과의 합의에 도달하지 못했기 때문이다. 이에 토마스 바슬리 어소시에이츠Thomas Balsley Associates는 이 공간의 썩 유쾌하지 못했던 과거를 감안하여, 주민들이 원하는 디자인을 반영하기 위해 프로젝트팀과 함께 주민들의 신뢰를 얻기 위한 과정을 거쳐야 했다. 여러 번에 걸친 토의를 바탕으로 진행된 신중한 의견수렴 과정을 통해 새로운 프로그램에 대한 동의를 이끌어 냈고, 큰 주제를 "앞으로 지역주민에 봉사하며, 접근이 쉬운 녹지 공원"으로 정했다. 이러한 일련의 과정이 진행되는 동안, 설계자는 이 독특한 모퉁이 공간이 자신만의 두드러진 색깔을 가질 수 있도록, 프로그램에 따른 요소들을 예술적인 형태와 색깔로 변화시킬 수 있는 자유와 격려를 보상으로 받을 수 있었다.

디자인의 큰 방향은 이러한 요소들을 통해, 도시의 체계와 격자, 그리고 사람의 상상력을 기만하는 도시의 질서들로 이루어진 모든 한계들로부터 벗어나도록 하는데 맞추어졌다. 소로들은 격자를 무시하며, 경사는 마치 공원과 같고, 나무들로 이루어진 살아있는 배경을 조성함으로써 공간에 대한 관심을 건물의 벽이 아니라 새롭게 창조된 나무들에 의한 경계로 전환시켰다. 이 밀집한 상록수 숲은 공원의 종방향으로 끝 부분까지 다다르고, 컬러풀한 리본모양의 곡선벽에 면하게 되는데, 이 벽은 5인치 간격의 조형물과 같은 파이프의 연속으로 이루어진 공간으로 나아간다. 상록수 경계와 대조되는 이 투과성의 벽은 공원 이용자에게 그 뒤에서 어떤 위험한 행동도 할 수 없다는 것을 명백하게 보여준다. 테라스를 둘러싸는 앉음벽과 벽들은 곡선을 이루고 컬러풀하며, 그 곡선은 다시금 격자의 느낌을 파괴한다. 키오스크는 원통형이고, 키오스크에 딸린 테이블과 의자들은 이용자들이 이용하는 동안 다른 차원의 편안함을 제공한다. 카페 키오스크와 전망테라스는 연중 내내 오픈되어 있고, 보행자의 느린 흐름을 유도한다.

이 공원은 조성 과정에서, 건전한 공공 오픈 스페이스 디자인 원칙과 의미있는 시민 리뷰를 거침으로써, 우리 도시 삶에 디자인의 창의성이 가져올 수 있는 드라마틱한 변형의 증거로서 자리하고 있다. 황폐하고 쓸모없던 공원이 탈바꿈된 전형으로 인정받고 있으며, 맨해튼과 다른 지역들의 수백의 공간들에 희망을 주고 있는데, 두 번째 기회를 통해 실패한 공간에 대해 실망하고 상처받았던 이웃들에게 충분히 도시적 삶과 자신감을 가져다 주었기 때문이다. 클라이언트가 토마스 바슬리의 업적을 기려 공원의 이름을 '바슬리 파크 Balsley Park' 로 개명한 것은, 뉴욕시 공공공간에 대한 기여의 답례로 이루어졌으며, 전례가 없는 일이었다.

Capitol Plaza

Location_ New York, NY, U.S.A
Lanscape Design _ Thomas Balsley Associates
Client _ Witkoff Properties
Consultants _ MEP Engineer - Rosenwasser / Grossman
Courtesy of Thomas Balsley Associates

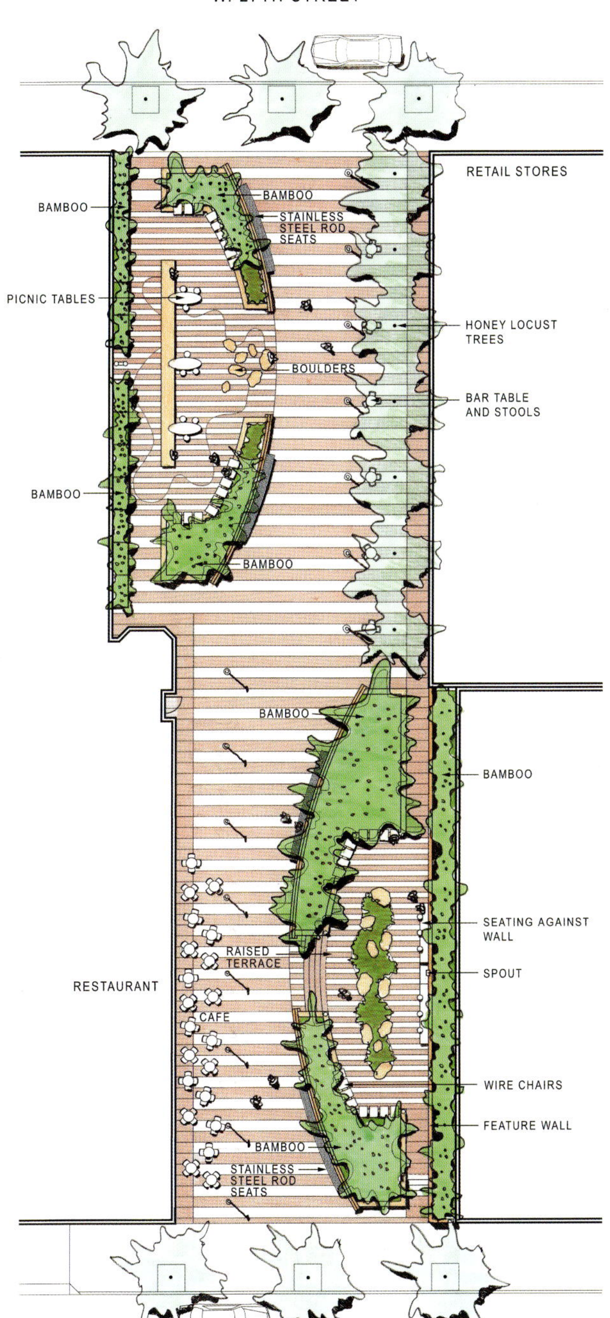

Plaza Plan

Capitol Plaza is located in the emerging residential neighborhood of Chelsea Heights amid weekend antiques markets and Flower District shops. This new public open space, which connects 26th Street and 27th Street just east of Sixth Avenue, features garden seating areas, a promenade, and cafes. In an area of Manhattan with too few public open spaces, Capitol Plaza's goal was to offer its young creative class and culturally vibrant constituency a place to pause, meet and socialize among lush bamboo groves and ornamental grass plantings, distinctive contemporary seating and adjacent cafes and shops -all in a synergistic composition that will ensure long term success and offer an artistically provocative design alternative to nearby Madison Square Park.

Curved, battered planter walls slice through the plaza, energizing its passage and organizing it into distinct areas with varying degrees of intimacy and enclosure. Innovative stainless steel furniture, such as bar tables with revolving stools, benches with attached cafe tables, and elliptical picnic tables, provide a multitude of seating options for its young diverse constituency. A 100-foot-long corrugated metal wall painted a vibrant orange and brightly lit in the evenings is intended to draw attention from Sixth Avenue's pedestrian and vehicular traffic. Elliptical cutouts animate the wall and reveal bamboo foliage behind helping to blur the line between the landscape and architectural treatments. One of the cutouts frames a stainless steel waterspout whose sounds add quality to the bamboo glade environment and reinforce its serenity.

© Michael Koontz

The northern shady area is dedicated to picnic tables, perimeter seating and a narrow, 100-footlong retail space that activates this end. The sunnier southern zone is raised slightly to provide a more intimate seating garden that is enhanced with the sounds of water and surrounded by bamboo glade. It overlooks and is animated by a flanking outdoor cafe that immediately became the neighborhood's favorite.

Scattered throughout downtowns in cities across the country are small urban spaces -some public, others privately owned- that have not lived up to their potential, but that have the power to enhance urban living by touching our daily lives in ways that larger destination parks can't. Capitol Plaza clearly and emphatically answers the question, "How can our profession make a difference in the city?"

With each small gesture like this, we collectively make city life so compelling that we help stem suburban sprawl and the destruction of our natural environment. This space is not a shallow gesture of form and color. Beneath its visual appeal lies a careful analysis of neighborhood dynamics and a deep understanding of the design principles that determine the success of small urban spaces and ultimately, of our cities.

© Michael Koontz

© Michael Koontz

캐피톨 플라자는 주말 벼룩시장과 플라워지구 상점들 사이에 있는 신흥 주거단지인 첼시 하이츠Chelsea Heights 안에 위치한다. 이 새로운 공공 오픈 스페이스는 26번가와 27번가를 연결하며, 6번 애비뉴의 동쪽에 면하고 있고, 정원의 앉는 공간과 산책로, 그리고 카페가 특징적이다. 공공의 오픈 스페이스가 매우 부족한 맨해튼 지역에 위치한 캐피톨 플라자의 목표는 젊고 창의적인 사람들과 문화적으로 민감한 지역주민들에게 다음과 같은 공간을 제공하는 것이다. 즉 이곳은 무성한 대나무 숲과 관상용의 지피 식재, 독특한 현대적 좌석, 그리고 인접한 카페와 상점들 가운데에서 휴식하고, 만나고 사회화되는 공간이며, 이 모든 상승효과를 가지는 조합들은 긴 안목에서의 성공을 보장하며, 근접한 매디슨 스퀘어 공원Madison Square Park의 대안으로서 예술적이며 자극적인 디자인을 제공한다.

곡선이며, 상부가 좁아지는 형태의 플랜터 벽은 플라자를 분할하며 통로에 에너지를 더하고 체계화함으로써, 플라자를 친밀함과 둘러싸임의 다양한 변화가 있는 독특한 공간으로 만든다. 회전의자가 달린 바 테이블, 카페 테이블이 부착된 벤치, 타원형의 피크닉 테이블과 같은 혁신적인 스테인리스 스틸 시설물은 젊고 다양한 지역주민들을 위해 다양한 앉는 방식을 제공한다. 100피트에 달하는 주름진 철제 벽은 선명한 오렌지색으로 칠해져 있고, 저녁이면 불이 밝혀지는데 이 벽은 6번 애비뉴로부터 오는 보행자와 차량의 주의를 유도하기 위한 것이다. 타원형의 작은 구멍은 벽을 생기 있게 하고 후면의 대나무 잎을 드러냄으로써 건축적인 처리와 조경 사이의 경계를 흐릿하게 만들도록 돕는다. 구멍들 중의 하나는 스테인리스 스틸 홈통의 틀이 되는데 이 홈통의 소리는 대나무 숲의 질을 높이고, 평온함을 더욱 강화해준다.

북쪽의 그늘진 지역은 피크닉 테이블, 주변 좌석과 이 지역을 활성화시키는 100피트에 이르는 좁은 공간으로 이루어진다. 해가 잘 드는 남쪽 공간은 보다 친밀하게 앉을 수 있는 정원을 만들고자 약간 높였으며, 물소리를 잘 들을 수 있고, 대나무 숲으로 둘러싸여 있다. 아래를 내려다 볼 수도 있는 이곳은 주민들이 즐겨 찾는 곳이 된, 노천카페의 측면에 위치함으로써 활성화되고 있다. 전국의 도시에 산재한 다운타운에 있는 작은 도시 공간들은 - 일부는 공공, 일부는 사유지인 - 그들의 잠재력대로 운영되지 못하고 있다. 그렇지만 그들은 대형 공원이 할 수 없는 다양한 방식으로 우리의 삶에 접속함으로써, 우리 도시 삶을 증진시키는 힘을 갖고 있다.

캐피톨 플라자는 명백하고 단호하게 질문에 답해준다, "우리 분야는 도시에 어떻게 차이를 만들 수 있는가?"

이러한 작은 각각의 몸짓을 통해 우리는 총괄적으로 도시의 삶을 칭찬하지 않을 수 없게 만듦으로써 우리 자연환경의 파괴와 교외지역의 팽창을 막도록 도울 수 있다. 이 공간은 형태와 색깔의 피상적인 행위로 이루어진 것이 아니다. 보여지는 매력 이면에는 주민들의 활동에 대한 주의 깊은 조사와 작은 도시 공간들의 성공을 결정짓는 디자인 원칙에 대한 깊은 이해가 있으며, 궁극적으로 우리 도시의 성공도 마찬가지이다.

© Michael Koontz

© Michael Koontz

©Erik Kvalsvik

Tide Point

Credited with turning a derelict old industrial site into the epicenter of Baltimore's Digital Harbor, this project is about transformation. Tide Point was the site of one of Procter & Gamble's main soap producing plants. Now the renovated buildings house high tech companies and institutions. Once, its boardwalk was off-limits to the public—now the site opens to the water and views beyond. Our role in the project included master planning, designing new construction, and landscape architecture.

Location _ Baltimore, Maryland, U.S.A
Size _ 13acres
Landscape Design _ W Architecture and Landscape Architecture, LLC
Courtesy of W Architecture and Landscape Architecture, LLC

©Janice Rettaliata

©Janice Rettaliata

©Barbara Wilks

©Barbara Wilks

The landscape of the site seeks create an environment that both draws neighborhood residents to the waterfront and also serves the needs of the office campus-creating flexible event spaces as well as places to meet, eat and rest. In the summer, the fog machine works to cool pedestrians, and hammocks and adirondack chairs dot the promenade and waterfront plaza.

We wanted to preserve the industrial character of the site, while also softening it enough to make it feel inviting. We broke up the scale of the site with smaller gardens of varying characters throughout. The centerpiece of the campus is a long wood promenade at the water's edge with one of the world's longest fog machines sending cooling clouds up through the planks on hot days. The misters are both a practical way to cool visitors during Baltimore's long hot summers, and also an exciting and visible way to build a presence for the site in Baltimore's active Harbor. At night, the mist can be lit with colored lights to create a dramatic presence. The site plan provides clear vehicular and pedestrian pathways and easements to Baltimore City for the public access to the waterfront. The project was certified as historic by the Maryland Historic Trust and the Department of the Interior.

While we have preserved much of the industrial character of the site, we have also integrated gardens of varying size throughout to soften the industrial hardness and encourage a sense of intimacy. The gardens are designed to each have a unique character and are meant to reflect different waterfront ecologies—an iris garden, a rain garden and a garden of marsh grasses. In our plantings, we chose plants native to the region and plants that would require minimal watering and fertilization to minimize environmental impacts and maintenance costs.

©Janice Rettaliata

©Janice Rettaliata

©Erik Kvalsvik

버려졌던 오래된 산업부지가 볼티모어 첨단 항구Baltimore' s Digital Harbor의 중심지로 변화된 이 프로젝트는 '탈바꿈transformation'의 사례로 잘 알려져 있다. 타이드 포인트는 Procter & Gamble의 주요 비누공장 중 하나였다. 이제 새롭게 재생된 빌딩들은 첨단 기술의 회사들과 연구소들로 사용되고 있다. 과거, 수변 산책로는 일반에게는 개방되지 않던 곳이었지만 이제는 물과 그 너머의 전망을 즐길 수 있도록 개방되었다. 이 프로젝트에서 우리가 맡은 역할은 기본계획과 새로운 건물의 설계, 그리고 조경이었다.

이 대상지에서 추구한 조경은 이웃 주민들을 수변으로 끌어낼 뿐만 아니라, 오피스 단지의 요구에도 잘 맞는 환경을 창조하는 것이었다. 즉, 유동적인 이벤트 공간이자 사람들이 만나고 먹고 쉬는 공간을 창출하고자 했다. 여름에는 안개 생성장치가 보행자를 시원하게 하고, 그물침대hammock와 아디론댁식의 의자adirondack chair가 보행로와 수변 광장 여기저기에 놓인다. 우리는 대상지의 산업적인 특징들을 보전하고 싶었기 때문에, 이런 특징들을 다소 완화시켰고, 이를 통해 방문객을 초대한다는 느낌을 가질 수 있도록 했다. 또한 구석구석에 다양한 특성을 가진 작은 정원들을 조성함으로써 대상지의 스케일을 세분화했다. 단지의 핵심은 수변에 있는 기다란 목재 산책로인데, 여름의 뜨거운 날에는 세계에서 가장 긴 안개 생성장치가 판자 사이로 시원한 안개를 발생시킨다.
그 안개 생성장치들은 볼티모어의 길고 뜨거운 여름동안 방문객들을 시원하게 하는 실용적인 방법일 뿐만 아니라, 활기찬 항구의 존재감을 형성하는 자극적이고 눈에 띄는 방식이기도 하다. 밤이 되면 안개에 다양한 색의 조명을 비춤으로써 드라마틱한 연출이 가능하다. 대상지의 계획은 차량동선과 보행자로, 그리고 시민들의 수변으로의 접근을 위한 볼티모어시의 지역권地役權을 명확하게 담고 있다. 이 프로젝트는 메릴랜드 역사재단Maryland Historic Trust과 내무부로부터 역사적인 작품으로 인정받았다.

우리는 대상지의 산업적인 특성들을 보존하는 가운데, 거부감을 완화하기 위해 다양한 크기의 정원들을 통합하였고, 친근감을 강화시켰다. 정원들은 각각 독특한 특징을 갖도록 디자인 되었고, '아이리스 정원', '비의 정원', '습지 초원의 정원' 처럼 수변생태를 반영하는 의미를 갖도록 했다. 식재방법으로는 지역의 자생종을 선택하였고, 최소한의 관수와 비료로 살아갈 수 있는 식물을 선정함으로써 환경에 미치는 영향과 관리비용을 최소화했다.

©Alison Cartwright

West Harlem Piers Park

Location _ New York City, NY, U.S.A
Size _ 9,800㎡
Landscape Architect _ W Architecture and Landscape Architecture, LLC
Client / Owner _ New York City Economic Development Corporation
Courtesy of W Architecture and Landscape Architecture, LLC

© Johannes Feder

© Alison Cartwright

The West Harlem Piers Park transforms a long and linear site into a public space that reconnects the Harlem community to the Hudson River. A narrow 69,000 sf parking lot has been expanded into a 105,526 sf park that re-imagines the threshold between city and river in a meaningful way. The West Harlem Piers Park is the culmination of a 30 year struggle by the community to regain their waterfront and renew their neighborhood through connection to the river.

AREA PLAN

The site, no wider than a tennis court between 129th and 133rd Streets along the Hudson River, had been cut off from the neighborhood by the adjacent streets and highway above. This point of access to the river, historically a natural cove between adjacent bluffs, and more recently an industrial port, had become a paved and fenced parking lot. The project intent was to create a continuation of the Manhattan linear Greenway along the river. Additionally, our goals included working with the client and the community to reconnect the residents to the water. Working closely with a strong group of community activists, the design team held a series of public meetings that resulted in a master plan for the area. The first phase of the plan was the creation of West Harlem Piers Park.

CONCEPT

SANDBAR FORMATION

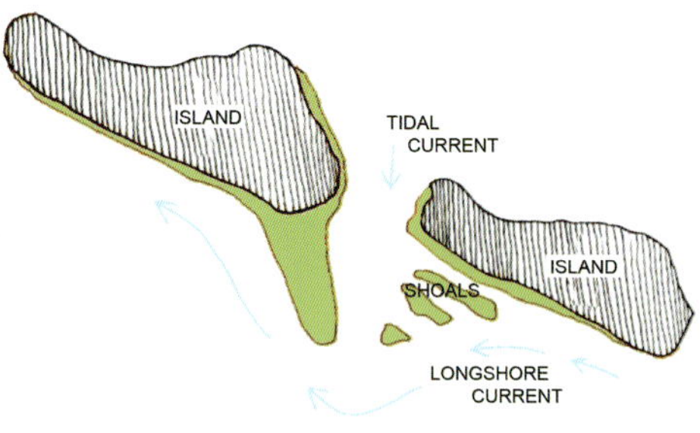

BUILT CONFIGURATION

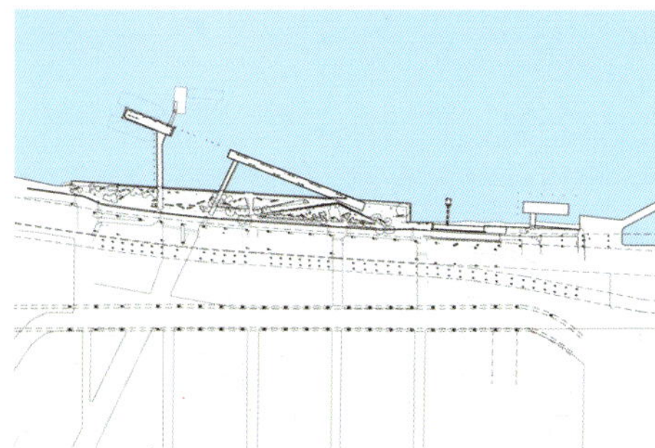

HARD EDGE

ERODED INTERFACE

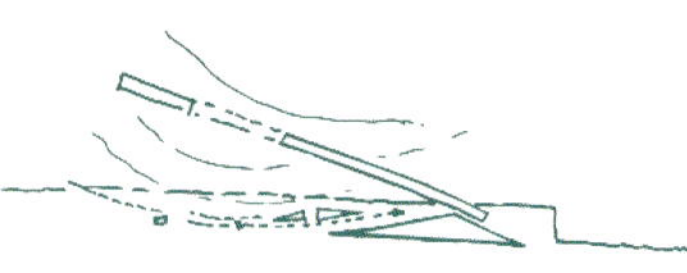

GRID RELATIONSHIP

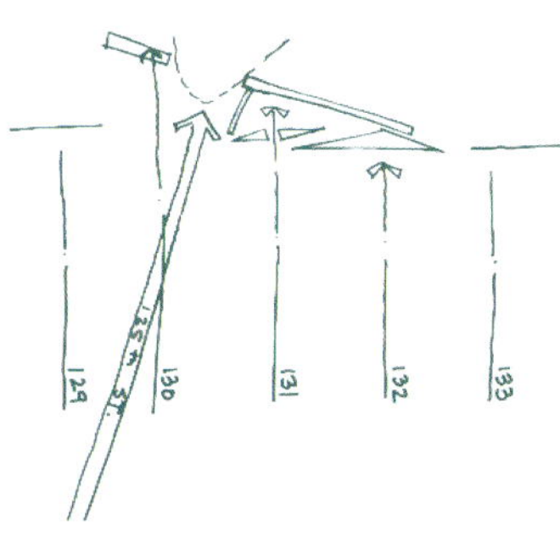

In order to create an effective gathering place, in addition to the linear connections required, the space of the park needed to expand. Several strategies were employed to increase the size of the space allocated for the park. The first was to literally expand it through the construction of new piers which were a key part of the community's vision. Historically piers aligned with the street grid, continuing out into the water. Looking for a new strategy, we used the model of sand bar formation to angle the piers. Connecting gangways created a loop of circulation out over the water, expanding the space. Another strategy was to expand the space by narrowing the adjacent roadway, eliminating unnecessary lane width. The reconfiguration of the street provided a 25% increase in park area. Finally, the park space was expanded visually though the extension of the diagonals created by the pier configuration into pathways that create long vistas. This diagonal structure also defines the two main areas of the park, an open cove and the shaded woodland.

©Johannes Feder

©Alison Cartwright

©Alison Cartwright

The cove is the terminus to 125th street, the "main street" of Harlem. It welcomes visitors into the park and is the main open gathering space. In earlier designs we had placed the piers in this area, but the community was firm in its vision of seeing open water down the 125th Street corridor. Incorporating this desire into the final design, the pier is broken at the cove and the connecting gangways follow the lines of the street out into the water. Other streets that terminate at the site create breaks and pathways that align with the street grid and create special places, like the jumping jewels water feature at 131st Street. Other elements in the cove like the benches, planters and raised lawns are arranged as if deposited by the receding tides.

The woodland provides an entrance to the park from 132nd St. The woodland rises from the street, then cascades down towards the water in continuous steps. These stairs provide seating and function as a shady overlook to the water. Other seating in the woodlands is created from scattered remnants of the original granite bulkhead which were replaced during construction. Pieces of the guardrail with inscriptions by the community created by artist Nari Ward are also arranged among the undergrowth, recalling past memories on this site.

A diverse program was necessary for the park to cater to its wide array of user groups. For community members a fishing pier provides a proper cutting and cleaning station that honors the long standing tradition of fishing in this area. A kayak float at the north end of the site encourages water related recreation and allows access to and from the river. For commuters, the transportation pier at the southernmost tip of the site will soon offer water taxi service and is configured for future ferry service as well.

For children, self activated fountains located between the two sloping "dunes" allows for water play in hot weather. For active recreation, the Manhattan bike trail follows the eastern edge of the site, honoring the community's desire for slow paced passive recreation to be located near the water like the required 12' waterfront shore public walkway which edges the water. The park also includes signage that documents the changing history of the waterfront.

긴 선형의 공간이었던 West Harlem Piers Park가 할렘지역과 허드슨강을 연결해주는 공공공간으로 거듭났다. 69,000평방피트의 좁다란 모양의 주차장은 105,526평방피트의 공원으로 확장되었고, 도시와 강을 연결해주는 의미있는 공간으로 변모되었다. 이 공원은 지역주민들의 수변공간을 되찾고자 하는 열망과, 강과의 연결을 통해 주변을 재생시키려는 노력으로 완성되었다.

허드슨강의 129번가와 133번가 사이에 위치한 이곳의 폭은 테니스장 보다 넓지 않은 편으로, 인접한 도로와 위로 지나는 고속도로에 의해 주변으로부터 단절되어 왔다. 강으로 접근하는 주요 지점인 이곳은 역사적으로는 인접한 험한 강변 사이의 자연적인 만이었고, 보다 최근에는 산업부두로서 사용되었으며, 포장 및 펜스가 둘러진 주차장으로 사용되었다. 이 프로젝트는 강을 따라 맨해튼의 선형 녹도를 연결하려는 목적으로 진행되었다. 더불어 우리는 지역주민과 수변을 다시 연결하기 위해 지역사회 및 클라이언트와 함께 일하는 것을 목표로 하였다. 지역의 자원 활동가 그룹과 긴밀하게 협업하면서 디자인팀은 여러 차례의 공공 미팅을 개최하였고, 이러한 미팅을 통해 지역을 위한 마스터플랜이 도출되었다.

적절한 모임공간을 창조하기 위해 선형 녹도의 연결에 더하여 공원 면적의 확장이 필요했다. 공원에 할당된 공간의 크기를 늘리기 위해 몇 가지 전략이 적용되었는데, 첫 번째 방법은 말 그대로 지역사회의 오랜 소망이었던 새로운 부두의 건설을 통한 확장이었다. 역사적으로 부두는 도로 그리드망과 정렬되고 수변으로 뻗어나간다. 새로운 전략에서 우리는 부두에 공간을 만들기 위해 모래톱의 형성 원리를 적용했다. 좁은 보도들이 물 위에서 연결되어 순환하며 공간을 확장하게 된다.
또 다른 전략은 불필요한 차로 폭을 없애고, 인접한 차도를 좁게 함으로써 공간을 확장하는 것이었다. 이러한 도로의 재배치를 통해 공원면적의 25%가 증가되었다. 결국 공원공간은 긴 전망을 창출하는 사선의 산책로를 부두에 배치함으로써 시각적으로 확장되었다. 또한 이 사선의 구조물은 공원의 두 주요 지역인 Cove와 그늘진 Woodland의 경계를 명확히 한다.

©Alison Cartwright

©Alison Cartwright

다양한 이용자 그룹에 적절히 대응하기 위해 공원에는 여러 가지의 프로그램이 필요했다. 낚시부두의 지역회원을 위해서는, 이 지역의 오랜 전통인 낚시를 할 수 있도록 적절하게 재단되고 깨끗하게 조성된 스테이션station이 필요했다. 대상지의 북쪽 끝에 있는 카약kayak장은 물과 관련된 레크리에이션을 가능하게 하고, 강으로 가거나 강으로부터 오는 접근을 쉽게 한다. 통근자를 위해서는 운송부두가 대상지의 최남단에 위치하는데, 곧 수상택시 서비스가 시작되고, 추후에는 페리 서비스도 개시될 예정이다. 어린이들을 위해서는 더운 날씨에 물놀이를 할 수 있도록, 두 개의 경사 "언덕" 사이에 자동 시동되는 분수가 설치되었다. 활동적인 레크리에이션을 위해서 대상지의 동쪽변을 따라 자전거 도로가 설치되었다. 또한 지역 사회의 느리고 수동적인 레크리에이션 욕구도 존중하여 수변 공공 산책로가 조성되었다. 공원에는 이 수변지역의 역사를 드러내는 사인물도 포함된다.

Copenhagen Harbour Bath

©JDS

Location _ Copenhagen, Denmark
Size _ 2,500㎡
Architect _ PLOT = BIG(Bjarke Ingels Group) + JDS(Julien De Smedt Architects)
Client _ The Municipality of Copenhagen
Completion _ 2003. 7
Photographer _ Nikolaj MØller
Courtesy of PLOT = BIG + JDS

©JDS

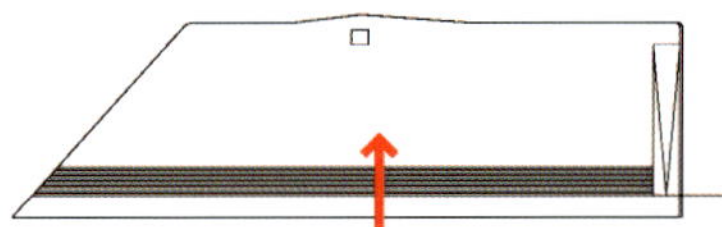

Continuity _ The Harbour Bath is the aquatic continuation of the Islandsbrygge harbour park. The Harbour Bath is a simple bended plate that connects the edge of the pier with the new harbour promenade and the water.

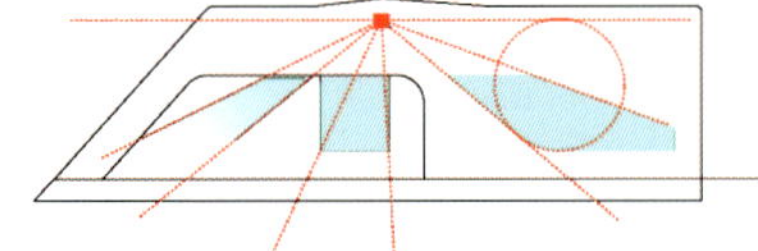

Safety _ The pools are laid out along radial lines extending from the center of the lifeguard tower, thus eliminating blind angles. The lifeguards can overview the entire bath from a single point. The diving pool expands in size to match the increasing jumping height. The minimum distance of 12 meters at the 5 m jump determines the geometry.

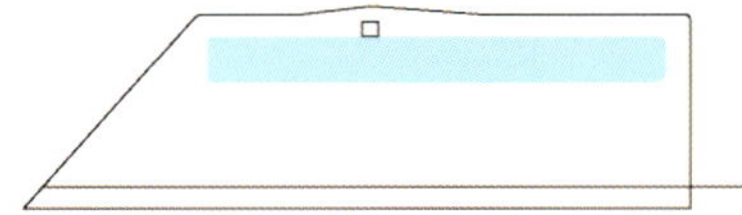

Swim _ A long linear swimmingpool strecthed along the entire length of the bath maximising the experience of free exercise. The pool is 86 meters long by 8 meters wide.

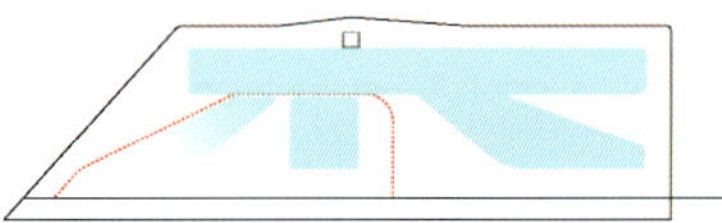

Play _ A paddle pool for children and the elderly facilitates the potential of the beach for play and chill out. Sloping from 0 to 0.3 to 0.6m of depth. A rectangular playpool of 1.2 m of depth is tailored for older children, ball games, the elderly and the disabled.

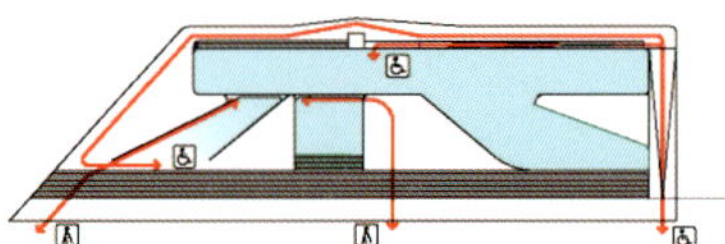

Accessibility _ A large ramp provides access to the entire facility incl. all pools. Strategically located handrails gives direct access to all pools for people with walking difficulties and the visually disabled.

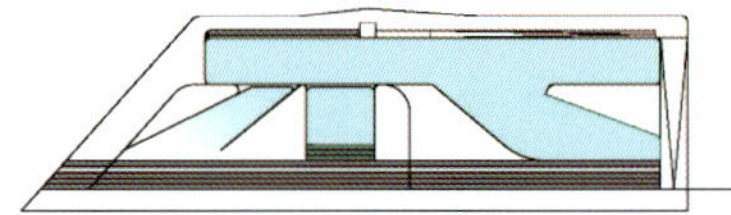

Harbourscape _ The design of the Harbour Bath has emerged by super-imposing the concerns for accessibility, lifeguards lines of sight and the different aquatic activities changing needs for land and water. The harbour bath appears as a bended wooden deck tipping on the edge between land and water.

Copenhagen's harbour is in the midst of a transformation from an industrial port and traffic junction to being the cultural and social centre of the city. The Harbour Bath has been instrumental in this evolution. It extends the adjacent park over the water by incorporating the practical needs and demands for accessibility, safety and programmatic flexibility. Rather than imitating the traditional Danish indoor swimming bath, the Harbour Bath offers an urban harbour landscape with dry-docks, piers, boat ramps, cliffs, playgrounds and pontoons. As a terraced landscape, the Harbour Bath completes the transition from land to water, making it possible for the citizens of Copenhagen to go for a swim in the middle of the city.

©JDS

©JDS

©JDS

Jyllands-Posten
Hvis De vil vide mere
©JDS

0,9 M

코펜하겐 하버Copenhagen' s Harbour는 산업항이자 교통 환승장이었던 곳이 도시의 문화적 · 사회적 중심지로 변화하는 핵심으로서의 의미를 지닌다. 항만 수영장Harbour Bath은 이러한 변화에 큰 도움이 되었는데, 접근성, 안전성, 그리고 계획에 따른 활용성을 위한 요구와 실용적인 필요성에 의해 인접한 공원이 수면위로 연장된 것이라 볼 수 있다. 전통적인 덴마크의 실내수영장을 모방하는 대신 항만 수영장은 건식 독dry-dock, 부두, 배모양 경사로, 절벽, 놀이터와 평저선平底船, pontoons 등의 시설이 있는 도심항구 풍경을 제공한다. 테라스화된 공간을 조성함으로써, 항만 수영장은 육지에서 물로의 전이를 완성하며, 코펜하겐의 시민들이 도심 한가운데서 수영하는 것을 가능하도록 했다.

Museo del Acero Horno^3 Monterrey

Surfacedesign inc. collaborated with Harari Arquitectura y Paisaje to transform a decommissioned blast furnace and a brownfield site into a modern history museum dedicated to the region's rich history of steel production. Borrowing from materials endemic to the site, innovative landscape design weaves together with modern architecture(Grimshaw Architects) to usher an old relic into the 21C. Environmentally sensitive technologies - such as green roofs and a storm water collection system - offer a new approach to the landscape while respecting the original context.

In 1986, the city of Monterrey, Mexico reclaimed an expansive 1.5ha brownfield site of a former steel production facility. Eleven years later, the site's decommissioned blast furnace has emerged as the Museo Del Acero Horno^3, the Museum of Steel, which serves as a new focal point for the region. Located at the center of the modern Parque Fundidora, which receives more than two million visitors per year, the Museo Del Acero Horno^3 narrates the story of steel production both to the generations who remember the history of the site and to younger visitors who may be unaware of the region's legacy.

Fundidora
ArcelorMittal
DEACERO

©Abigail Guzman Tamex

©Abigail Guzman Tamex

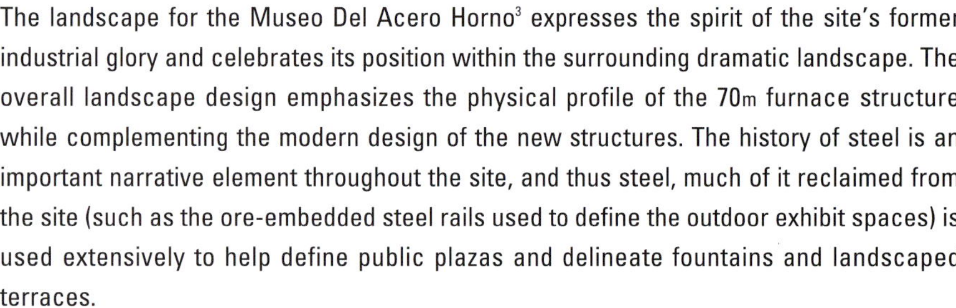

The landscape for the Museo Del Acero Horno[3] expresses the spirit of the site's former industrial glory and celebrates its position within the surrounding dramatic landscape. The overall landscape design emphasizes the physical profile of the 70m furnace structure while complementing the modern design of the new structures. The history of steel is an important narrative element throughout the site, and thus steel, much of it reclaimed from the site (such as the ore-embedded steel rails used to define the outdoor exhibit spaces) is used extensively to help define public plazas and delineate fountains and landscaped terraces.

Large, free-formed steel objects unearthed during excavation were incorporated as stepping stones and other features. The design approach melds industrial site reclamation—and the adaptive re-use of on-site materials—with ecological restoration through the use of green technologies.

All of the storm water runoff within the site's boundaries is treated in a series of on-site treatment runnels. These surround the exhibition areas and reinterpret the former industrial canals that once moved steel production by-products within the site. Aquatic plants and wetland macrophytes bio-remediate and treat storm water before it enters an underground cistern where it is stored for dry season irrigation.

Two water features are integral to the narrative of the project, while helping to define and locate the public space adjacent to the museum. In the main esplanade, the steel plates that formerly clad the exterior of the main hall were repurposed into a stepped canal over which water cascades. The 200m long feature alludes to the tracks used daily to train in the thousands of tons of raw materials that were off-loaded in this location, and serves as a visual connection to the rain garden in the landscape beyond. At the museum's entrance, the stepped canal culminates in the misting fountain, a grid of rocks visibly embedded with ore. This trompe l'oeil evokes the caustic heating process once used to extract ore, but instead of steam it generates a cooling mist that blows over the plaza—a pleasant surprise for visitors in Monterrey's hot and arid climate.

The use of green roofs (extensive and intensive) over the museum - which comprises the largest such roof system in Latin America - helps to reduce the visual impact of the new buildings. The existing furnace rises from this newly created ground plane. On the higher roof, a variety of drought-tolerant sedums have been arranged according to the structural roof patterns of the new architecture, and are contained by what appears to be a floating steel disk. A circular viewing deck allows visitors to take in the expanse of surrounding regional landscape, including the distant Sierra Madres, which are echoed in the roof's mounded shape. Below, a less constrained meadow of tall grasses - an abstraction of the native landscape - creates a connection to the landscape's pre-industrial context both functioning as a bioremediation for degraded soil and increasing thermal benefits for the new structure.

Principals of sustainability are at the core of the landscape design of the Museo Del Acero Horno[3]. By thoughtfully repurposing found industrial artifacts and incorporating new green technologies that work in concertwith the architecture and the greater landscape, the designers have created an outdoor exhibition space that interprets the area's historic uses while celebrating artistic opportunities for the future.

Location _ Monterrey, Mexico
Size _ 275 acres
Landscape Architects _ Surfacedesign Inc.+ Harari arquitectos
Client _ The City of Monterrey, Mexico
Architect _ Grimshaw Architect
Associate architect(s) _ Oficina de Arquitectura, S.A. de C.V
Courtesy of Surfacedesign Inc.

Surfacedesign은 기능을 상실한 용광로 공장과 그 주변 지역을 강철 생산과 관련된 역사기념박물관으로 탈바꿈하기 위해 Harari Arquitectos와 협력하여 프로젝트를 진행하였으며, 건축은 Grimshaw Architect에서 담당하여 21세기형의 모던한 스타일의 구조로 완성하였다. 주변 구조물들의 자재들은 현지에서 많이 활용하였으며, 비를 저장할 수 있는 우수처리시설이 적용된 옥상녹화시설을 포함한 다양한 요소들은 이곳을 현대적인 정원설계에 신선한 시도를 한 작품으로 인식되도록 하고 있다. 부지 전체에는 과거 활발했던 공업의 정신을 기리기 위해 강철의 역사와 느낌을 볼 수 있도록 철로 제작된 갖가지 구조물들을 설치해 두었다. 또한 부지의 경계를 구분짓기 위해 작은 수로를 만들어 물이 흐르도록 하였는데, 건기에는 물을 저장하여 사용할 수 있는 구조로 설계되어 있다. 박물관 입구에는 철광석으로 제작된 프레임의 블록형태 구조물들을 연속적으로 배열하였고, 사람들이 다닐 수 있는 길은 철판으로 특수제작하여 설치하였다.

© Paul Riveria

© Abigail Guzman Tamex

TECATE

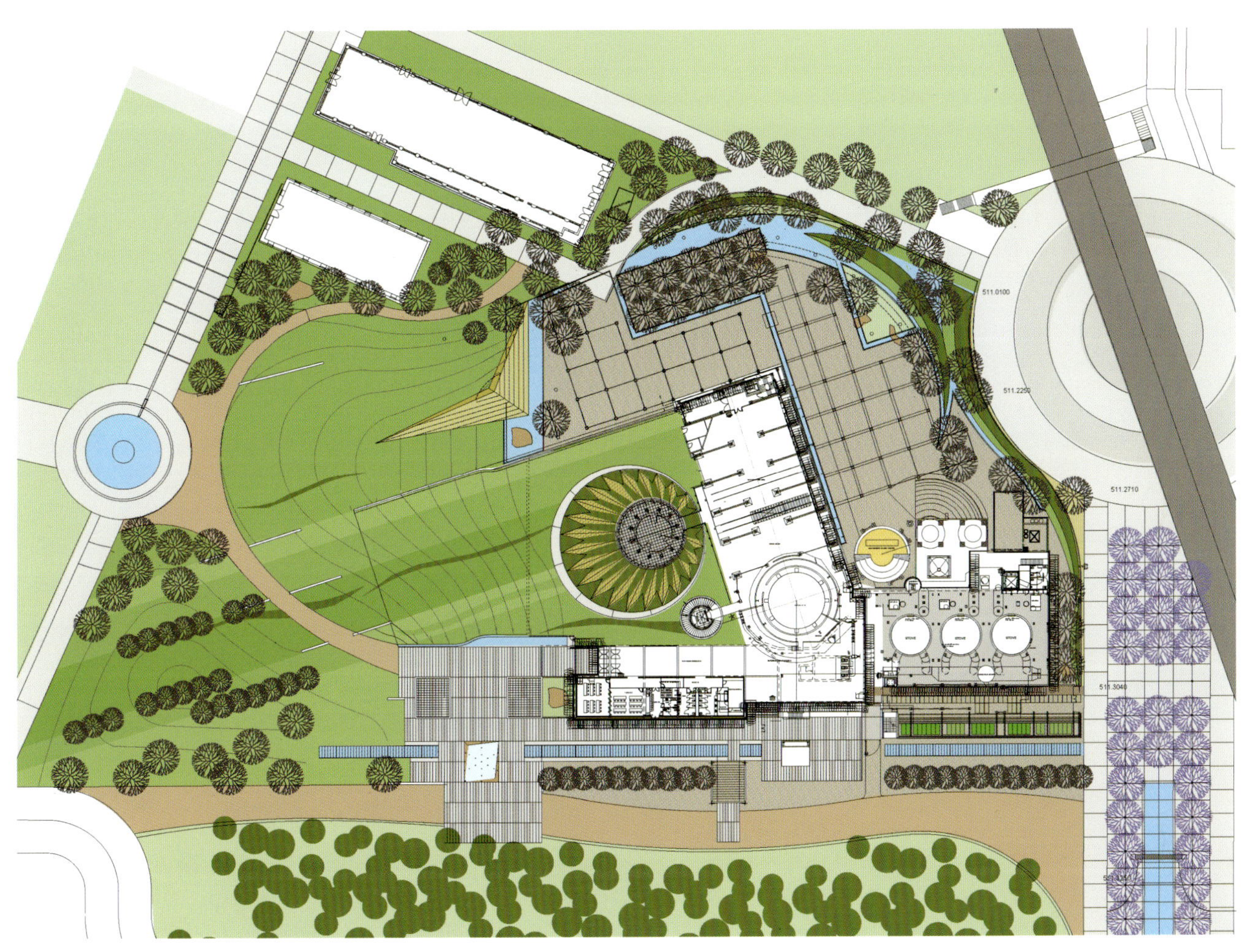

200미터 정도의 길이로 제작된 철로모양의 구조물은 실제 기차가 다녔던 곳인데, 정원과 물의 조화가 이루어진 시각적 연관성을 지닌 공간으로 재탄생 하였다. 철로모양으로 제작된 구조물 끝에는 철광석 그대로를 모아두고 스팀이 나오도록 했는데, 이는 박물관 내부에서 작동되는 냉각시설에 의해 나오는 수증기로서, 방문객들에게 특징적인 볼거리를 제공하고 있다. 그 밖에도 라틴 아메리카에서 가장 큰 우수처리시설이 도입된 옥상녹화시설을 갖추고 Native Grass와 Sedum들로 꾸몄으며, 이 공간은 새롭게 특화 조성된 조경시설로 경관에 큰 영향력을 주고 있다.

SASAKWA HILL LODGE

Location _ Serengeti, Tanzania, Africa
Landscape Architect _ WATG + The Really Useful Landscape Company
Architect _ WATG
Associate Architect _ MD Consultancy
Client/Owner _ VIP Safari Club
Interior Designer _ Interiors Designs Company Ltd.
Photography _ Singita Grumeti Reserves
Courtesy of WATG

This flagship hilltop resort with its wide wrap-around verandas offers awe-inspiring views over the private 350,000-acre Grumeti Game Reserve(ideal for viewing the annual migration of wildebeest, zebra and gazelle). Sasakwa's free-standing cottages 'designed in the style of a turn-of-the-century East African ranch home' feature wood paneling, antique furnishings and infinity-edge pools set in private gardens.

The concept was inspired by a desire to merge with the built and natural environments; the colors, materials and building heights reflect the landscape. The goal of the master plan was to give the land an untouched feel. Trees were carefully relocated—no mature trees were cut down—and most of the plants used in the design are indigenous.

During construction, two quarries were set up from scratch, and local villagers were trained to cut and finish the stone. A "factory" was created on site, where the workers produced their own concrete blocks. Transportation impacts were diminished significantly, and the local economy benefited. The contractor also trained local villagers for various trades such as bricklaying, plastering and carpentry. Some of these workers are now employed as part of the lodge's maintenance team.

Sustainability에 중점을 둔 마스터 계획 및 건물 설계 초안은 기존에 존재하는 자연환경에서 그 아이디어를 가져왔으며, 자연에 서식하는 경관의 이미지에 색상이나 건물군을 조화롭게 어울리도록 최대한 고려하였고, 현존하는 자연 및 경관을 최대한 덜 파괴하며 자연과 조화롭게 개발하는데 역점을 두었다. 또한 부지내 아름다운 수목들을 피해 건물들을 설계했으며, 문제가 되는 부분의 수목들은 자르기 보다는 조심스럽게 옮겨 그 가치를 보존토록 했다. 대부분의 조경 재료들은 현지에 자생하는 식생 재료들로 조경 재료의 현지화에 중점을 두어 물이 귀한 지역에서 관수 시설이 필요 없도록 절수가능한 계획을 했다.

7동의 카테이지 및 빌라 건물들(스파, 레스토랑, 야외 수영장 등)은 세기초의 동아프리카의 랜치 홈 스타일에서 영감을 받아 설계했으며, 나무 패널이나 고전적인 가구들 및 인피니티 에지 수영장들도 이러한 설계영향을 받았다.

프로젝트 건설 시행 단계에서는 공사 부지내 두 군데의 채석장을 설치하고 현지 주민들을 교육시켜 공사에 인부로 채용하는 등 철저히 현지화에 중점을 두었으며, 소규모의 콘크리트 공장을 세워 현지 주민들로 하여금 프로젝트에 필요한 콘크리트 블럭들을 직접 생산토록 하여 주민들에게 경제적 이익이 돌아가게 했다. 또한 이러한 부지내 시설물들은 야생자연구역내에 건설 관련 교통량을 현저히 감소시켰고, 현지에서 교육받고 인부로 일한 주민들은 공사 완공 후 정직원으로 채용되어 시설물 관리 등의 임무를 맡게 되었다.

이 프로젝트는 초기 단계부터 철저히 Sustainability development에 촛점을 맞추어 계획 · 설계 · 시공이 되었으며, 자연과 조화를 이루며 자연 환경 파괴의 최소화와 현지화에 성공한 프로젝트로서 현지 주민의 적극적인 지지를 받았고, 경제화에 성공된 모델로서 The World' s 100 Ultimate Escapes, Travel+Leasure의 Global Vision Award, 15 to Watch 등에 이름을 올렸다.

LOST DO

Lost Dog Wash Trailhead

Situated at the southern gateway to the City of Scottsdale's McDowell Sonoran Preserve, the Lost Dog Wash Trailhead has established a new standard for sustainable arid region design, including strategies for planning, preservation and construction. The project balances the various needs of the Preserve visitors and methodologies to protect the fragile desert ecology, including minimization of site disturbance, utilization of onsite materials, solar power, composting toilets, rainwater and gray-water harvesting. The Lost Dog Wash Trailhead is the third of nine planned Preserve Access Areas designed by the award winning team of Weddle Gilmore Architects and Floor Associates, landscape architects, and the project now serves as the model for all future Preserve Access Areas and desert sensitive design in general.

© Bill Timmerman

Located in north Scottsdale, Arizona, the McDowell Sonoran Preserve consists of mountains, heavily vegetated alluvial washes and large tracts of undisturbed Sonoran Desert that create natural open space corridors linking adjacent communities, the Tonto National Forest and the Maricopa County Regional Park. When completed, the McDowell Sonoran Preserve will encompass nearly 57 square miles (147km^2) or approximately one-third of Scottsdale's total land area, creating one of the largest urban preserves in the United States. The McDowell Sonoran Preserve is intended to be maintained in as natural a state as possible. As such, the Preserve Access Areas are envisioned to be demonstration projects for sustainable desert design practices and are intended to explore new technologies and develop new methodologies that promote and contribute directly to sustainable design practices.

The Preserve Access Areas allow visitors to experience the Preserve through hiking, horseback riding, mountain biking, nature studies, bird watching, scenic viewing, picnicking, rock climbing and more. The 7-acre Lost Dog Wash Trailhead site is located at the convergence of two popular multi-use trails at the foot of the McDowell Mountains, allowing for access to more than 17 miles of recreational trails within the Preserve.

The program for the Lost Dog Wash Trailhead project included new trailhead facilities, outdoor classroom areas and trail improvements for pedestrian and equestrian uses, each being served by separate parking areas. The facilities are home to various educational programs that include guest lectures, field-based classrooms and guided interpretive hikes that introduce participants to local area history and instill appreciation for the varied flora and fauna of the Sonoran Desert.The surrounding desert context was the touchstone for all site and architectural improvements within the project.

Extensive studies and analyses were conducted by the design team which included archeological assessments, topographic surveys, slope and hydrology analyses, vegetation survey, soil studies, wildlife and habitat surveys and native plant inventories. These studies informed all design decisions from initial design through the project's construction phase. All program elements were carefully sited between existing desert washes to minimize environmental and visual impacts and to integrate into the existing landscape. Significant natural areas with smaller wash corridors were left undisturbed and protected throughout construction to allow drainage while preserving large swaths of desert vegetation.

Early conceptual studies were frequently reviewed and proposed development limits were repeatedly fine-tuned allowing for maximum preservation of native trees and cacti stands. Where development required disturbance, the native desert pavement, trees, cacti, shrubs and natural artifacts were salvaged, stockpiled, and reinstalled on site as part of the planting design. In total, more than 1,000 specimen plants were salvaged and reintroduced to the site as part of the revegetation effort.

The planting concept is derived entirely from the site itself. As part of the site analysis process, Floor Associates conducted extensive field surveys to determine which plant species were indigenous to the specific areas of the site, including cataloging of plant densities and plant associations. In the desert, plant species, densities and associations can vary widely within even relatively small sites due to such environmental influences as drainage patterns, soil characteristics, exposure and elevation. This data was used to "re-vegetate" the site to replicate pre-construction conditions. Re-vegetation efforts mirrored the symbiotic plant relationships of the surrounding Sonoran desert and drew from native plant densities and growth patterns cataloged prior to construction. The focus on these patterns and densities helped to 'blur the edges' of natural and recreated landscapes, as well as to preserve and protect wildlife habits. In addition to the native plants, the top four inches of native soil and native "desert pavement" cobble within the limits of disturbance were salvaged and stockpiled for later redistribution. This allowed the project's top-dressing to seamlessly integrate into the adjacent coarse and rugged desert floor.

The desert cobble textures allow the establishment of plant communities by providing a protected place to take root. To minimize site disturbance, the construction process was carefully monitored by the design team. This included locating temporary fencing at the edge of the construction envelope to limit activities to designated areas; designating staging areas for materials and equipment and restoring these areas upon project completion; and developing a stringent construction recycling program.

Over seventy-percent of building materials were of local origin or directly supplied from the site in the form of native soils and screened rock materials. The parking lot areas and vehicular drives were constructed of stabilized decomposed granite, in lieu of traditional paving materials, to increase percolation and minimize surface runoff and heat gain. Stabilized granite parking areas and trail surfaces utilized screened site soils to match the adjacent colors and textures of the desert floor.

The walls of the trailhead structures are rammed earth, utilizing soil from onsite excavations.

The structures' metal roof panels and steel framing members were left in their unfinished raw state and allowed to naturally rust and patina, thus avoiding the use of paint or other potentially toxic finishes. This natural approach was extended to include the site furnishings, such as light bollards, trash receptacles and vehicular wheel stops. This thoughtful use of natural indigenous materials effectively integrates the amenities into the native site landscape.

The Trailhead lavatories use composting toilets that provide an annual savings of 20,000 gallons of water compared to conventional systems and eliminated the need for installation of municipal sewer lines. The rooftop panels of the main trailhead structure house a 2,000-watt solar array which allows the Trailhead to be self-sufficient and independent from the municipal electric grid. The array powers LED lighting and a drinking fountain chiller. The project is essentially "off the grid" with no utility connections, with the single exception being potable drinking water.

Rainwater from the roof of the main structure and gray-water from its lavatory sinks and drinking fountains is collected and stored in a 4,000-gallon underground cistern. This gray-water is reused by the drip irrigation system as supplemental water to new landscape plantings. A two-year timetable was set to wean new plantings from irrigation, thus relying solely on natural rainfall. A temporary irrigation system utilizing potable water was used for initial establishment, particularly for the dry-scattered seed mix, and has since been phased out, as initially planned. The collected gray-water will continue to provide supplemental water during severe summer months to offset the long-term drought the region is currently experiencing.

The Lost Dog Wash Trailhead project exemplifies the principals of sustainable arid region design practices and has received a number of design awards from such entities as the American Society of Landscape Architects and the American Institute of Architects. The design team of Weddle Gilmore Architects and Floor Associates has recently completed the design of the next Preserve Access Area, known as Gateway, which will open to the public next year. Each project is unique in both the specifics of their sites and their programmed uses, however, each project successfully pushes the envelope of sustainable design and construction practices to create beautiful demonstrations on how to properly and appropriately build in the desert.

대상지는 산들과 초목이 자라고 있는 충적토로 구성된 부지, 그리고 근접한 The Tonto National Forest와 The Maricopa County Regional Park를 이어주는 길게 뻗은 자연도로가 형성되어 있었고, The McDowell Sonoran Preserve이 차지한 지역은 거의 57평방마일에 해당되었는데, 이는 Scottdale 전체부지의 3분의 1을 차지하는 면적이며 미국에서 가장 큰 보호구역이기에 가능한 자연 그대로의 모습을 유지하려 노력했다고 한다. Lost Dog Wash Trailhead는 하이킹이나 승마, 산악자전거, 조류관찰, 경관감상, 피크닉, 암벽등반 등을 즐기려는 사람들이 주로 찾는 곳으로 다양한 기능을 할 것으로 생각된다.

The Lost Dog Wash Trailhead의 총괄적인 계획에 있어서 야외 학습장으로서의 기능, 보행자와 말을 타는 사람들이 이용하는 데 있어서의 편리함, 공원과 주차장의 구분 등이 고려 되었다. 부지 곳곳에는 하이킹의 기본교육, 지역의 역사소개, 소노란 사막Sonoran Desert의 동식물 분포 등에 관한 시설물들이 설치되어 교육적 요소로 활용되고 있다.

주변환경에 대한 연구와 분석은 설계팀에 의해서 이루어 졌으며, 주로 고고학적인 평가, 지형분석, 경사 및 수문학적 분석, 토양 및 야생동식물 관찰 등이 대부분이었다. 연구결과는 프로젝트의 부지양상 및 전반적인 설계의 정확성을 구축하는데 많은 도움이 되었으며, 설계팀은 주변 현황을 고려하여 시각적인 효과를 두드러지게 하고, 지역 특성상 침식가능성이 높은 지역에 대해서는 조심스럽게 작업을 진행하였다고 한다. 또한 중요한 지역은 그대로 보존하여 활용하였고, 초원의 생태계를 유지하는데 큰 역할을 하고 있는 수계는 가급적 보호하고자 노력하였다.

© Christopher Brown

야생식물들은 최대한 보존하고 개발이 필요한 지역의 경우 야생식물들을 그대로 채취하여 다시 식재하는 과정으로 설계를 변경해 나갔는데, 이곳에는 약 1천여 종 이상이 번성하고 있었기 때문에 전반적인 프로젝트의 컨셉은 바로 자연적인 부지 그 자체만으로도 가치가 있었다. 따라서 자연 그대로의 부지특성을 살리고 기존의 야생식물을 재활용하는 계획이 적용되었으며, 이러한 재활용의 의미는 're-vegetate' 라는 표현으로 소노란 사막지대의 상징적인 요인이 되었다고 한다. 재활용에 대한 부분은 토양도 마찬가지여서 기존의 토양은 4인치 두께의 자연적인 도로로 재활용하였다.

건축물 및 시설물 제작에 있어서도 70% 이상은 현지에서 활용할 수 있는 토양 및 석재를 재활용 하였으며, 주차장은 부식된 화강암을 사용하여 삼투압 효과를 증대시키고 지면의 열을 줄이는 효과를 도모했고, 화강암과 산길 지면 색채의 적절한 조화는 자연스러운 분위기를 연출하고 있다. 구조물을 둘러싼 외벽은 다져서 굳힌 흙으로 만들었고, 지붕 판넬은 철골 프레임으로 형성된 구조물 위에 설치함으로써 도료를 사용하지 않고 자연스럽게 녹이 생기도록 하였다. 조명볼라드나 휴지통, 재활용 수거함 등도 자연스러움을 강조하기 위해 건축물과 같이 조화를 이루도록 제작하였다.

Site plan

우수저장시설에 연간 2만갤론 정도의 물을 보관하여 시설물 내 화장실에서 사용하고 있어 별도의 수도관 설치는 필요하지 않았으며, 또한 지붕에는 태양광을 모을 수 있는 집열판을 설치하여 저장된 2천와트의 전력이 LED조명과 음수대를 돌리는 원동력이 된다. 지붕을 타고 내려오는 빗물이나 음수대, 시설물 내부에서 사용하고 남은 물들은 정화처리하여 약 4천갤론 정도의 물을 보관할 수 있는 지하탱크에 저장되는데, 이 물은 주변에 산재된 초목들의 공급원으로 활용이 된다.

The Lost Dog Wash Trailhead 프로젝트는 자연환경을 재활용한 좋은 사례를 제시할 뿐 아니라 미국조경가협회 ASLA와 미국건설협회The American Institute of Architects 등으로 부터 다수의 디자인상을 수상함으로써 그 가치를 인정받았다.

Nicolai School Centre

Location _ Kolding, Denmark
Size _ 3,800㎡
Landscape Design _ Arkitekt Kristine Jensens Tegnestue
Collaborators _ Dorte Mandrip Arkitekter
Client _ The Municipality of Kolding
Design period _ 2003~2007
Completion _ 2007. 10
Courtesy of Arkitekt Kristine Jensens Tegnestue

JULIUS & ESTHER

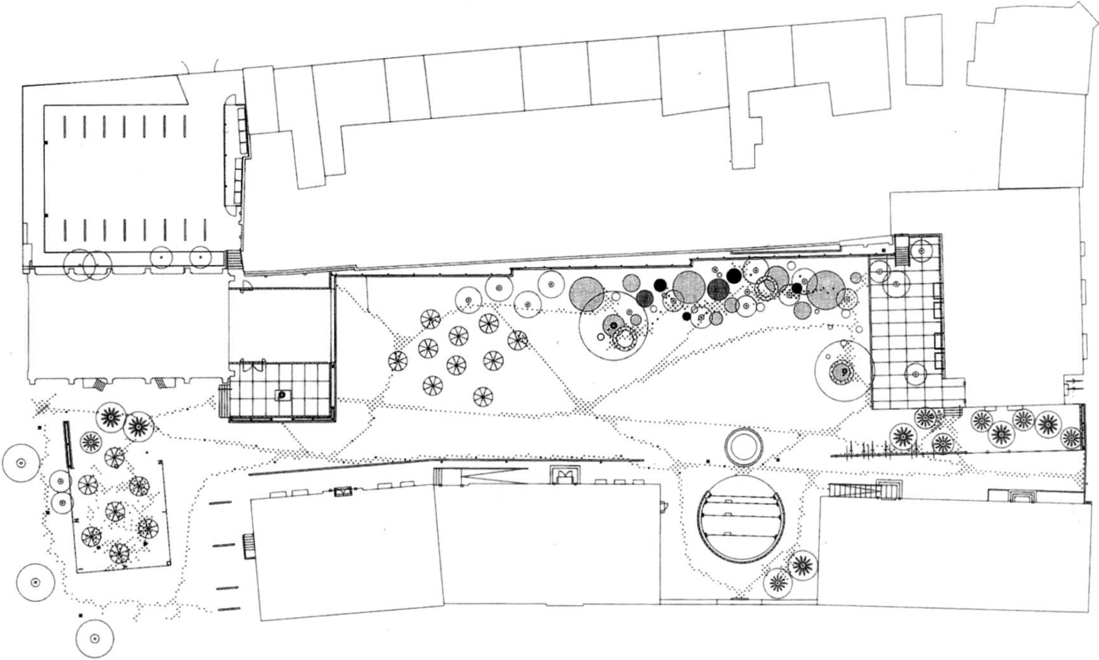

St. Nicolai School Centre from the 1850's is a continuous reincarnation of everything from the traditional Danish school with gyms and schoolyards. With five art houses as the new function the possibility of a new typology of creative and dynamic culture centre emerges. The project evolves with the idea of the schoolyard as the centre, joining the different functions together, an active free space in a more cheerful schoolyard, than the historical version. A yard, creating a base of the different house programs, movies, literature, house of children, craftsmanship and music.
Along the west side a continuous wall of corten steel is built along two terraces and a stage/ tribune also made in corten steel. Furthermore different outdoor activity spaces are added to the schoolyard: The Garden of Tales, The City Garden, The Market Square and the Plaza, opening up towards Katrinegade. The tilting terrain in the schoolyard is prevailed and a new enclosing surface of black asphalt is laid out, on which patterns are drawn with thermo plastic and street paint. The schoolyard is designed with a big white drawing on a new layer of black asphalt- as a new drawing board. The project development has been theme based and five groups of users have during the design process participated in concretizing the programming and the future schedule of utilization and maintenance.

Since 1890 St. Nicolai has been a school complex. The schoolyard as well as all the buildings had since then been worn down, unfit for the modern school life. Therefore the whole building complex with the schoolyard and the five buildings was to be turned into a cultural place, renovated by The Municipality of Kolding and the Realdania Foundation.
The complex is an almost 3,400m² big schoolyard with five different school buildings from various periods of time. Build from 1890 until late 1930, all made in different styles and with either red or yellow brick or plaster. The courtyard has a long rectangular shape almost 100m by 35m with a small terrain slope in the middle. Along one side of the courtyard to the west there was an ugly condemned fence and planted area that ran for almost 70 meters. The buildings had all various stairs and extensions in poor state. The courtyard was used only as a parking lot and the whole state was literally a messy backyard site.

The aim for the site has been to turn the whole area into a cultural complex, each house having different purposes as follows: House of literature, House of films, House of crafts, House of children's culture and House of music and the schoolyard should be created so that it would benefit all of the dynamic features from the houses. Hereby we harvested following programs: a cinematic place as an outdoor scene also able to feature small concerts or plays, places for exposition and markets, places for children to play and everyone else to stay gather, run and just be in the sun. The aim of intervention by an architectural point of view: On the city level or structural codex we did two things to full fill the yard on both sides: A new wall almost 70 meters long and a new structure in the row of houses on the other side. Our aim was to make a new concept for a modern school yard, one that was a lot livelier, full of joy, participation and through feeling of dynamic purpose.

The new schoolyard now consists of various situations and elements that communicates with the five building programs; a entrance area when coming from Blæsbjergade, a garden for children's play and relaxing on the grass, a terrace for the cinema cafes guests, a big circular stage for summer cinema screenings and theatre plays or concerts, a area to have markets, a small courtyard for drinks and talks when visiting the house of music and a square for multiple use.
Having a differentiated building complex the aim for the project has been to bind the complex together as a hole. Therefore the selection of materials has been essential. The main material used is corten steel, asphalt and white thermoplastic combined with pale pink magnolia trees and bright blue, yellow and red elements in the garden.
The asphalt serves as a blanket covering the whole schoolyard, the bright white thermoplastic creating links between the houses. The white dots challenges one to walk from dot to dot or merely creating ones one by drawing directly on the asphalt. The corten steel creates, with its orange colour, an inclosing wall that links the various buildings together. But also creates areas for recreation like the terrace and the stage. The blue, yellow and red in the garden are elements for playing such as a sandbox and water basins.

The new schoolyard now consists of various situations and elements that communicates with the five building programs; a entrance area when coming from Blæ sbjergade, a garden for children's play and relaxing on the grass, a terrace for the cinema cafes guests, a big circular stage for summer cinema screenings and theatre plays or concerts, a area to have markets, a small courtyard for drinks and talks when visiting the house of music and a square for multiple use.

Having a differentiated building complex the aim for the project has been to bind the complex together as a hole. Therefore the selection of materials has been essential. The main material used is corten steel, asphalt and white thermoplastic combined with pale pink magnolia trees and bright blue, yellow and red elements in the garden.

The asphalt serves as a blanket covering the whole schoolyard, the bright white thermoplastic creating links between the houses. The white dots challenges one to walk from dot to dot or merely creating ones one by drawing directly on the asphalt.

The corten steel creates, with its orange colour, an inclosing wall that links the various buildings together. But also creates areas for recreation like the terrace and the stage. The blue, yellow and red in the garden are elements for playing such as a sandbox and water basins.

The new schoolyard has not got a central space but has many points of view - a polyfocal space. Is has become a place for various activities link to the themes of the buildings, a public urbane space that now invites one to participate.

1850년대부터 성 니콜라이 학교 Nicolai School Centre는 체육관과 운동장을 포함한 전통적인 덴마크 학교로서 가졌던 모든 것을 지속적으로 재생, 변화시키고 있다. 이곳은 새로운 기능을 가지게 될 5개의 예술극장과 함께 창의적이고 역동적인 문화단지로 거듭나, 새로운 상징이 될 가능성이 높아졌다.
이번 프로젝트는 역사적인 해석에 치중하기 보다는 서로 다른 기능의 결합, 보다 즐거운 운동장을 위한 자유로운 공간을 조성함으로써 공간의 중심으로서의 운동장이 되는 개념을 발전시켰다. 즉, 영화, 문학, 어린이, 장인정신 및 음악과 같은 서로 다른 건물 프로그램의 본거지로 창조되는 운동장인 것이다.
서쪽을 따라 연속되는 코르텐강의 벽은 두 개의 테라스와 무대 옆으로 조성되었고, 강단도 코르텐강으로 제작되었다. 그에 더해 동화의 정원, 도시 정원, 시장광장과 Katrinegade 쪽으로 개방되는 플라자 같은 다양한 외부 활동공간들이 운동장에 더해졌다. 운동장의 경사진 지형은 극복되었고, 새로운 검은색 아스팔트의 포장이 펼쳐졌으며 열가소성수지와 페인트로 문양이 그려졌다.
운동장은 검은색 아스팔트 포장층을 새로운 화판으로 삼아 대형의 흰색 그림이 포함되어 설계되었다. 이번 프로젝트의 전개는 테마를 기반으로 했고, 5개의 사용자 그룹은 이용 및 관리의 향후 계획과 프로그램을 구체화시키는 디자인 프로세스에 참여했다.

1890년부터 니콜라이는 학교 단지였다. 그때부터 사용된 건물과 운동장은 낡아서 현대의 학교생활에는 맞지 않았다. 결국 운동장과 다섯 개의 빌딩을 포함한 전체 단지는 The Municipality of Kolding and the Realdania Foundation에 의해서 리노베이션 됨으로써 문화적인 장소로 거듭나기에 이르렀다. 이들은 1890년부터 1930년대 후반까지 붉은색과 노란색 벽돌 또는 벽토를 재료로 이용해 서로 다른 스타일로 건설되었다.
운동장은 100m×35m 크기의 긴 직사각형 형태로, 중앙에 작은 경사지형이 있었다. 안마당 서측의 한편을 따라 흉측한 펜스가 있었고 식재지역은 70m에 달했다. 건물에는 상태가 좋지 않은 다양한 형태의 계단과 연장된 부분이 있었고, 안마당은 거의 주차장으로만 사용되었으며, 전체적인 상태는 말 그대로 지저분한 뒷마당 같았다.

이 대상지에서 목표로 삼은 것은 전체 지역을 문화단지로 조성하는 것이었으며, 각각의 건물은 문학의 집, 영화의 집, 장인의 집, 어린이 문화의 집, 음악의 집과 같은 서로 다른 목적을 지니도록 했다. 그리고 학교 운동장은 반드시 각 건물의 역동적인 특징의 장점을 살릴 수 있도록 조성되어야 했다. 이를 위해, 작은 공연이나 연극을 할 수 있는 외부무대로서 영화적인 공간, 전시나 시장을 위한 공간, 어린이들이 뛰어놀거나, 모든 사람들이 함께 모이거나, 뛰거나 그저 태양 아래 있을 수 있는 공간과 같은 프로그램을 선정했다.
건축적인 관점에서는 도시의 수준 또는 건축 법령에 따라 운동장을 완성시키기 위해 양쪽 편에 각각 70m에 달하는 새로운 벽과 다른 편에 열 지어 서 있는 건물군에 새로운 구조물을 세우는 등 두 가지를 실행했다. 우리의 목표는 현대적인 운동장을 위한 새로운 개념을 만드는 것이었다. 즉 보다 생기 있고, 즐거움과 참여를 만끽하고, 역동적인 느낌을 부여하기를 기대했다.

새로운 운동장은 다양한 상황과 요소들로 구성되었고, 입구 지역, 어린이들의 놀이와 잔디 위에서의 휴식을 위한 정원, 시네마 카페 방문객을 위한 테라스, 여름의 영화상영·연극·콘서트를 위한 대형 원형무대, 시장이 열리는 공간, 음악의 집을 방문할 때 이야기 하거나 가벼운 음료를 마실 수 있는 작은 안마당, 그리고 다양한 목적을 위한 광장 등 다섯 개의 건물 프로그램과 연계된다. 뚜렷이 구분되는 건물들로 이루어진 단지였기 때문에, 이번 프로젝트의 목적은 단지를 하나로 묶는 것이었다. 그래서 재료의 선택은 매우 중요했다.
사용된 주요한 재료는 코르텐강과 아스팔트, 담홍색 목련과 결합된 흰색의 열가소성수지, 그리고 정원에 사용된 밝은 파랑, 노랑, 붉은색의 요소들이다. 아스팔트는 전체 운동장을 덮는 재료로 작용하며, 밝은 하얀색 열가소성수지는 각 건물들 간의 연결을 만들어준다. 이 하얀색 점들은 사람들이 점에서 점으로 걷도록 하거나, 단지 아스팔트 위에 직접 그림으로써 만들어보도록 유도한다. 오렌지색 코르텐강은 다양한 건물들을 하나로 연결하며 둘러싸는 벽이 되어 준다. 이외에도 테라스나 무대 같은 레크리에이션을 위한 공간도 코르텐강으로 조성되었다. 정원에 쓰인 파랑, 노랑, 그리고 붉은색은 모래상자, 수반 같은 놀이를 위한 요소로서 사용되었다.

새로운 운동장은 집중된 중앙 공간을 갖지는 않지만 여러 개의 뷰포인트를 가진 다초점의 공간으로 볼 수 있다. 즉 이곳은 건물들의 주제를 연결시켜 주는 다양한 행태를 위한 공간이자, 사람들의 참여를 끌어 들이는 도심 공공공간으로 거듭난 것이다.

Gudbrandsjuvet - Viewing platforms & bridges

Location _ Gudbrandsjuvet, Norway
Landscape Design _ Jensen & Skodvin Architects
Architect _ JSA
Client _ Roads dept
Courtesy of JSA

The main platform is constructed by 25mm laser cut steel sheets, cantilevered like a bridge around the cliff, hung in each end. The railing has a geometry that allows it to be continuous even with very different security requirements from place to place. The large inward curve allows the tourists to securely lean out over the deadly waters. The bridges are made from different materials according to what is most appropriate at each site. The platform at the parking side is made from prefabricated elements of concrete, like a bicycle chain, an element that is connected in the corners but rotated in the angle that will fit the site. This was appropriate at this site because cantilevering prefabricated elements had obvious advantages economically and practically. A related geometric concept is used for the service center.

주 전망대는 레이저로 커팅된 25mm 판상의 강철로 제작되었으며, 한쪽 끝이 절벽쪽으로 연결된 켄틸레버(외팔보) 다리의 형태로 되어 있다. 난간은 장소마다 요구되는 매우 다양한 안전강도를 모두 만족시키면서도 서로 연결이 가능한 기하학적인 모양으로 설치되어 있다. 안쪽으로 크게 휘어지도록 하여 관람객이 계곡의 실감나는 경관을 바라보기 위해 상체를 내밀더라도 안전하도록 설계되었으며, 교량들은 각 장소마다 가장 적합한 서로 다른 재료들을 사용하여 제작되었다. 주차장 측의 전망대는 조립식 콘크리트로 제작되었는데, 마치 자전거 체인처럼 각 조각들이 끝단에서 연결되고 있지만, 대상지에 맞추어 코너에서는 순환된다. 재료들이 대상지에 적합했던 이유는 무엇보다 캔틸레버식으로 만들기 위해서는 조립식 조각들이 경제적이고 실용적인 장점들을 갖기 때문이다. 이러한 기하학적 개념은 서비스센터에도 적용되었다.

Erman Residence Garden

In the summer of 2005, Surfacedesign were commissioned by Mark Erman, an early-rising hedge-fund manager, to design a back yard for his newly built home in the Noe Valley neighborhood of San Francisco. Seeking to transform the fairly typical, 25-foot wide by 50-foot long rear space, the client, who lives alone in the bottom flat of the two-condo building, had only two programmatic requests : that the garden be easy to maintain, and that it contain a hot tub for unwinding at the end of the day.

Location _ The Noe Valley, San Francisco, California, U.S.A
Landscape design _ Surfacedesign inc.
Client _ Mark Erman
Landscape Contractor _ 5 Elements Design
Photography _ Jeremy Harris
Courtesy of Surfacedesign inc.

At first glance, the lot's challenges were easier to spot than its potential. The developer had thoughtfully provided a carpet of sod (nowgone to seed) and a water feature (broken). More daunting was that in order to maximize every square foot, he had dug the garden into the existing grade. Defined by a 14-foot retaining wall on the north side, a ten-foot wall to the west, and the house itself-a four-story elevation that fairly looms over the rear-the garden felt far smaller and more constrained than it actually was. Thus, the goal was to not simply insert a prefab spa within the existing landscape, but to find ways of creating movement and life in the garden and to provide opportunities for experiencing nature with carefully selected and placed plant materials.

In order to maximize the journey from the house to the spa, the designers began by organizing the garden into three primary zones. Immediately off the rear of the house a stone patio was expanded to create a place for outdoor dining that visually (and practically) extends the interior, which opens fully to the outdoors. From there one embarks along a path comprised of parallelogram-shaped stone slabs to the middle zone, a 20-foot tall grove of bamboo that has been under-planted with a dense carpet of mondo grass, an assortment of self-naturalizing spring flowering bulbs, and winter-blooming hellebores. This zone also acts as a delicate veil to the most secluded portion of the garden. Defined at its threshold by a diagonal Japanese boxwood hedge whose crisp edges mimic the geometry of the paving, the back of the yard contains the hot tub, which is flanked by three Japanese snowdrop trees that rise sculpturally from a field of decomposed granite.

With the inclusion of the hot tub, often considered something of an eyesore, the designers embraced the challenge of putting "banal" elements to new use. Encasing the prefab spa is an 8ft square, steel-framed cover made of eco-sourced ipe wood, that appears almost to be inserted into the platform extending from the rear fence. When the tub is in use, this movable cover slides on tracks to become a waterside deck for sunbathing and toe dangling. It then zips along on stainless steel wheels forty feet through the garden and up to the terrace, where it doubles as a dining table—the literal interpretation of movement in the garden. Pausing in the middle, it becomes a three-dimensional frame for the plantings, which fit deftly within its volume. Even when the hot tub is uncovered, the steel plate that is fixed to its front continues to partially screen the spa from view while reflecting the tonality of dark fence.

Also imparting a sense of movement to the space is the side fence, which conceals the original concrete wall mass and its oppressive verticality while adding a feeling of craftsmanship to the generally austere space. Variously sized off-the-shelf redwood planks were stained black and intermittently canted inward—adding visual interest even as its inherent horizontal quality lengthens the appearance of the garden. As evening approaches, lights placedbetween the fence and the wall glow like lanterns through the gaps, casting interesting shadows and softly illuminating the space. An outdoor grill embedded in the fence in the front of the garden is used at al fresco gatherings in warmer months.

Through a series of budget-conscious interventions, the designers created a garden that met the programmatic requirements of the client within a verdant space that is as low maintenance as it is pleasurable—from winter, when the hellebores are in bloom, to summer, when the snowdrop trees are smothered in blossoms and rain their circular patterns of petals onto the ground.

2005년 여름, 클라이언트인 Mark Erman 씨가 샌프란시스코의 노 밸리에 장만한 그의 집 정원을 새롭게 조성해달라는 요청으로부터 프로젝트는 시작되었다. 이 정원은 25피트 너비, 50피트 길이의 공간이었다. 14피트에 이르는 펜스가 서쪽방향으로 설치되어 있으며, 집은 4층으로 뒤뜰이 잘 보이는 구조로 배치되어 있어, 정원은 실제 보다 더 작고 아담하게 보일 수 있도록 설계되어 있었다. 설계자들은 스파가 놓여진 곳부터 집까지 3가지 중요한 요소들로 정원을 구분하였다.

우선, 스파 뒷편에 있는 저녁식사를 즐길 수 있는 공간까지 석재 테라스Stone Patio를 연결하여 실내외 인테리어의 구분을 시각적으로 좀 더 확대하였으며, 그 이동공간의 가운데에는 규칙적으로 배열한 넓은 석판을 두었다. 20피트 높이의 작은 대나무숲을 조성하고, 부분적으로 잔디류Mondo Grass를 심어 봄의 새로운 생명력과 겨울의 크리스마스 트리와 같은 이미지를 연출할 수 있도록 했다. 그 외에 설계자들은 정원의 평범함에 도전하는 무언가 새로운 것을 첨가하고 싶어했는데, 그 대안으로 스파시설물을 덮을 수 있는 8평방피트의 정사각형 덮개구조를 만들어 철로를 이용해서 뒤쪽 펜스까지 이동할 수 있도록 하였다.

이 구조물이 설치됨으로써 정원은 입체적인 구조를 띄게 되었으며, 주변을 둘러싼 어두운 색의 펜스들은 보다 차분하고 위요된 은은한 분위기를 연출하게 되었다.

펜스와 벽 사이에는 조명시설이 설치되어 야간에는 그 틈새에서 빛이 새어나옴으로써 그림자와 빛의 은은한 효과가 공간전체에 고즈넉한 분위기를 선사하고 있다.

Baroda Garden

© Tom Bonner

Location _ Beverly Hills, California, U.S.A
Size _ 5 acres
Landscape Architect _ Rios Clementi Hale Studios
Original Architect of house _ A. Quincy Jones
Renovation / Restoration Architect of house _ Rios Clementi Hale Studios
Client _ David Bohnett
Courtesy of Taylor & Company

© Tom Bonner

Landscape has the ability to create a powerful sense of place. In 1955 the architect A. Quincy Jones worked with landscape architect Garrett Eckbo to design a home for the actor, Gary Cooper. Their idea was to create a modern, indoor-outdoor lifestyle with views and connections to the magnificent surrounding trees and Arroyo. Modernist geometries were continued from the architectural forms of the house out into the site, interior materials were brought outside, and landscape materials, such as stone, were brought inside.

It was this sculptural modern architecture, the beauty of the surrounding site, and the ability to inhabit both, that inspired the current owners to purchase the property. Throughout the past several years, two additional adjoining properties have been purchased and combined into the current 5-acre garden under the design guidance of Rios Clementi Hale Studios. Rios oversaw significant changes to grading, circulation, hardscape, and planting thatunified the site and created a comprehensive set of experiences for the owners and their guests.

The inspiration for this garden starts with the modern forms of the house and the transparent way the building sits on the site. The modernist garden vocabulary of geometric forms and a minimal plant palette set the stage for the entry and approach to the house. The back lawn garden culminates with the merging of the modern garden into the surrounding naturalistic setting. Down in the arroyo and on the adjoining newer properties the natural landscape becomes intensified, engulfing the visitor in its exaggerated scale.

The gardens transform according to exposure and site conditions. Shade gardens, succulent gardens, and a variety of other plant ecosystems are on array throughout the overall experience of the site. A stroll through the gardens will transport the visitor's memory to a variety of locales. It is the strong linkages between place and experience, architecture and garden, that distinguish this garden.

조경은 강한 공간감을 창조하는 힘을 갖고 있다. 1955년 건축가인 퀸시 존스A. Quincy Jones와 조경가인 가렛 에크보Garrett Eckbo는 연기자인 게리 쿠퍼Gary Cooper의 집을 설계하기 위해 함께 일했다. 그들의 아이디어는 모던할 뿐만 아니라 전망이 좋은 실내외 겸용의 생활공간을 창조하는 것이었는데, 특히 주변의 훌륭한 숲과 계곡으로 연결되기를 기대했다. 모던하고 기하학적인 조합은 주거건물의 건축 형태로부터 대상지로 연결되었고, 인테리어 재료들은 외부공간에 놓였으며, 석물 같은 조경요소들은 실내공간으로 반입되었다.

조각품처럼 모던한 건축, 대상지 주변의 아름다움, 그리고 거주하기 좋은 장점이 현재의 소유주가 이 토지를 구입하는데 결정적인 역할을 했다. 소유주는 지난 몇 년 동안 두 개의 인접한 토지까지 구입하였고, 리오스 클레멘티 해일 스튜디오Rios Clementi Hale Studios(이하 리오스 스튜디오)의 디자인 지침 아래 현재와 같은 5에이커 면적의 정원으로 포함시켰다. 리오스 스튜디오는 대상지 공간을 통합하고, 소유주와 방문객들의 포괄적인 체험을 위한 공간을 창조하기 위해 현저한 지형의 변화, 동선의 순환, 포장 공간hardscape, 식재 등을 두루 검토하였다.

이 정원에 대한 착상은 주택의 모던한 형상과 건물이 대상지에 잘 보이도록 자리 잡은 방식으로부터 출발하였다. 기하학적 형태와 최소한의 식재 색조palette로 표현되는 모던한 정원언어는 입구와 주택의 접근로 부분에 반영되었다. 뒤편의 잔디정원은 이 현대식 정원을 대상지를 둘러싸고 있는 자연적인 배경과 통합되는 절정을 보여준다. 계곡의 아래, 그리고 새로운 토지 부근의 자연적인 풍경은 점점 증폭되며, 방문자들은 확장된 규모에 빠져들게 된다. 정원은 방위와 대상지의 조건에 따라 변형되었으며, 그늘정원shade garden, 다육식물정원succulent garden, 그리고 다양한 식물 생태계를 총체적으로 경험할 수 있도록 구석구석 배치되었다. 정원을 거닐다 보면, 다양한 장소에 대한 추억이 떠오를 것이다. 이곳은 장소와 경험, 건축과 정원의 강력한 결합을 표현하고 있으며, 이것이 바로 이 정원을 독특하게 만들어 주고 있다.

© Tom Bonner

West Village Townhouse Garden

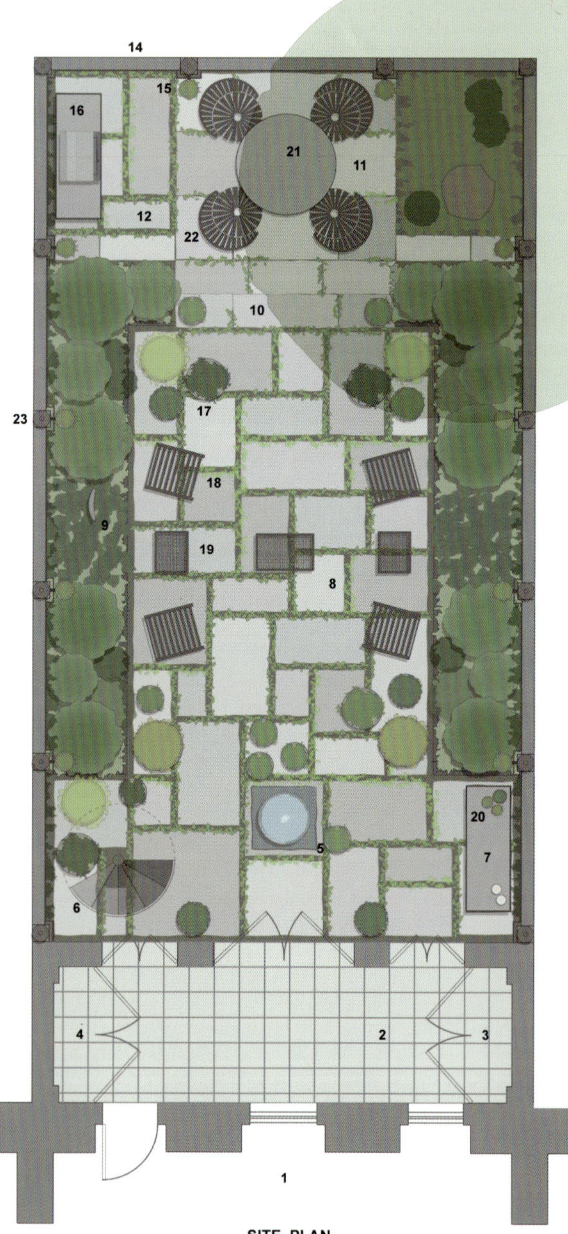

Location _ 520 Eighth Avenue, 21st Floor, New York, U.S.A
Landscape Design _ Sawyer/Berson Architecture & Landscape Architecture, LLP
Client _ Anonymous
Lighting Design _ William Armstrong Lighting
Structural Engineering _ Robert Silman Associates, PC
Courtesy of Sawyer/Berson

For a townhouse restoration in Manhattan's historic landmark district of the West Village, two gardens were created: one facing the street, and the other at the rear of the townhouse. The layout and style were designed to complement the nineteenth century Italianate brownstone architecture, and to meet the client's desire for a tasteful period garden that would be functional for entertainment and relaxation. Materials, plants and furnishings were all selected to create a beautiful garden setting with luxuriant green and provide for a comfortable outdoor living space.

LEGEND

1 TOWNHOUSE
2 LOGGIA
3 STORAGE CABINET FOR CHAIR CUSHIONS
4 UTILITY CABINET FOR TRANSFORMERS AND IRRIGATION CONTROLLER
5 PEDESTAL FOUNTAIN
6 SPIRAL STAIR TO KITCHEN ABOVE
7 SERVING CONSOLE
8 SEATING AREA
9 BRONZE SCULPTURE 'MINISTER OF WORDS'
10 STEPS UP TO DINING/BBQ AREA
11 DINING AREA
12 BBQ AREA
13 EXISTING NORWAY MAPLE
14 LATTICE FENCE
15 BRACKET MOUNTED POT (10 TOTAL)
16 BBQ CONSOLE
17 IMPRUNETA TERRACOTTA POTS
18 IRON CHAIRS
19 IRON SIDE TABLE
20 ZINC POTS
21 BLUE STONE DINING TABLE WITH BROWN STONE PEDESTAL BASE
22 ANTIQUE DINING CHAIR
23 FINIAL ON CAP OF LATTICE FENCE WALL (14 TOTAL)

A specimen Dr. Merrill Magnolia is featured in the front garden flanking the entry steps to the parlor floor of the townhouse. The Magnolia is under planted with sequential displays of crocus, dwarf iris, then white tulips, and finally New Guinea impatiens with white caladiums and ferns showing above. These plantings are choreographed to the seasonal progression ending in early fall. Window boxes and gray Impruneta terracotta pots placed at the parlor front entry are planted with variegated English ivy and white caladiums, and echo the seasonal plantings in the garden below. An ornate iron gate of the period provides entry from the sidewalk to the garden space, and to the ground floor of the building. The gate is anchored at each side by a restored cast iron railing. Antique bluestone paving bordered with milled brownstone curbing bisects the garden, and a large gray Impruneta terracotta pot planted with a dwarf Japanese boxwood provides the terminus to the gate axis.

The rear garden is seven hundred square feet, measures twenty feet wide by thirty-eight feet in depth, and is enclosed by the recent addition of the loggia and kitchen above, and the new painted mahogany lattice fence at the rear and two sides. The lattice fence panels and lattice doors at the loggia openings are backed with bronze colored acrylic mirrors. Thorndale ivy was planted to soften the architecture of the perimeter fencing, and is maintained to provide approximately forty percent coverage, subtlety revealing the surprise of the mirrors behind the lattice.

The main element of the rear garden is the intimate seating space surrounded by lush plantings in the beds and strategically placed terracotta pots. The lattice paneled fence provides the backdrop for the space, and the overhead canopy of the maple provides a sense of privacy from the windows of the neighboring townhouses. The space is then punctuated with the close placement of a bronze sculpture entitled "Minister of Words" by Melissa Zink. The sculpture is anchored to a pedestal that includes a hidden stainless steel turntable engineered to provide both an anchorage and viewing access to all sides of the sculpture.

The view and sound of water cascading from the pedestal fountain adds a cooling and a meditative sensation from within the seating area. Up the steps from the seating area is the dining space appropriately flanked by a BBQ console. Moon lighting is provided for the seating area by fixtures strapped onto branches within the highest reaches of the Norway maple. The dining table, sculpture, pedestal fountain, fence panels and the trunk of the Norway maple are all illuminated to varying degrees, and are circuited separately for desired night lighting affects. An electronic dripirrigation system supplements the watering of the beds, planted paver joints and the pots.

Furniture for the seating area and the consoles were designed to echo the form of an antique iron chair found in a Southampton shop, and are comfortable and functional accommodations for entertainment activities. The dining chairs are restored antiques. Fabrics for all the chair cushions were chosen to compliment the plant palette as well as for durability.

맨해튼 역사경관지구 내의 웨스트 빌리지 타운하우스 정원 복원 작업을 통해 2개의 정원이 새로 조성되었다. 정원의 전체적인 레이아웃과 스타일은 19세기 이탈리아 풍을 가진 부유계급층의 건축물과 조화를 이루도록 조성하고, 발주처의 요구에 따라 오락과 휴식공간으로서의 다양한 용도로 사용할 수 있도록 새롭게 설계되었으며, 자재, 식물, 소품 등 모든 요소들은 풍부한 녹음과 편안한 외부공간을 지닌 아름다운 정원을 위해 엄선되었다

맨해튼 역사경관지구 내의 웨스트 빌리지 타운하우스 정원 복원작업을 통해 2개의 정원이 새로 조성되었다. 정원의 전체적인 레이아웃과 스타일은 19세기 이탈리아 풍을 가진 부유계급층의 건축물과 조화를 이루도록 조성하고, 발주처의 요구에 따라 오락과 휴식공간으로서의 다양한 용도로 사용할 수 있도록 새롭게 설계되었으며, 자재, 식물, 소품 등 모든 요소들은 풍부한 녹음과 편안한 외부공간을 지닌 아름다운 정원을 위해 엄선되었다.

정원 입구에서부터 타운하우스의 응접실에 이르는 계단까지는 태산목이 열식되어있다. 태산목의 하부에는 크로커스*Crocus*와 꼬마붓꽃*Dwarf Iris*, 하얀색 튤립과 뉴기니 임파첸스*New Guinea impatiens*가 아메리카산 관엽식물의 일종인 칼라디움, 양치식물 등과 함께 순차적으로 식재되었다. 이들 화초류들은 대개 초가을이 끝날 무렵까지 계절의 변화를 보여줄 수 있도록 계획되었으며, 창문의 플라워박스와 응접실 입구의 회색 테라코타 화분에도 서양담쟁이덩굴*English Ivy*과 하얀색 칼라디움을 식재하여, 정원의 계절별 자취를 보여줄 수 있도록 했다. 화려한 문양의 철제 게이트는 타운하우스 정원을 둘러싼 철제 난간과 연결되어 있으며, 고풍스러운 청회색 사암 보도와 갈색사암 연석이 정원을 양분하고, 축의 정점에는 회양목을 식재한 회색의 대형 테라코타 화분을 두었다.

후정의 면적은 약 700평방피트로, 20피트의 폭에 38피트의 길이이다. 정원은 최근 조성된 로지아(한쪽이 트인 주랑)와 부엌 위쪽, 그리고 뒤쪽과 두 면에 새롭게 칠해진 마호가니 격자모양 펜스로 둘러싸여 있다. 펜스와 문들은 청동색으로 색이 입혀진 아크릴 거울로 장식이 되어있으며, 펜스의 대략 40퍼센트 정도를 덮고 있는 쏜데일 담쟁이덩굴은 건축적으로 부드러운 느낌을 주면서, 격자 하부의 거울과 함께 신비로운 느낌을 줄 수 있도록 하였다.

후원은 깔끔히 배열된 테라코타 화분과 푸른 숲이 우거진 화단으로 둘러싸인 휴식공간이 특징이다. 격자무늬 판넬로 제작된 펜스는 정원의 배경이 되며, 단풍나무로 제작된 캐노피는 이웃 타운하우스들의 창문에서 볼 수 있는 시선을 막아줌으로써 개인적인 프라이버시를 방해하지 않도록 고안되었다. 한 켠에는 정원 어느 각도에서든 볼 수 있도록 설치된 Melissa Zink의 청동상 'Minister of Words' 가 놓여있다.
캐스케이드의 물소리와 경관은 시원스러움과 함께 사색을 즐길 수 있도록 하고 있으며, 저녁 식사를 위한 공간에는 바비큐를 위한 시설이 조성되었다. 야간에는 노르웨이산 단풍나무잎 사이로 달빛이 스며들어와 고즈넉한 분위기를 연출한다. 설치된 소품들은 사우스햄튼샵에서 구매한 고풍적이고 매력적인 가구들로서 휴게공간을 편안하면서도 오락적인 공간으로 만들고 있으며, 공간에 놓인 의자와 쿠션 및 직물들도 정원의 전체적인 컬러에 맞도록 선택하였다.

ela annals 1

INTERNATIONAL DESIGN COMPETITION IN KOREA

GWANGGYO LAKESIDE PARK | MAGOK WATERFRONT | GANGBUK LARGE PARK | CENTRAL OPEN SPACE IN MULTIFUNCTIONAL ADMINISTRATIVE CITY

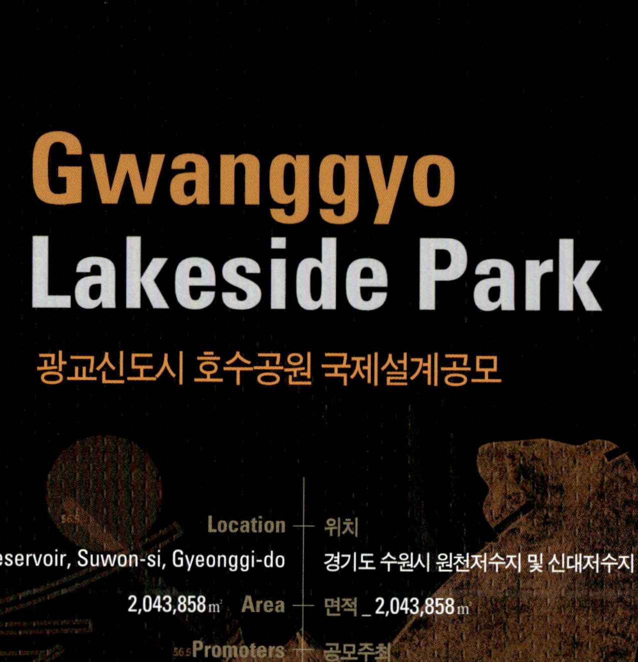

Gwanggyo Lakeside Park

광교신도시 호수공원 국제설계공모

Location	위치
Woncheon & Sindae reservoir, Suwon-si, Gyeonggi-do	경기도 수원시 원천저수지 및 신대저수지 일대
2,043,858㎡ **Area**	면적 _ 2,043,858㎡
Promoters	공모주최
Governments of Gyeonggi-do, Suwon-si, Yongin-si, Gyeonggi Urban Innovation Corporation	경기도, 수원시, 용인시, 경기도시공사
Schedule	공모일정
Announcement of Design Guideline 2008. 8. 13	설계지침제공 2008. 8. 13
Exploration of Field 2008. 8. 25	현장답사 2008. 8. 25
Final Submission 2008. 11. 26	작품제출 2008. 11. 26
Announcement of Winner 2008. 12. 8	당선작발표 2008. 12. 8

In December, 8, 2008, International Design Competition for Gwanggyo Lakeside Park, Suwon, Korea lead by Governments of Gyeonggi-do, Suwon-si, Yongin-si, and Gyeonggi Urban Innovation Corporation has announced the result of awards. To make up a world level Lakeside Park as a focal point of natural and cultural of Gwanggyo new town, 8 companies located at local and foreign which submitted a the request for qualification(RFQ) for participation got a designated invitation.

The prizewinner, 'Urban Reservoir', was awarded by Synwha Consulting, '8Landscapes' as 2nd by Field Operations and Group Han, and 'Ribbon 55 & Community 76' as 3rd prized winner awarded to Seo-Ahn Total Landscape.

The first prizewinner design emphasized that the role of Lakeside Park as a green heart and a center of urban infrastructure should be activated a dynamic life in urban living and suggested a new paradigm of Lakeside Park on a basis motive of unique value and culture memorization, using a word, 'Urban Levee' & 'Dumbung'.

2008년 12월 8일 경기도와 수원시, 용인시, 경기도시공사가 주최한 '광교신도시 호수공원 국제설계공모' 의 심사결과가 발표되었다. 광교신도시의 자연적 · 문화적 거점으로 성장할 수 있는 세계적 수준의 호수공원을 조성하는 것을 목표로 진행된 이번 공모전은 1단계 국제지명초청 방식으로 진행되었으며, 국내 조경설계 공모로는 처음으로 참가의향서를 제출한 업체를 대상으로 초청작가가 선정되었다.

국내외 8개 업체가 지명초청작가로 선정된 가운데 당선작으로는 1등에 신화컨설팅이 제출한 'Urban Reservoir' 가, 2등에 필드 오퍼레이션스와 그룹한이 제출한 '8 Landscapes' 가, 3등에 조경설계 서안이 제출한 'Ribbon 55 & Community 76' 이 선정되었다.

1등작은 도시의 외곽에서 머물렀던 호수를 도시의 중심이자 푸른 심장으로, 더 나아가 도시의 다이나믹한 삶으로 작동하는 호수공원의 역할을 강조해, 땅이 가진 고유한 가치와 문화적 기억을 모티브로 'Urban Levee' 와 '둠벙(물이 마르지 않는 작은 웅덩이)' 이라는 단어를 사용해 호수공원의 새로운 패러다임을 제시하고 있다.

1 **1st Prize** URBAN RESERVOIR
2 **2nd Prize** 8 LANDSCAPES
3 **3rd Prize** RIBBON 55 & COMMUNITY 76
4 **Submitted Works** SERENE VITALITY
TWO LAKES
GWANGGYO LAKESIDE PARK
GWANGGYO PIER LAKESIDE PARK
JOURNEY OF WATER
5 **Jury Comments** _ Niall Kirkwood

※수상작 외 초청작 수록순서는 국내 업체는 가나다 순, 해외업체는 알파벳순 정렬임

| **Invited Designer** |

Synhwa Consulting | Field Operations + Group Han Associates | Seo-Ahn Total Landscape | Balmori Associates, Inc
WRT(Wallace Roberts & Todd, LLC) | GROSS MAX. | Stan Allen Architect | GAONE Design Group

1st Prize

URBAN RESERVOIR

Synhwa Consulting (주)신화컨설팅
Partners _ Park Kee Seong 박기성 + Kim Do Kyong 김도경 + Hong Hyoung Soon 홍형순

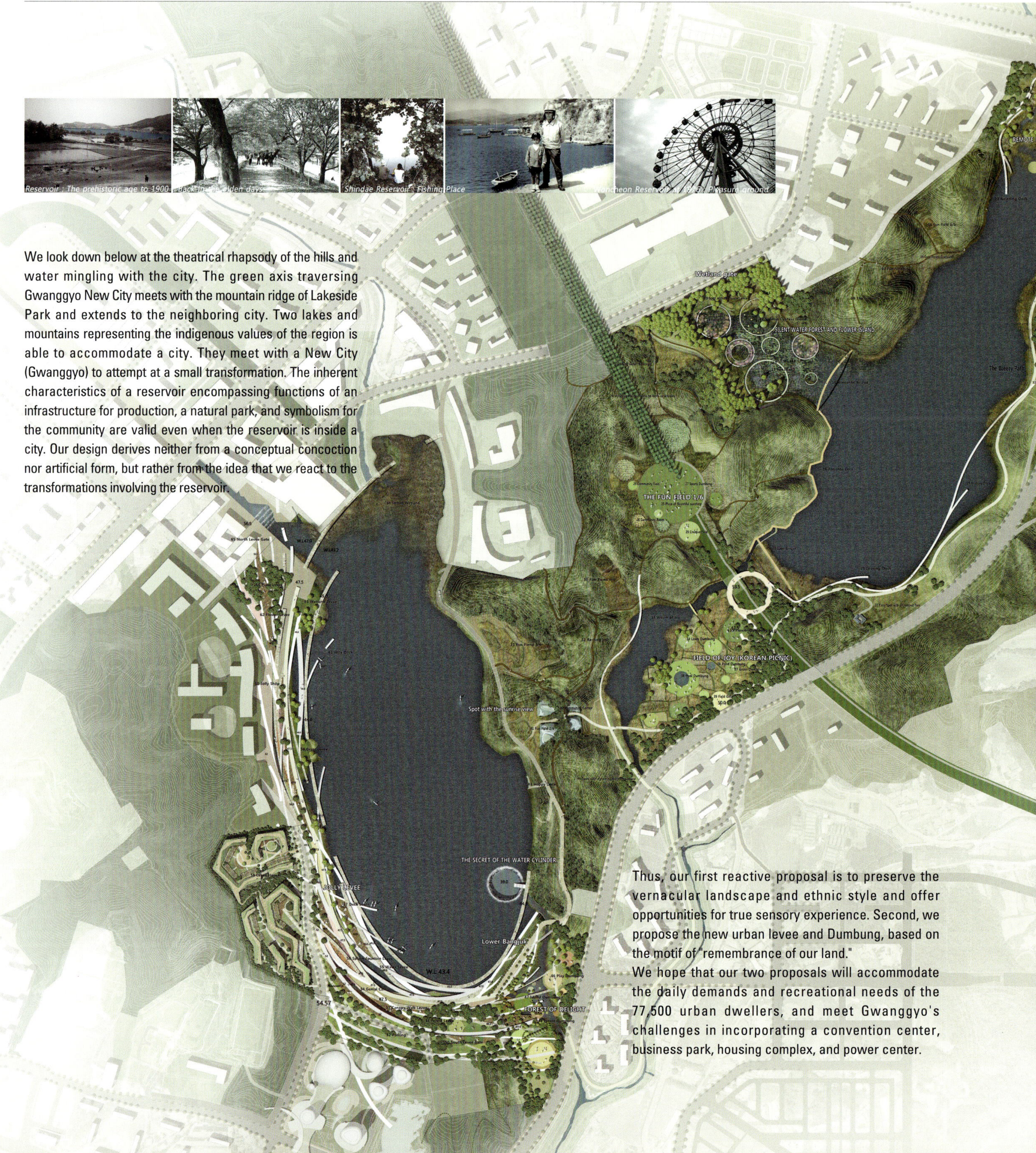

We look down below at the theatrical rhapsody of the hills and water mingling with the city. The green axis traversing Gwanggyo New City meets with the mountain ridge of Lakeside Park and extends to the neighboring city. Two lakes and mountains representing the indigenous values of the region is able to accommodate a city. They meet with a New City (Gwanggyo) to attempt at a small transformation. The inherent characteristics of a reservoir encompassing functions of an infrastructure for production, a natural park, and symbolism for the community are valid even when the reservoir is inside a city. Our design derives neither from a conceptual concoction nor artificial form, but rather from the idea that we react to the transformations involving the reservoir.

Thus, our first reactive proposal is to preserve the vernacular landscape and ethnic style and offer opportunities for true sensory experience. Second, we propose the new urban levee and Dumbung, based on the motif of "remembrance of our land."
We hope that our two proposals will accommodate the daily demands and recreational needs of the 77,500 urban dwellers, and meet Gwanggyo's challenges in incorporating a convention center, business park, housing complex, and power center.

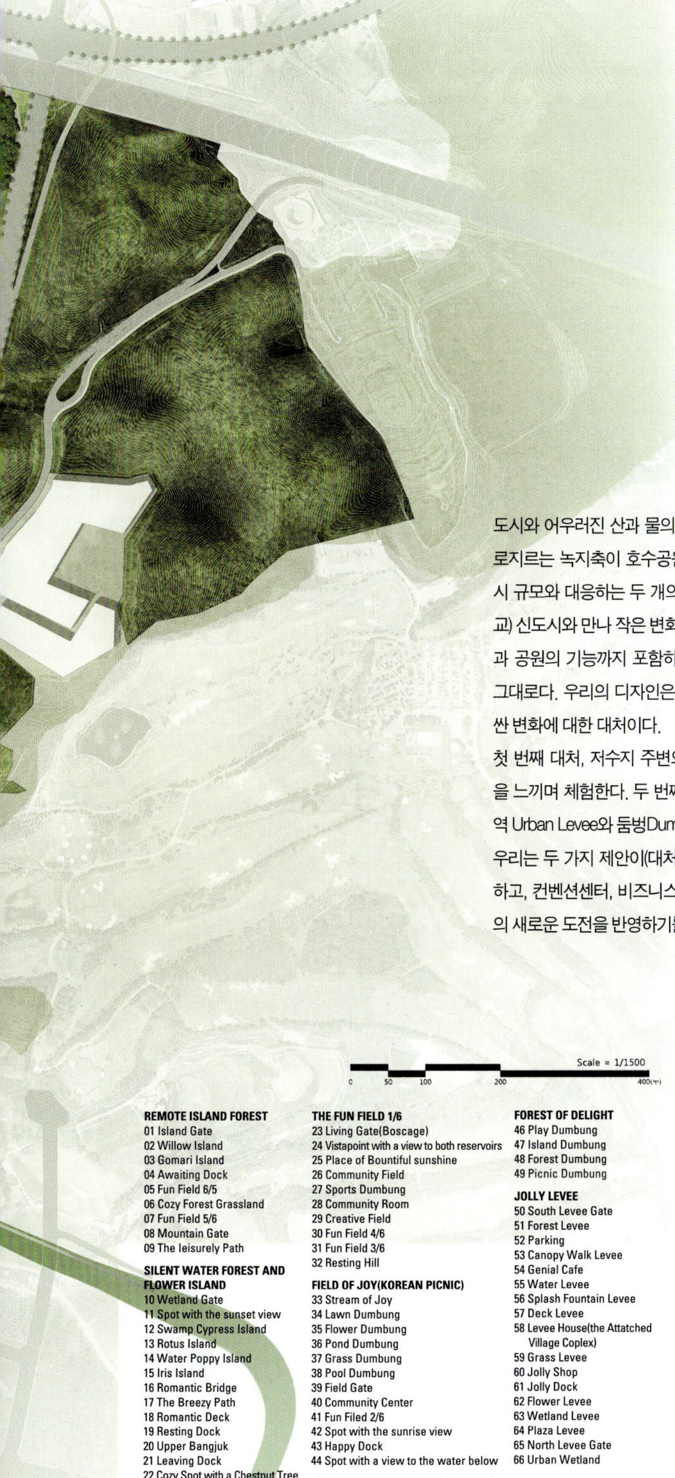

도시와 어우러진 산과 물의 극적인 랩소디를 내려다본다. 광교 도시를 가로지르는 녹지축이 호수공원의 능선과 만나고 인근 도시로 이어진다. 도시 규모와 대응하는 두 개의 호수와 산은 지역의 고유성을 유지한 채, (광교) 신도시와 만나 작은 변화를 시도한다. 생산적인 인프라, 공동체의 역할과 공원의 기능까지 포함하는 저수지의 본질적인 의미는 도시 안에서도 그대로다. 우리의 디자인은 고안된 의미나 형태가 아니라 저수지를 둘러싼 변화에 대한 대처이다.

첫 번째 대처, 저수지 주변의 토속적 풍토 경관을 지키고, 구체적인 감각을 느끼며 체험한다. 두 번째 대처, 땅의 기억을 모티브로 하여 새로운 영역 Urban Levee와 둠벙Dumbung을 제안한다.

우리는 두 가지 제안이(대처가) 77,500명의 도시민의 일상과 여가를 수용하고, 컨벤션센터, 비즈니스파크, 연립주택단지, 파워센터를 포함한 광교의 새로운 도전을 반영하기를 희망한다.

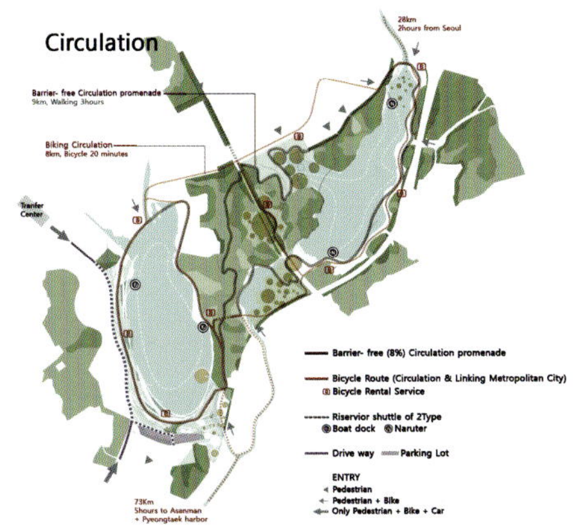

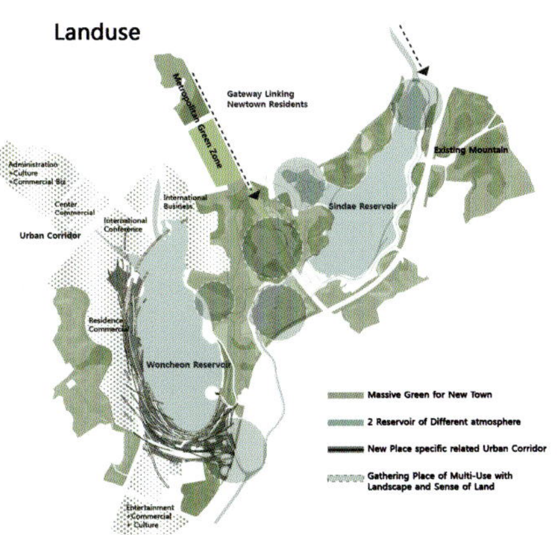

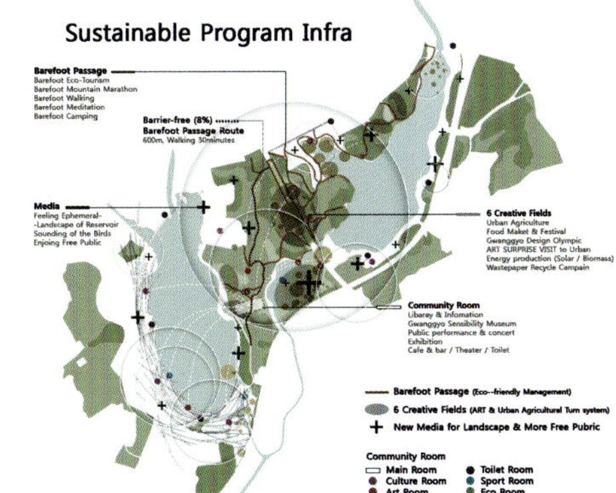

Vernacular Landscape & Ethnic Style

Preserve the vernacular landscape and ethnic style, thereby offering opportunities for true sensory experience...

The low-lying hills and farmland covering 11,282,521㎡ will transform into an urban town. Only the hills touching the reservoir and its mirror image reflected off of the water will hold the memories of the surrounding land. Our idea of a vernacular landscape is not a physical piece of land subject to development. It is rather that something which elicits true sensory experience through contrasts between light and dark and cold and warm, colors, textures, sounds, metaphors, reflections, contemplations, and meditations.

저수지를 둘러싼 11,282,521㎡의 나지막한 산과 농경지가 도시로 바뀐다. 이제 저수지와 맞닿은 산만이 물에 비친 실루엣과 함께 이 땅의 기억을 말한다. 우리의 토속경관은 개발의 대상이 되는 물리적 대지가 아니라, 밝고 어두움, 차갑고 따뜻함, 색과 질감, 소리, 은유, 반영, 관조와 명상으로 구체적인 감각을 느끼며 체험되는 대상이다.

광교산과 연결되는 녹지축과 원천저수지 동쪽, 신대저수지 주변의 산은 토속경관으로 보존되며 이곳에 산과 물이 주는 소리와 색, 질감과 향기, 빛과 바람의 구체적인 체험을 짓는다. 감성의 자극은 해독을 원하는 도시민을 씻겨 줄 것이다.

Ideas from the Indigenous Value and Cultural Memories of the Region

Proposing a new urban Levee and Dumbung

The new areas, derived from the motif reflecting the indigenous values and cultural memories of the region, is where nature and city meet and humans and all other forms of life coexist. They are the urban levee and Dumbung, both extensions of the local values and each serving the needs of the city. The overlapping and/or sharply contrasting areas of the new and old instill renewed excitement. The urban levee and Dumbung give new meaning to the city. We propose to create the urban levee as a beach and passageway to the city and the Dumbung as a new type of resting place for the urban dwellers.

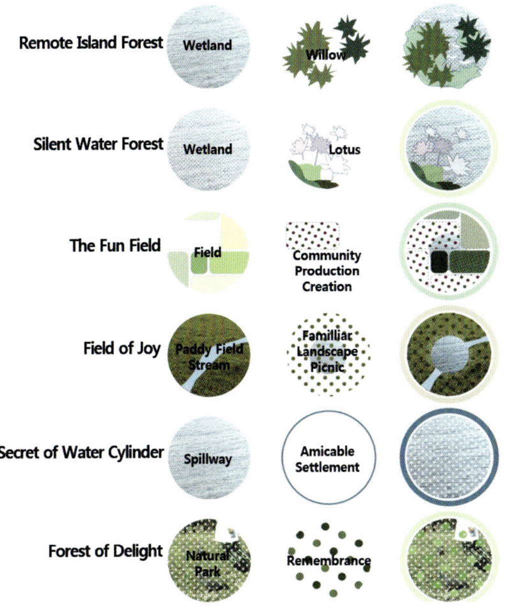

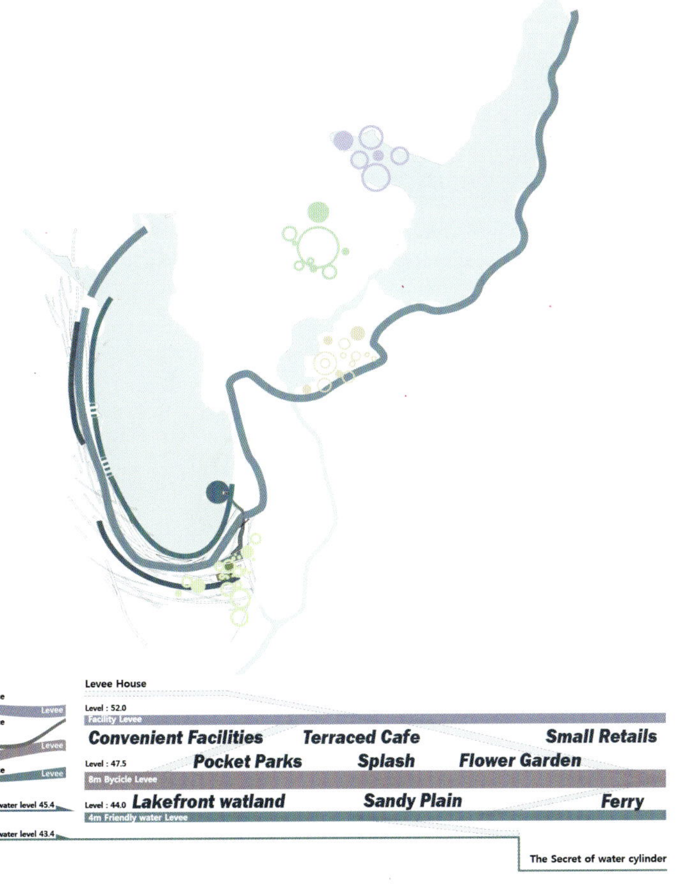

지역의 고유성과 문화적 기억에서 출발한 새로운 영역은 자연과 도시의 접점이며, 모든 생명과 인간이 공존하는 장이다. 새로운 영역인 Urban Levee와 둠벙은 지역의 가치와 연장선에 있으며, 도시의 요구를 수용한다. Levee와 둠벙은 기존 경관과 중첩, 대조됨으로써 환기를 주며, 도시안에서 새로운 의미들로 발견된다. 도시와 연결되는 통로인 Urban Levee와 새로운 도시민을 받아줄 수 있는 장소 Dumbung을 제안한다.

Emotional Narratives

Pleasures of Emotional Impression

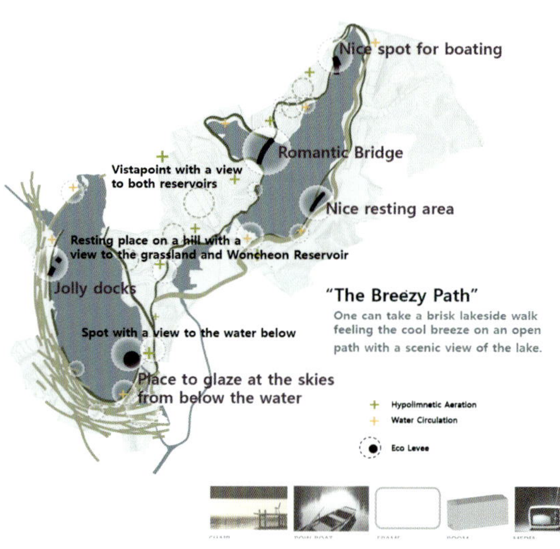

Rhapsodies of Ephemeral Landscape

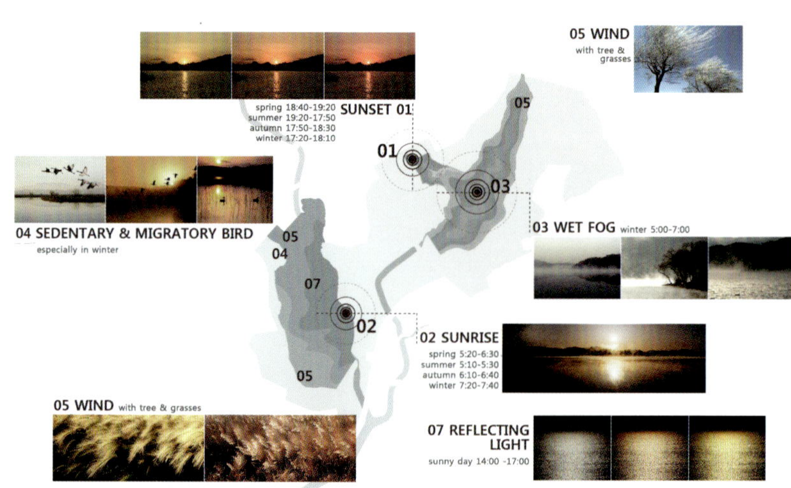

Emotional Matrix

Management Narrtives

This provides the opportunity to get closer to nature as visitors will voluntarily participate in the ecological management of the Park.

he six creative fields in the middle of a silent forest temporarily transform into a bustling arts park. Acts of creation using nature's time and space as the material combined with fun things that people do will complete the Park's arts program that we are proposing.

Campaigns for energy conservation, use of solar energy, and use of recycled paper are all part of the "Less Co2" Project.

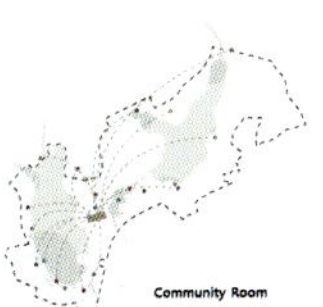

Urban Levee

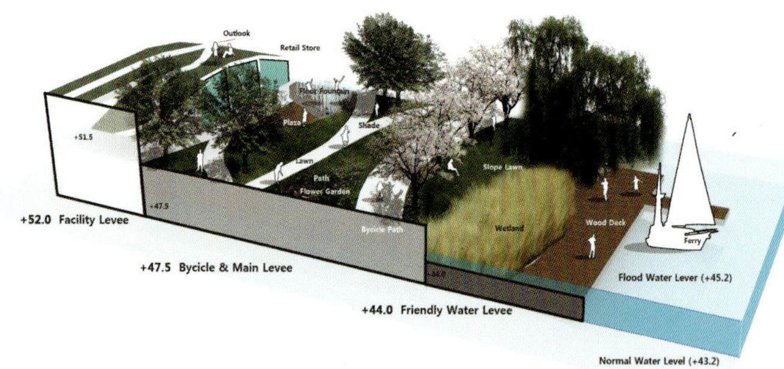

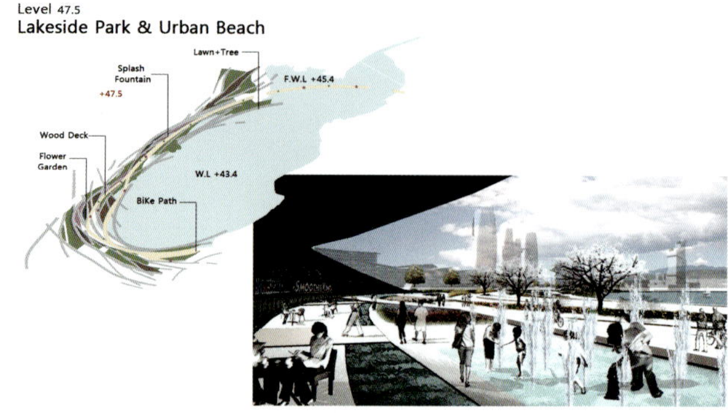

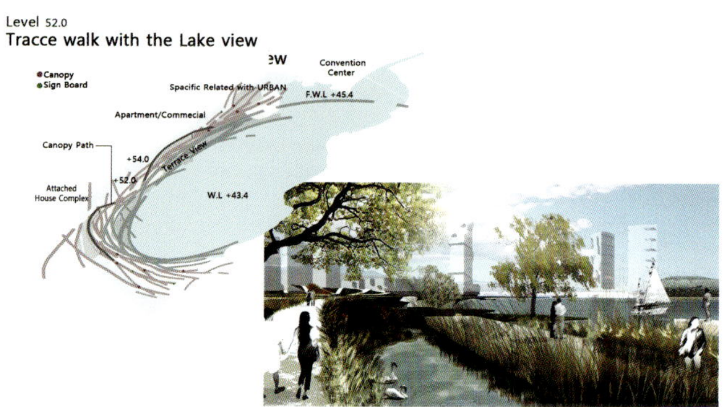

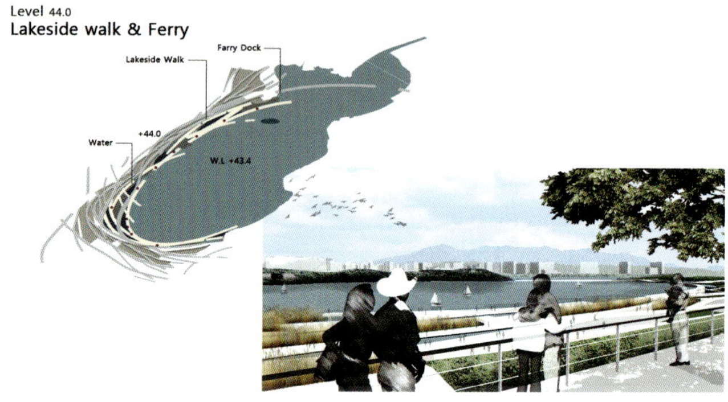

Ecological Narratives

Barefoot Mountain

맨발로 산을 걸으며 녹음이 짙은 곳에서 오감을 체험하다

생태자연도에 언급된 4가지 유형의 산림의 특성에 맞는 복원방법을 제시하고, 등급에 따른 휴넌제를 적용한 이용 및 보존 시스템을 마련한다. SECONDARY FOREST와 ARTIFICIAL PLANTING FOREST의 이용과 보존시기를 정하여 내부 산책로들을 부분별로 사용한다. 보다 건강해진 숲속에서 맨발로 흙을 밟고 걸으며 오감을 체험한다.

Existing Forest

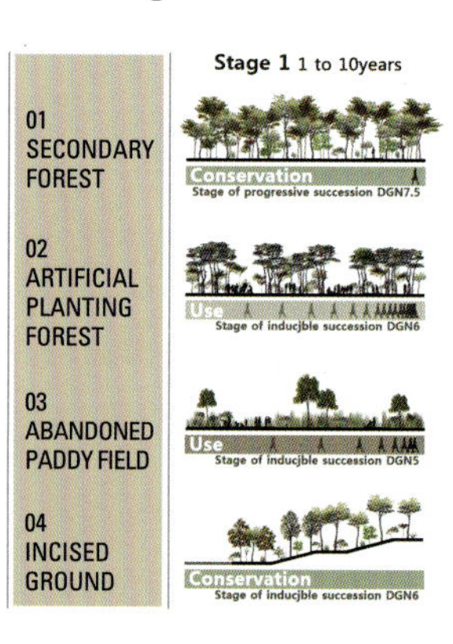

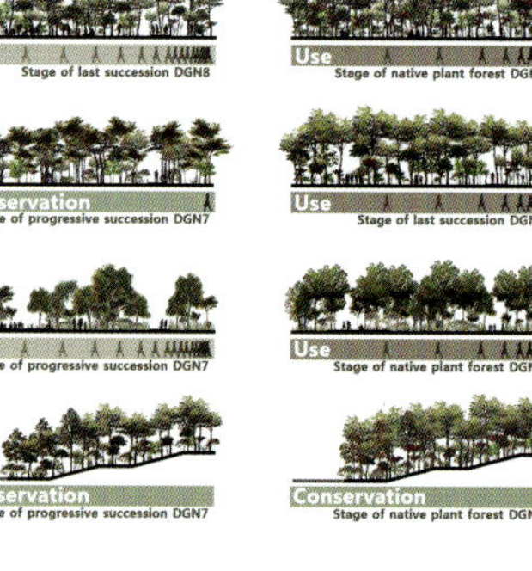

	Stage 1 1 to 10years	Stage 2 10 to 20years	Stage 3 20 to 30years
01 SECONDARY FOREST	Conservation Stage of progressive succession DGN7.5	Use Stage of last succession DGN8	Use Stage of native plant forest DGN9
02 ARTIFICIAL PLANTING FOREST	Use Stage of inducjble succession DGN6	Conservation Stage of progressive succession DGN7	Use Stage of last succession DGN8
03 ABANDONED PADDY FIELD	Use Stage of inducjble succession DGN5	Use Stage of progressive succession DGN7	Use Stage of native plant forest DGN8
04 INCISED GROUND	Conservation Stage of inducjble succession DGN6	Conservation Stage of progressive succession DGN7	Conservation Stage of native plant forest DGN8

Water Mate

Water quality is most strongly affected by the type of relationship it shares with neighboring people, streams, mountains, and cities. Only when the relationship is two-way and friendly will nature maintain a balance and recover the ability to purify itself.

물은 인접한 사람, 하천, 산, 도시와 이웃하며 어떤 관계를 맺고 있는가가 수질에 가장 큰 영향을 받는다. 서로 주고 받는 유익한 친구사이가 될 때 자연은 균형을 찾고 자정능력을 회복한다.

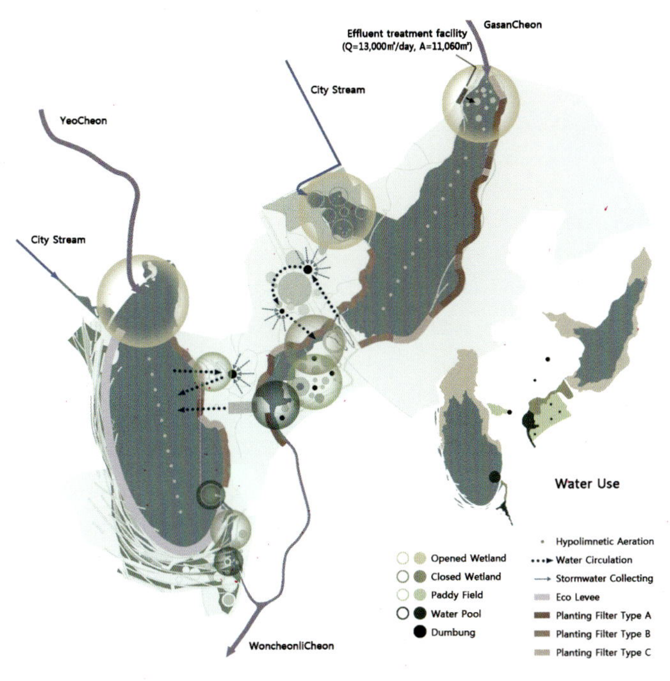

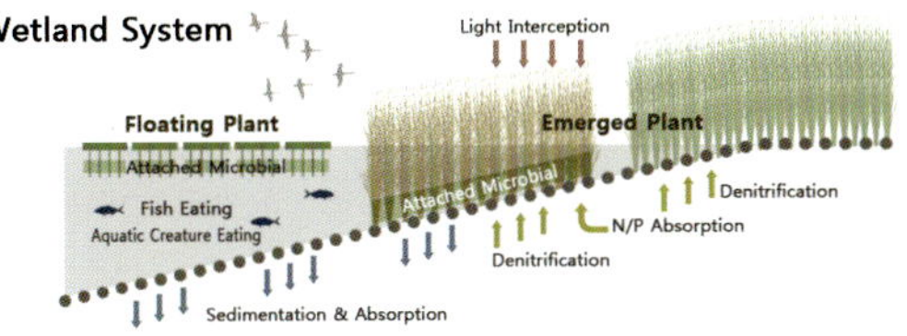

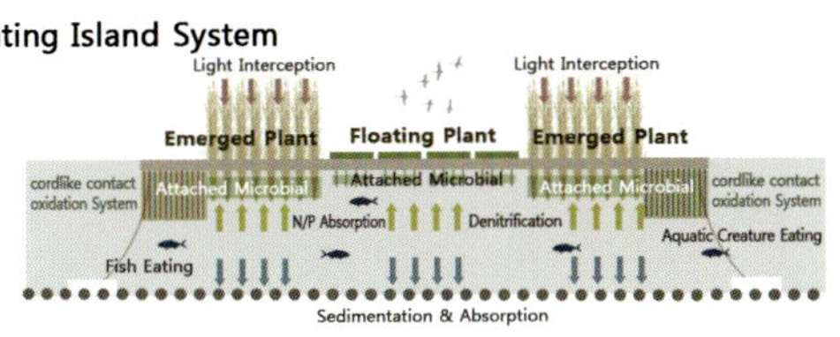

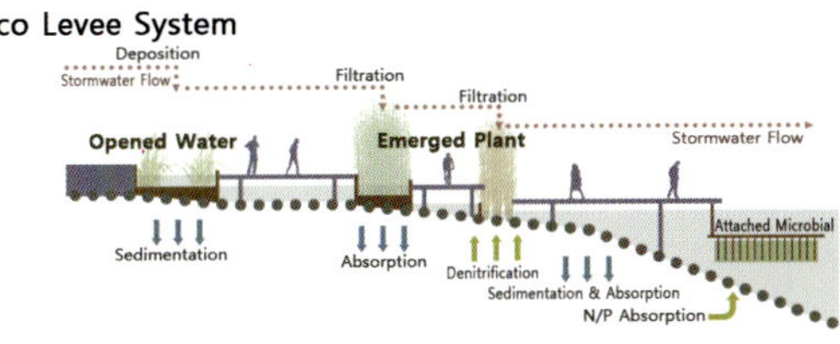

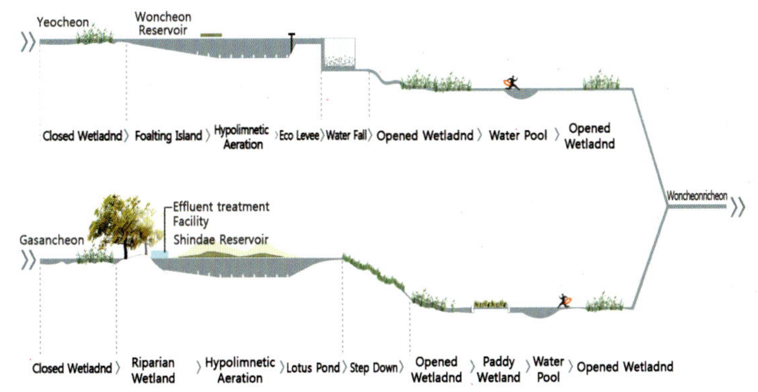

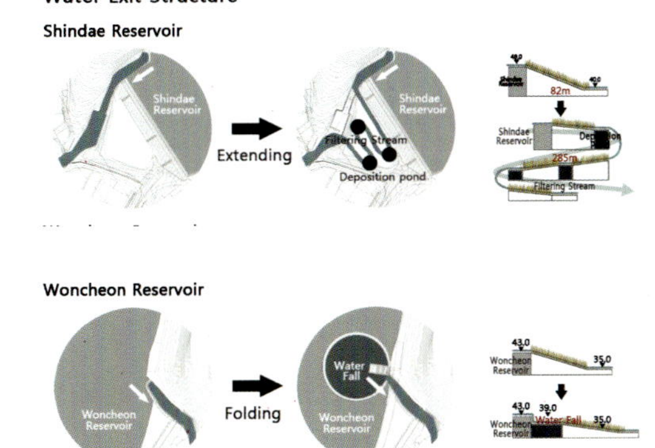

Planting Filteration

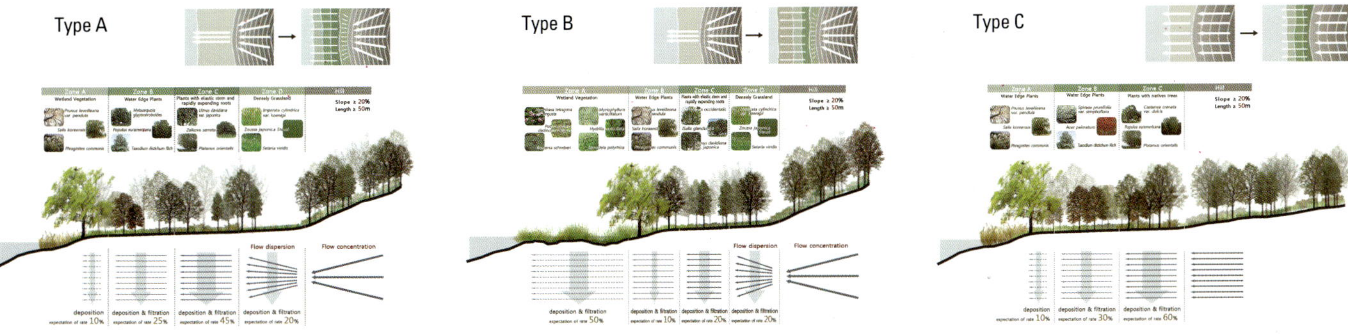

Remote Island Forest

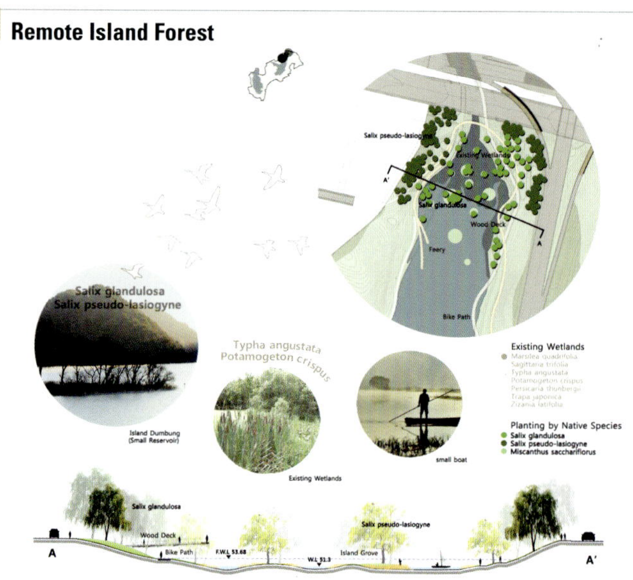

Silent Water Forest and Flower Island

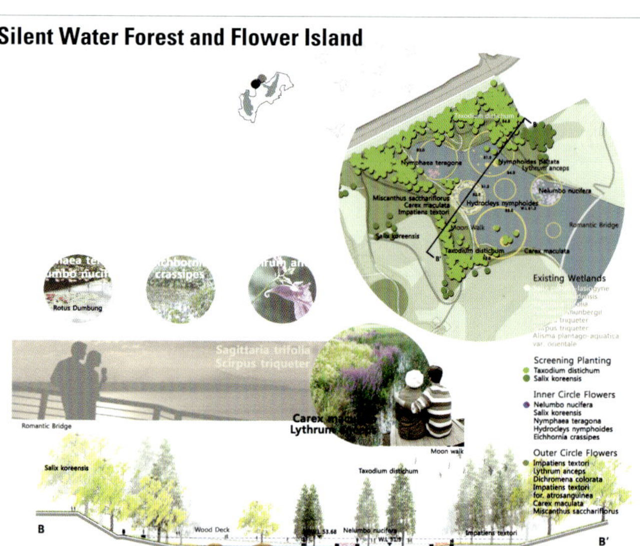

Forest of Delight

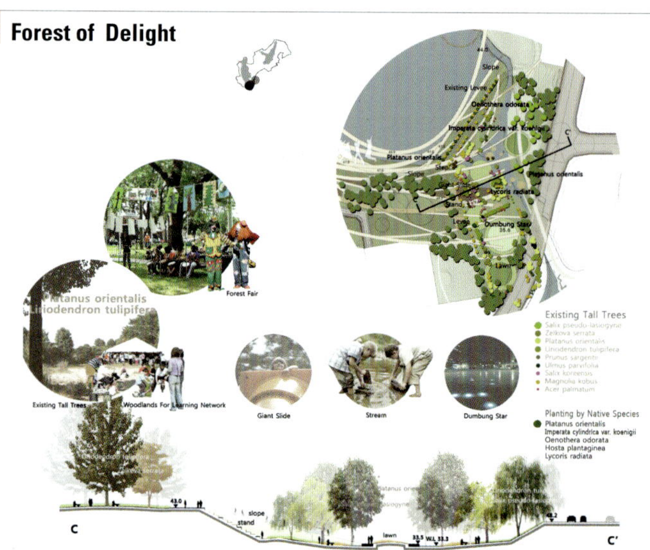

Field of Joy(Korean picnic)

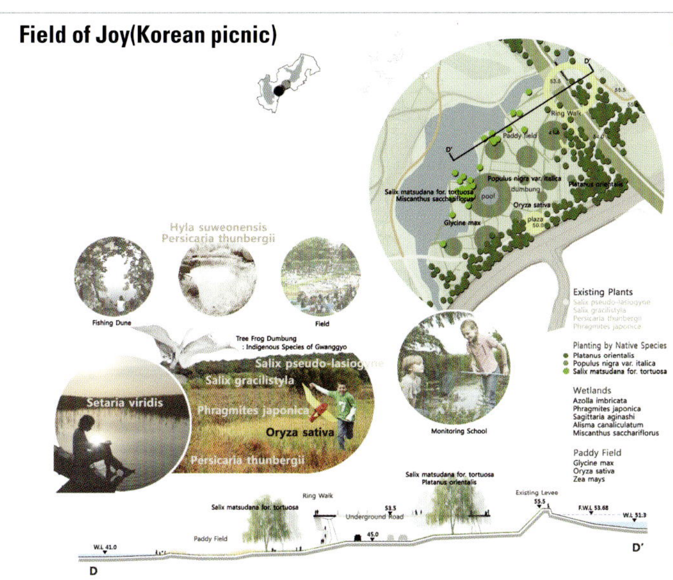

The Fun Field

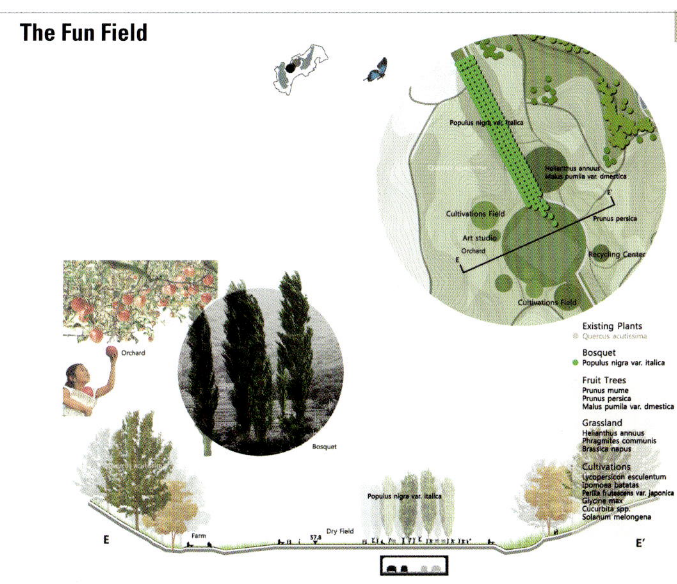

Secret of the Water Cylinder

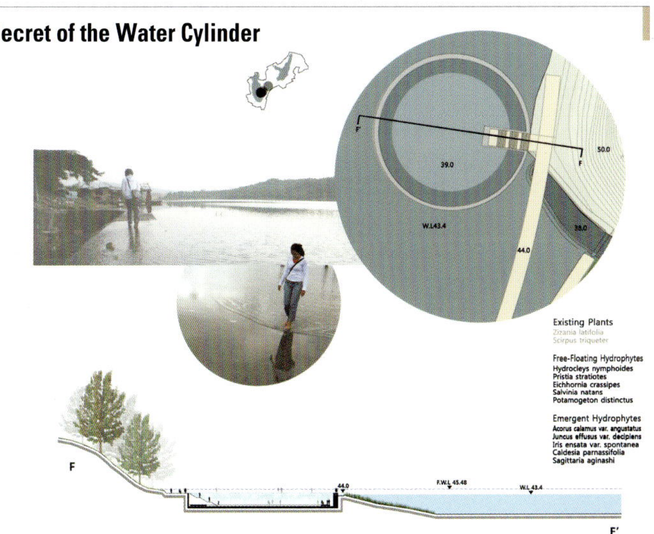

2nd Prize

8 LANDSCAPES

Field Operations + Group Han Associates (주)그룹한 어소시에이트
Partners _ HaeAhn Architecture 해안건축 + H Associates + Buro Happold

8 LANDSCAPES

field operations + group han

with
haeahn architecture
h associates
& buro happold

Experiental Beauty

Each of these landscapes offers a unique and special experience for the future park visitors, connecting them to nature, time and place. Instead of an over-programmed theme park, the design approach here seeks to conserve and re-present the most powerful and dramatic aspects of Lakeside nature, creating paths, trails, boardwalks, terraces, gardens, scenic vistas and overlooks, lakeside beaches, wetlands and woodland settings - all for people to stroll, jog, cycle, congregate and access the many diverse places in the park.

8경에 해당하는 각각의 경관은, 자연과 시공간을 연계하여, 장차 호수공원을 찾게 될 이용자들에게 독특하고 특별한 경험을 제공할 것이다. 마치 테마파크처럼 과도한 프로그램이 도입된 공원을 지양하고, 수변의 가장 강력하고 드라마틱한 특징을 보존하고 재현하는 것에서부터 디자인을 시작하고자 한다. 그것은 호숫가의 자연, 산책로와 오솔길, 테라스와 정원, 아름다운 풍경과 근사한 전망, 호숫가와 같은 것이며, 이용자들은 공원 내의 아주 다양한 장소에서 산책을 즐기거나 조깅을 하거나 자전거를 타거나 함께 모여서 공원을 즐기게 될 것이다.

Suwon Eight Spectacle

SNOW ON GWANGGYO MOUNTAIN

FOG IN PALDAL MOUNTAIN

WILLOWS ON THE BANK

AZALEA MOUNTAIN WITH PINES AND BIRDS SINGING

LOTUS ON NORTH POND

SUNSET ON WEST LAKE

WATER CASCADE

NIGHT SKY (WAITING FOR MOON RISING AT DRAGON POND)

Lakeside Edges & Places

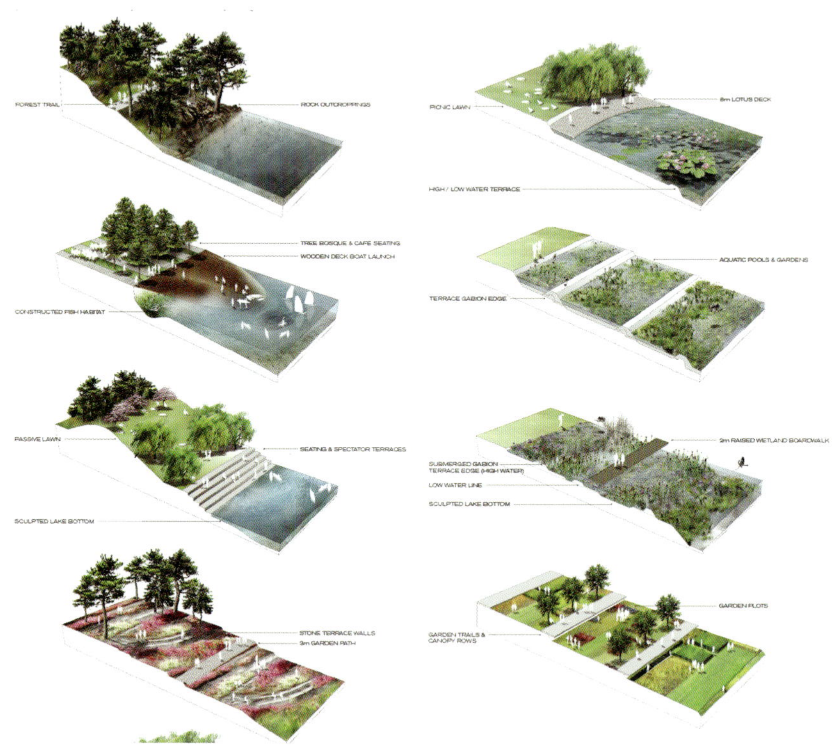

Eight Landscapes

MOUNTAIN FOREST

LOTUS POND

WONCHEON & SINDAE LAKES

SPRINGWATER TERRACES

WILLOW PARK

WATER CASCADES & CATTAIL WETLANDS

AZALEA GARDEN TERRACES

GARDENS & FIELDS

Four Strategies

First, the Mountain Forest

Our design proposes a massive reforestation program, beginning with clearing of underbrush and removal of invasive and dead plants, and continuing with underplanting of more diverse species and age structures. Oaks, Maple and Beech, with some Pine, will showcase the south-facing slopes, while more Pine and Hemlock will characterize the shadier, wetter north-facing slopes. This will allow for an interesting effect throughout the seasons, while highlighting north-south orientation within the site.

덤불가지들을 정리하고 병충해를 입었거나 고사한 수목을 제거하고, 다양한 종류의 하부식재와 다년초 식재를 통해, 다시 풍성한 숲을 가꾸고자 한다. 남측 경사면에는 떡갈나무, 단풍나무, 너도밤나무, 소나무 등을, 그늘이 많고 습한 북측 경사면에는 소나무와 헴록 등을 식재할 계획이다. 이런 시도는 대상지 내의 남북방향에서 계절에 따라 흥미로운 효과를 줄 것이다.

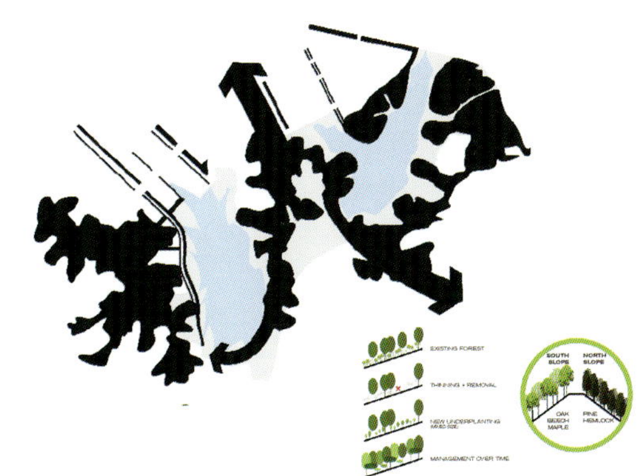

Second, the Lakes

The lakes will first need to be dredged of all contaminating material, and upstream sources will need to be controlled in terms of watershed water quality. Once this is done, we prose various measures be constructed as part of the park to ensure continual water quality improvement.

우선 수질정화를 위해 호수 바닥의 더러운 오염물질을 준설하고, 호수로 흘러들어오는 상류의 수원을 관리해야 한다. 이러한 작업이 마무리되면, 지속적으로 수질 관리가 유지될 수 있도록 공원 내에 다양한 방법을 제안할 것이다.

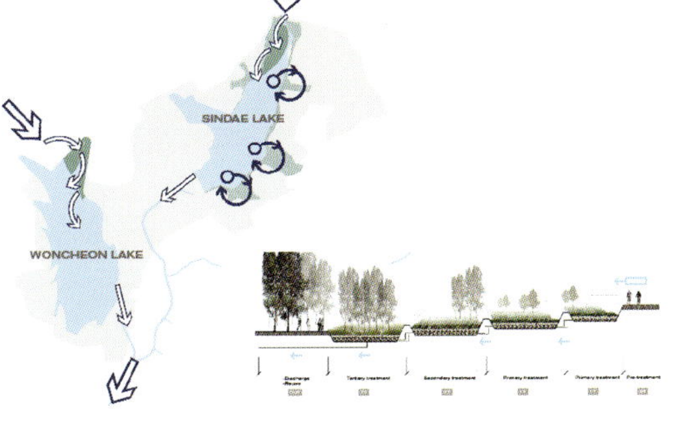

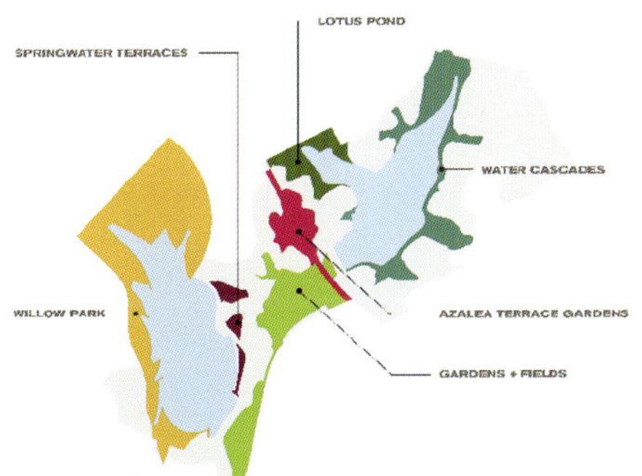

Third, the Lakeside Landscapes

In addition to the mountain forest and the two lakes, six other landscapes define the Park. These are all lakeside landscapes, the places where most people can stroll around and gain access to the lakes themselves.

숲과 두 개의 호수, 그리고 6경(6landscapes)이 공원의 특징을 단적으로 나타내준다. 이것들은 호숫가 풍경과 여러 장소들로, 많은 사람들이 호숫가를 찾아 둘러볼 수 있는 곳에 위치해 있다.

Fourth, the Circulation

Once inside the park, the visitor can access over 30 kilometers of interconnected paths and trails without having to cross any vehicular roads. Some of these paths can be specially designated for specific uses, such as cycling and roller-blading.

일단 공원에 들어서면, 대략 30킬로미터가 넘는 동선을 접할 수 있다. 이 사람을 위한 동선은, 차도의 간섭을 받지 않도록 꾸며진다. 그리고 이 동선 중의 일부는 자전거나 인라인 스케이트 이용자를 위해 조성될 것이다.

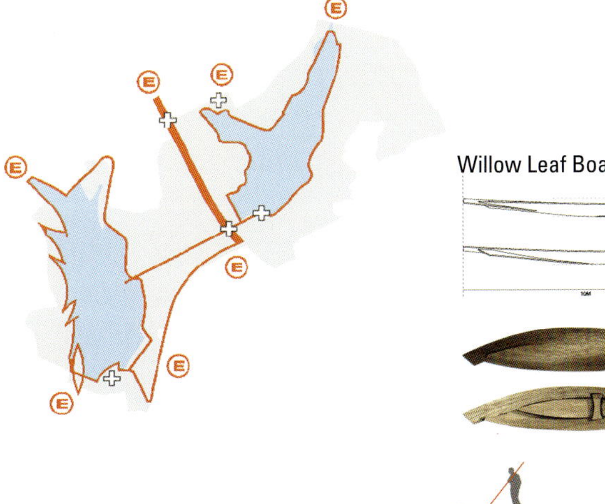

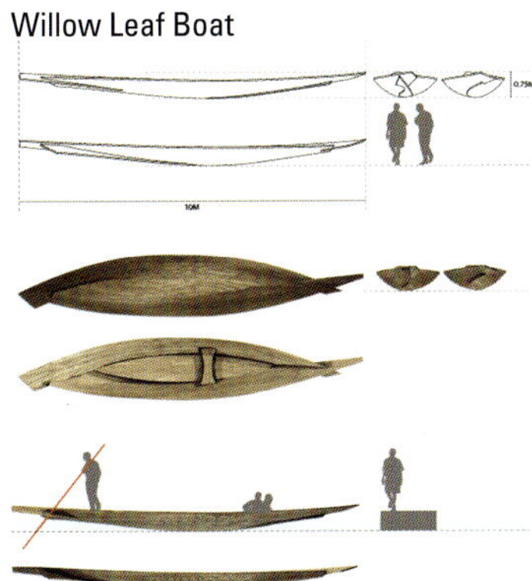

GWANGGYO NEW CITY
WILLOW PARK AND WATERFRONT
ZONE C
15 minute walk
6 minute kayak paddle
GRAND FOUNTAIN TERRACES
SEATING TERRACES
RECREATION LAWN
WILLOW PARK
CHERRY TREE GROVE
AMPHITHEATER
WADING TERRACE
CHERRY GROVE
SHADE PROMENADE
PUBLIC BOAT LAUNCH
SPECTATOR LAWN

LAKESIDES
Public access to the water's edge is maximized throughout the design, with many novel ways to approach, access and experience the water.
NIGHT-TIME WATERFRONT
SHADED PROMENADES
NEW BEACHFRONT
KAYAK & BOAT LANDINGS
LAKESIDE GARDENS
WETLANDS & BOARDWALKS

Zone C Development

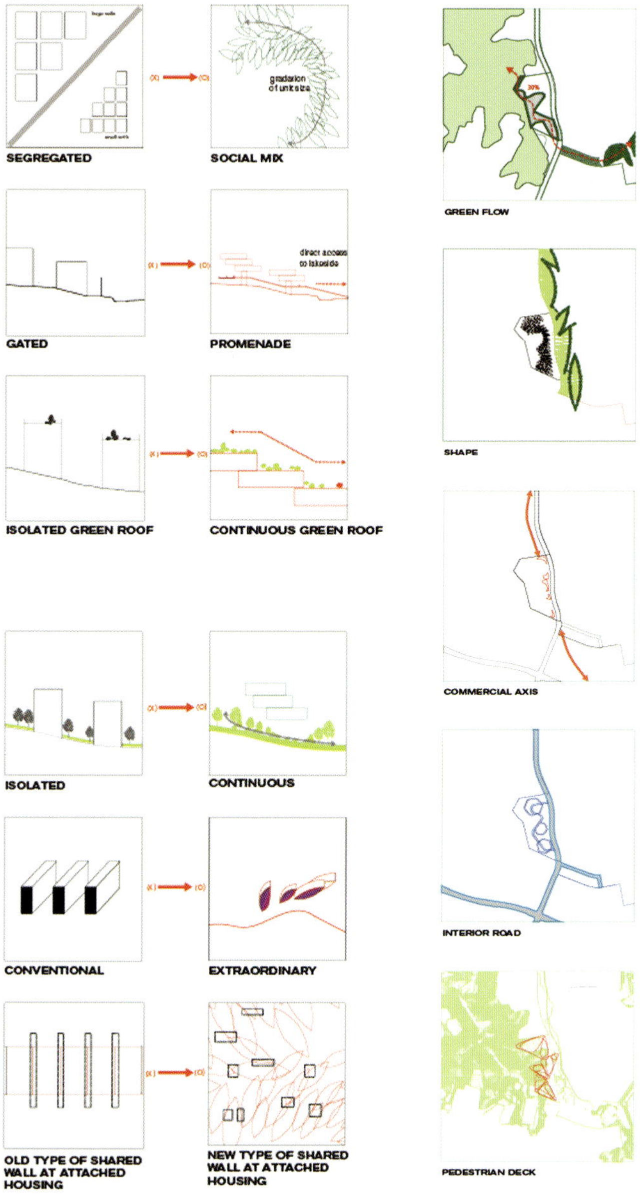

SEGREGATED
SOCIAL MIX
gradation of unit size
GATED
PROMENADE
direct access to lakeside
ISOLATED GREEN ROOF
CONTINUOUS GREEN ROOF
ISOLATED
CONTINUOUS
CONVENTIONAL
EXTRAORDINARY
OLD TYPE OF SHARED WALL AT ATTACHED HOUSING
NEW TYPE OF SHARED WALL AT ATTACHED HOUSING
GREEN FLOW
SHAPE
COMMERCIAL AXIS
INTERIOR ROAD
PEDESTRIAN DECK

Massing Concepts

PROGRAM
valley
ridge
valley
ridge
valley
FITTING
VIEW OPTIMIZATION

MOUNTAIN FOREST

WILLOW PARK AND WATER

AZALEA GARDEN TERRACES

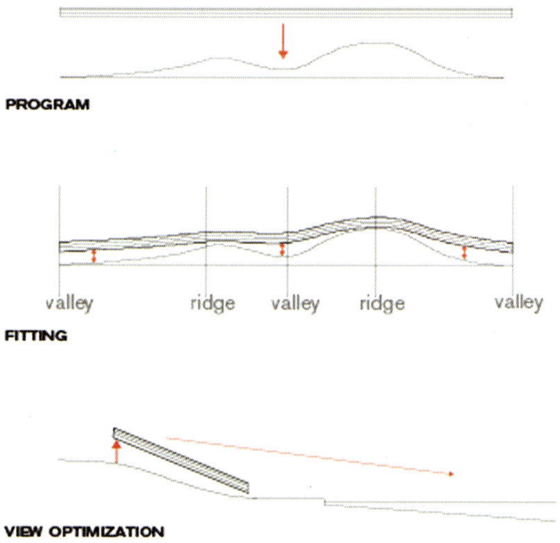

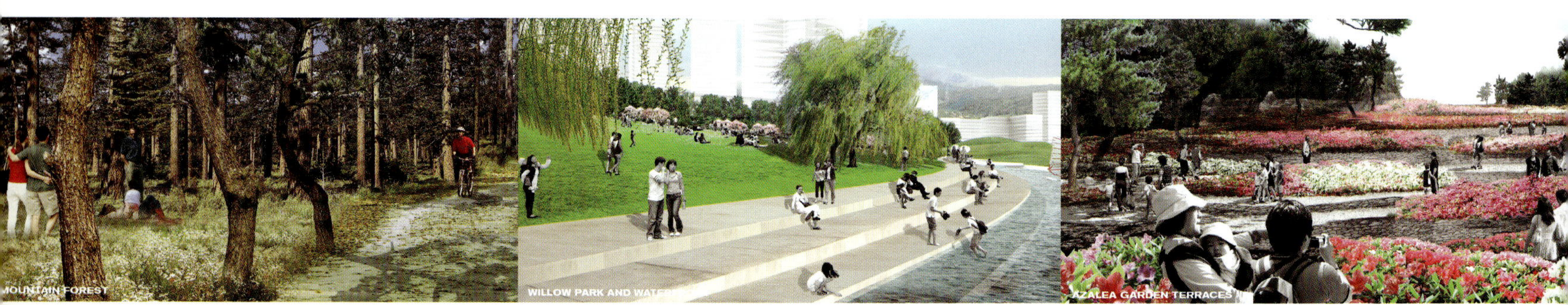

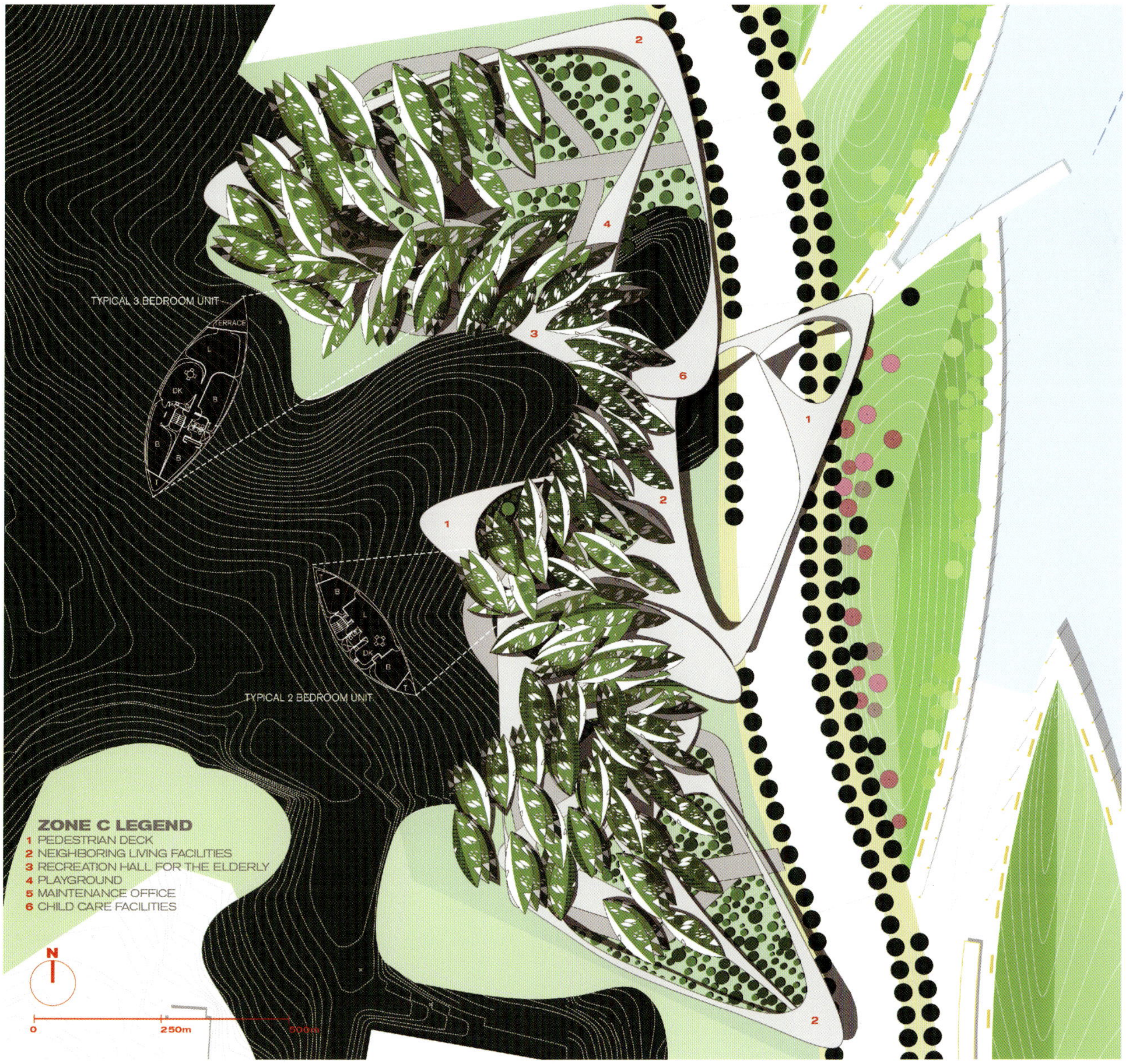

TYPICAL 3 BEDROOM UNIT
TERRACE
TYPICAL 2 BEDROOM UNIT
ZONE C LEGEND
1 PEDESTRIAN DECK
2 NEIGHBORING LIVING FACILITIES
3 RECREATION HALL FOR THE ELDERLY
4 PLAYGROUND
5 MAINTENANCE OFFICE
6 CHILD CARE FACILITIES
N
0
250m
500m

AT WONCHEON LAKE

SPRINGWATER TERRACES

LOTUS GARDEN

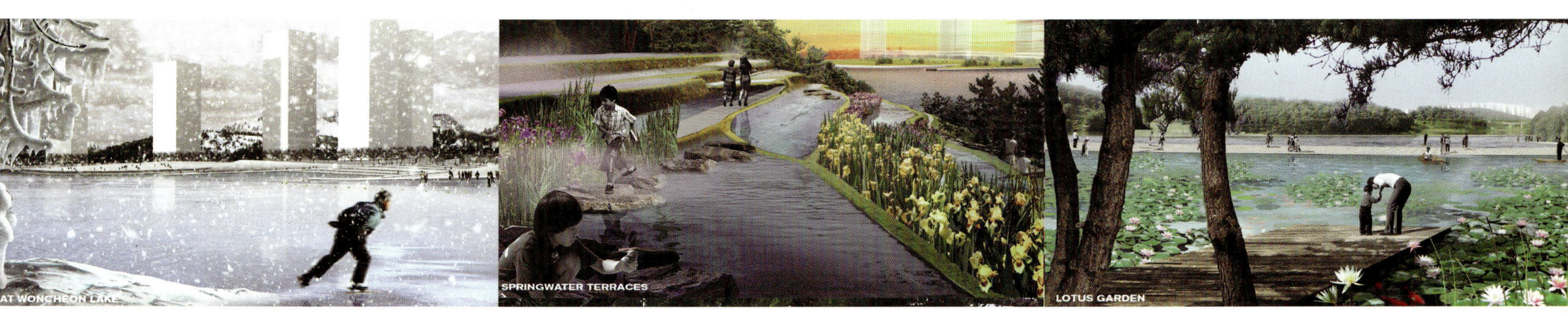

3rd Prize

RIBBON 55 & COMMUNITY 76

SeoAhn Total Landscape조경설계 서안(주) | **Partners** _IBI+D.G.OZ+Gartenlabor

Two Lakes. One Park.

This is the story of the reunion of two lakes, Shindae and Woncheon, who were born of a small river channel many years ago. The two lakes, were like twins, separated at birth by a mountain which kept them apart for many, many decades. The two reservoirs grew to become two lakes, and each one developed its own character. The first lake, Shindae, preferred the country - it thrived on being surrounded by nature. This lake flourished with natural beauty, changing with the seasons, alive with the rugged character of the wilderness. Yet there were times when Shindae desired the culture, the diversity, the animation and the excitement of the city that it knew was located on the other side of the mountain. The second lake, Wonchon, loved the city. It was filled with people, with smells and sights, and sounds that contributed to the hustle and bustle and made the city come alive in the morning and remain lively until late in the evening. The sounds were like music, celebrating the life and culture of the people. Yet, there were also times when Wonchon longed for the peace and solitude of nature - the quietude that existed on the other side of the mountain. This is the story of the reunion of two lakes. The Gwanggyo Lakeside Park has brought together the two lakes into one park. It is a park that strikes a perfect balance between nature and urbanity.

이 이야기는 오래전 작은 물줄기로부터 시작된 신대호와 원천호의 결합에 대한 것이다. 두 개의 호수는 수십년동안 산의 지세에 의해 나뉘어 있었으나, 마치 쌍둥이처럼 닮아 있다. 두 개의 저수지는 두 개의 호수로 변화해가면서, 각각의 특징을 지니게 되었다. 첫 번째 호수인 신대호는 풍부한 자연환경에 둘러쌓여 전원적인 특징을 갖게 되었고, 원생적인 특징이 남아있어, 자연의 아름다움, 사계절에 따른 다채로운 변화가 풍부한 곳이다. 그러나 신대호는 지리적 위치 때문에 도시의 문화, 다양성, 오락성과 활기를 품기에는 더 많은 시간이 필요해보였었다. 두 번째 호수인 원천호는 신대호와 달리 도시적인 특성을 보다 많이 갖고 있다. 아침부터 사람들의 활기와 거기에 스며있는 빛과 소리와 향기가 가득하고, 저녁 늦게까지 그런 활기가 유지되는 곳이다. 마치 음악처럼 사람들의 삶과 문화를 축복하는 곳이라 할 수 있는 것이다. 그렇지만 원천호는 신대호가 도시적인 특성이 부족한 것처럼, 지리적 위치로 인해, 자연의 고요함을 맛보기엔 어려움이 있었다. 우리는 두 가지 다른 성격을 가진 호수를 하나의 공원으로 결합시켜, 자연과 도시 사이의 완벽한 균형을 추구하고자 한다.

Three Strategies

Intergrating the Ecosystem — Environment Strategy

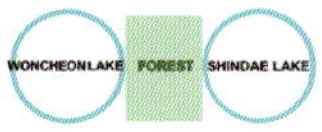

WONCHEON LAKE FOREST SHINDAE LAKE

task 1. HYDROLOGY - showcase water purification system
task 2. VEGETATION - enhance forest and lake edge
task 3. ECOSYSTEM - strengthen connection between hydrology & vegetation

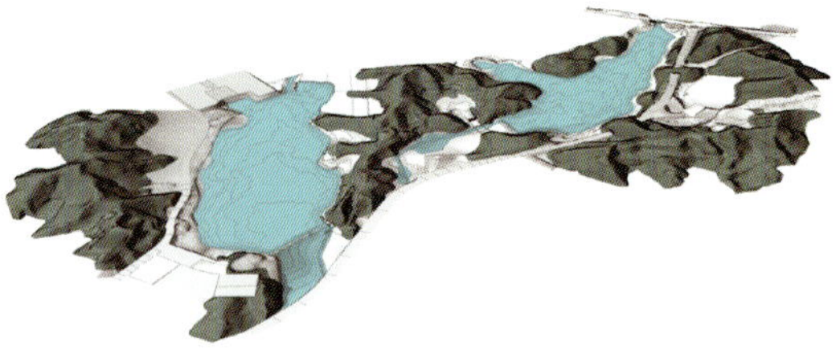

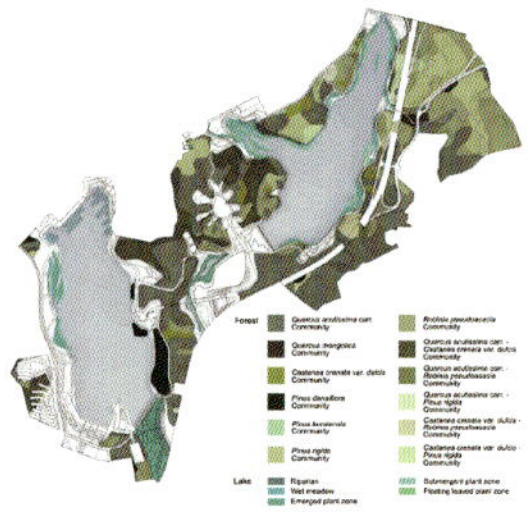

Hydrology

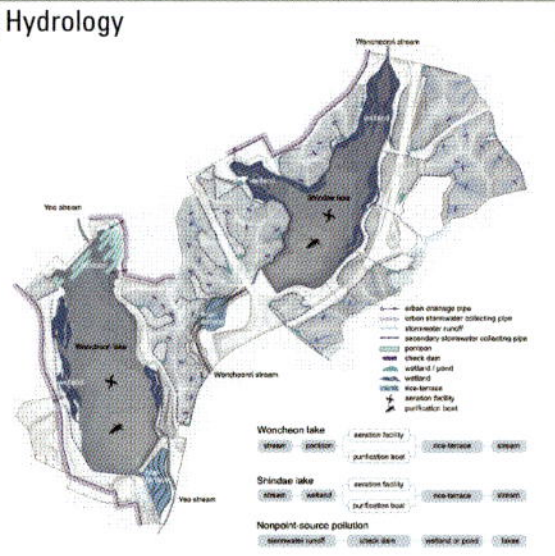

Vegetation

Ecosystem

Ribbon 55 - Environment Strategy

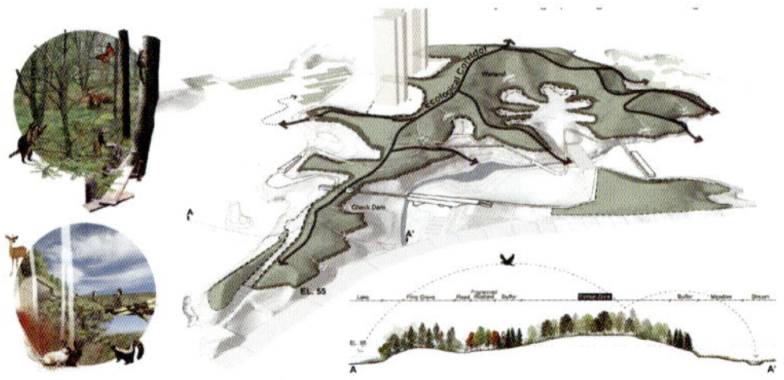

Linking the Lakes — Program Strategy

THIRD NATURE CULTURE 55 FIRST NATURE

task 1. Define the spaces and character
task 2. Combine and Streamline programs
task 3. Link spaces and programs

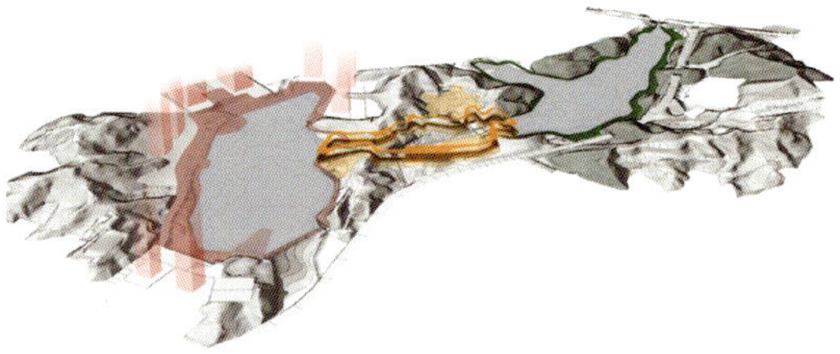

Define space and character

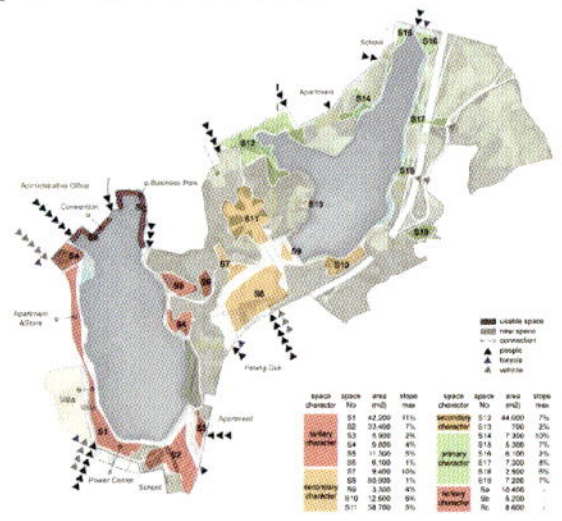

Combine and Streamline Programs

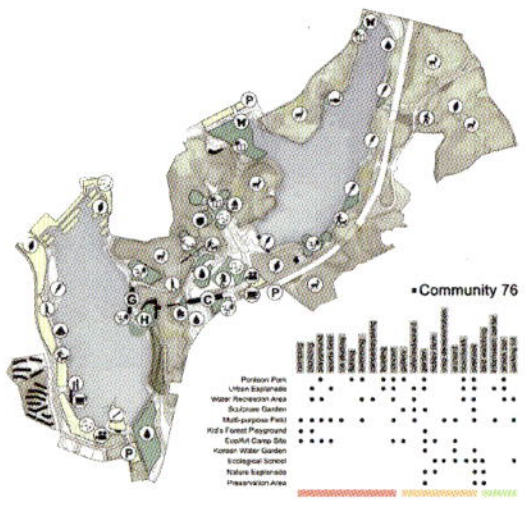

Link Space and Programs

Ribbon 55 - Program Strategy

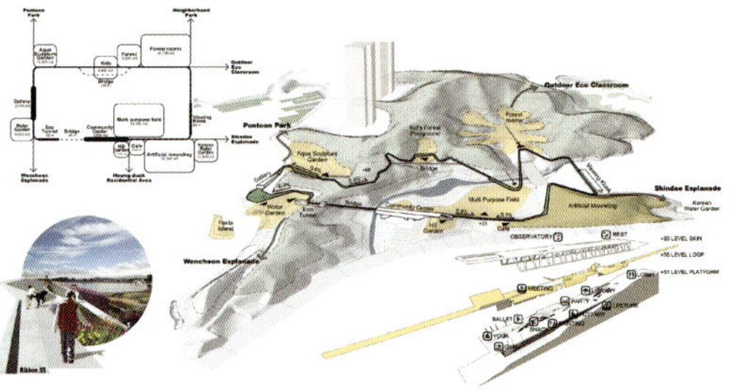

Quilting the Landscape — Landscape Strategy

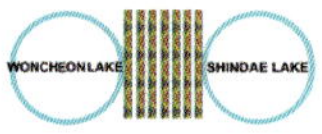

WONCHEON LAKE SHINDAE LAKE

task 1. Define dominant landscape features & view axis
task 2. Color the site : Forest & Lake
task 3. Landscape Narrative - frame the view, control the view

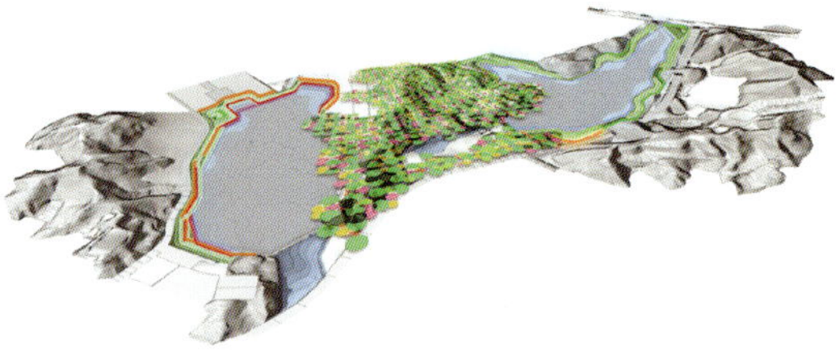

Dominant Landscape Features & View Axis

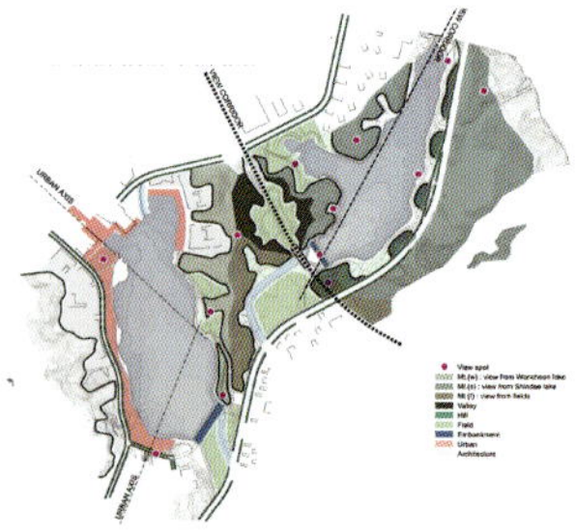

Color the Site

Landscape Narrative

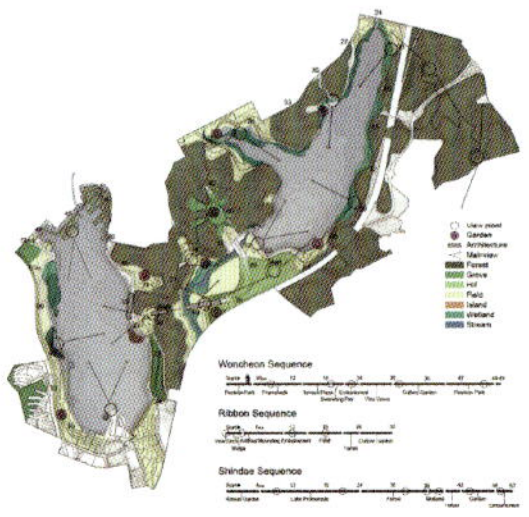

Ribbon 55 - Landscape Strategy

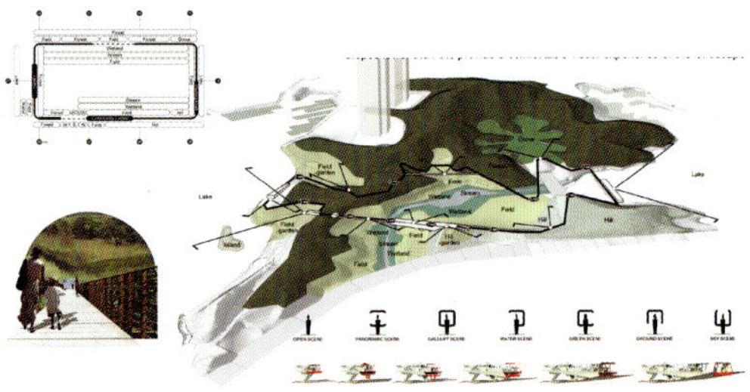

GREEN PARKING
RIBBON 55
PARKING UNDER DECK
GREEN PARKING
LEGEND
01. PONTOON PARK
02. URBAN ESPLANADE
03. LAKE TERRACE PLAZA
04. FLOATING PIER
05. RICE TERRACE WATER PURIFICATION
06. FESTA ISLAND
07. WATER GARDEN
08. AQUA SCULPTURE GARDEN
09. KID'S FOREST PLAYGROUND
10. TUNNEL BRIDGE
11. FOREST ROOMS
12. ARTIFICIAL MOUNDING
13. COMMUNITY CENTER
14. MULTI-PURPOSE FIELD
15. RICE TERRACE WATER PURIFICATION
16. RICE TERRACE/SKATING RINK(WINTER)
17. KOREAN WATER GARDEN
18. NATURE ESPLANADE
19. FOREST TRAIL / LOOKOUT
20. FOREST REGENERATION AREA
/ FOREST EDUCATION PROGRAM
21. WATER PURIFICATION PLANT
22. COMMUNITY FARM
23. FISHING PIER
24. OUTDOOR ECO-CLASSROOM/WETLANDS DISPI
25. PARK INFO CENTER
26. MANDAEROO(VIEWING KIOSK)
27. CONNECTION TO NEIGHBORHOOD PARK
A. LAKESIDE CAFES
B. ATTACHED HOUSING COMPLEX
C. BOUTIQUE HOTEL
D. LAKESIDE GALLERY

Sustainable System in the Park

Sustainable Urbanism

task1. Secure wind tunnel connecting Gwanggyo mountain, lake and Heung-duk area

task2. Reduce Heat island effect with encouraging wind tunnel effect

task3. Air purification - park functioning as an air filter for new town

Mt. Kwanggyo
Mountain_18.0℃
Building_25.9℃
Pontoon_17.7℃
Wetland_17.7℃
Lake_18.9℃
Heating Souce (New town)
Cooling Source (Green space, Lake)
CO2-Fixation
CO2-Fixation
Air filter
clean breeze
Two Lakes

H_2O Treatment System

system1. Wetland / Two ponds system

system2. pontoon system

system3. Water quality improvement facilities

system4. Rice-terraces / Meander

Carbon Offset Park

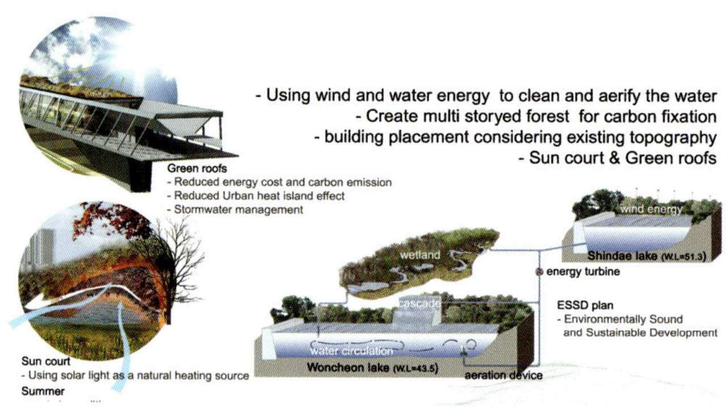

1. Wetland / Two ponds system

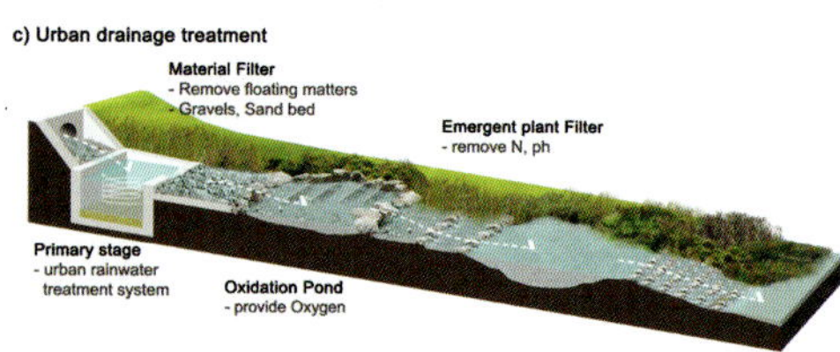

2. Pontoon system

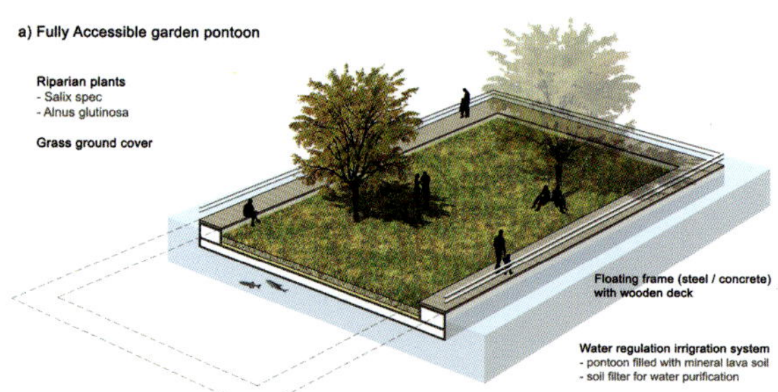

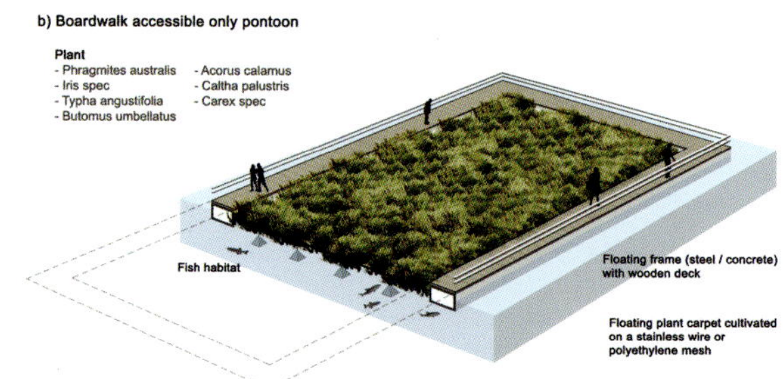

3. Water quality improvement facilities

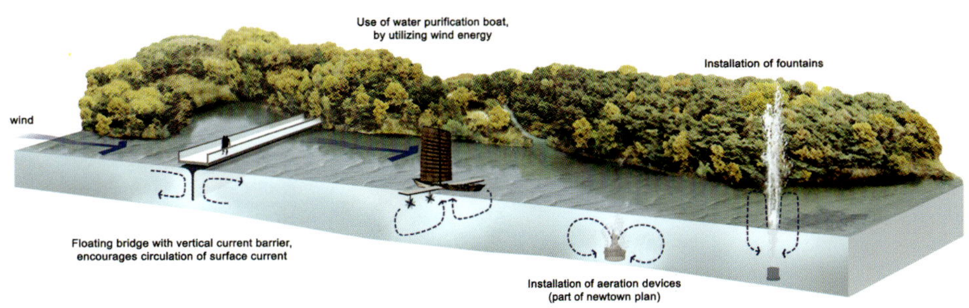

4. Rice-terraces / Meander

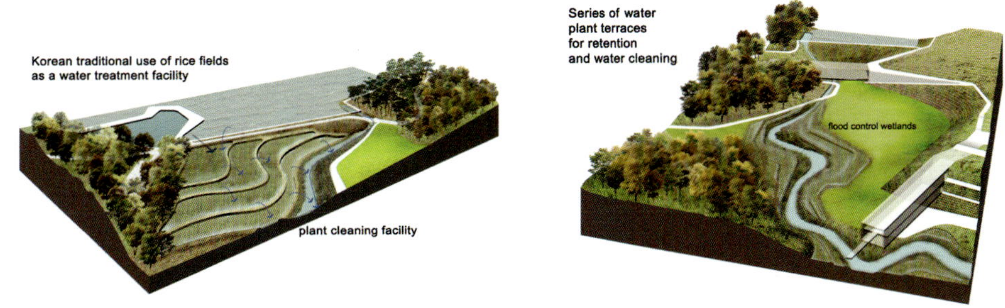

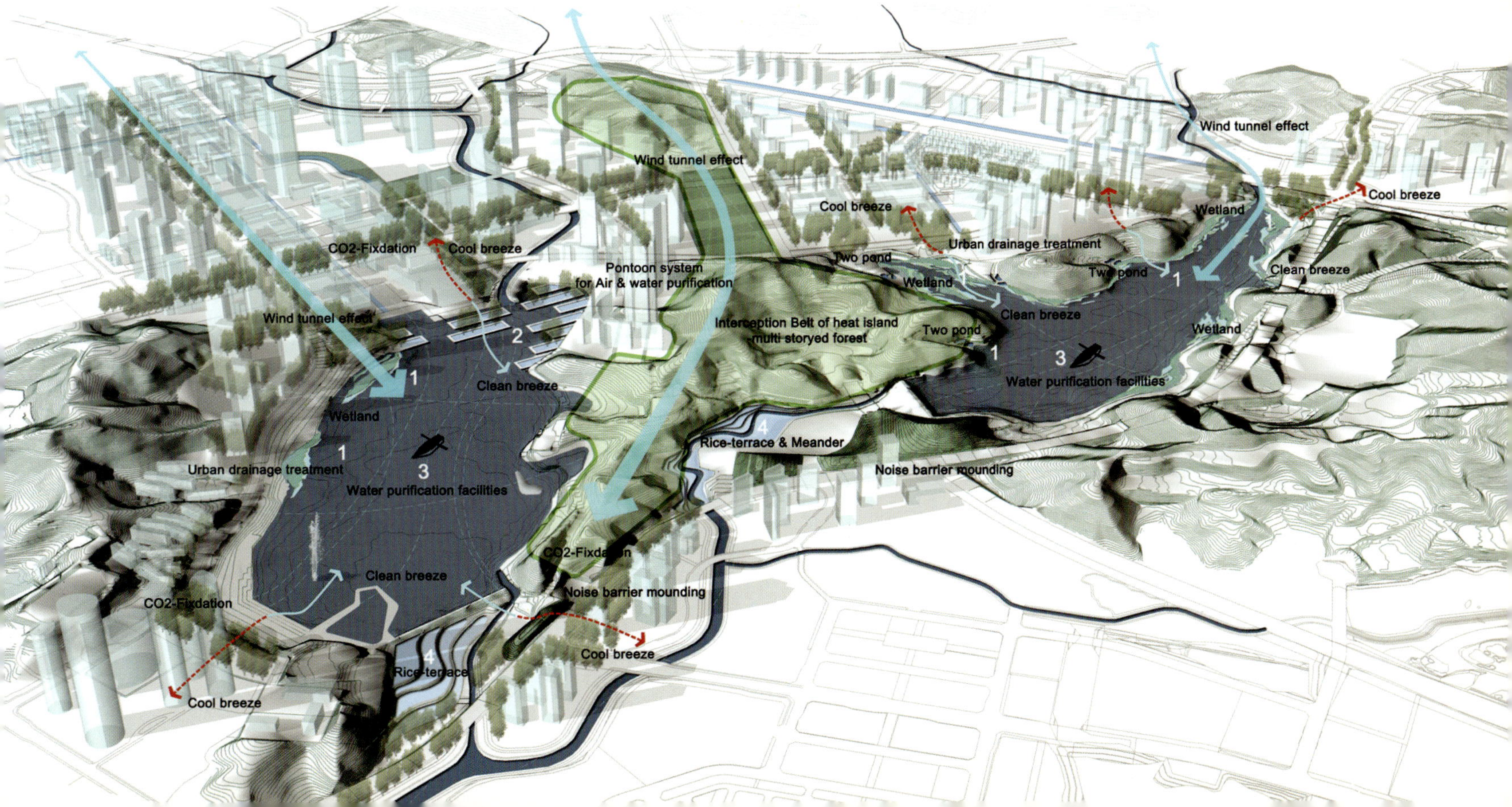

Pontoon park

A number of floating structures anchored near the new city center is creating an amphibious park of accessible and non accessible green surfaces. The pontoon park is a meeting point between urban and aquatic life. It is at the same time a place for leisure and a facility to help to rehabilitate a good water quality through plant and soil filtering. The pontoons surfaces offer new habitats for birds, fish and other animals which can easily be observed by visitors.

Pontoon Park

Urban esplanade

The Urban esplanade not only connects convention & business center to the Power center and also access to the lake for the residential area. The open lawn gentle sloped to the lake provides relaxed space for enjoying the scenery of lake and hills. Boardwalks provide the unique impression of walking on the water. In addition, the café and restaurant on the floating deck attracts and draws visitors to the water's edge.

The esplanade is a ribbon-like pathway that has been designed to create urban, creative and cultural experiences. It has a cadence or a rhythym which is like the streets and blocks of the city. At every block there is a pulse, where the pier meets the water and each one is celebrated by design, through a fountain, a sculpture, a stage, or a lighting event. All together, in the evening, they are coordinated to music, like a dance that celebrates the city and the culture.

Urban Esplanade

Lake terrace plaza & Floating pier

At the southern end of the Urban Esplanade in the Lake Terrace Plaza, where the terraces of a sloped lawns and reflecting ponds are structured into straight lines – seating platforms that step down to meet the water and allow for gatherings of larger groups. String quartets and jazz festivals animate the terraces as groups gather to celebrate one major gathering space along the waterfront.

Floating pier - With all this water, purified and cleansed, it would be a shame if you could not stop and swim on a hot day. The swimming pier allows for an experience of swimming in a natural pool, which is physically at the water level of the lake. Kids run and splash in the many interactive tidal pools which shoot water high up into the sky, creating squeals of laughter from children and adults alike. One could also sun bathing on the lounge chair or sand bathing in the floating sand box.

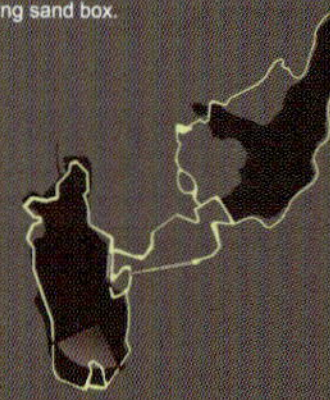

Lake Terrace Plaza & Floating Pier

Festa island & Boutique Hotel

Landscapes are the natural setting for major life-changing events in people's lives. The setting of a wedding island, with natural rockery, with topiaries, with fountains and with ceremonial horticulture is one that will become recognized throughout the city and the country. The wedding island will have a seasonal celebration of colour throughout the year as the blossoms, the fall colours and the shapes of the ornamental trees become the setting for many photographs to be remembered.

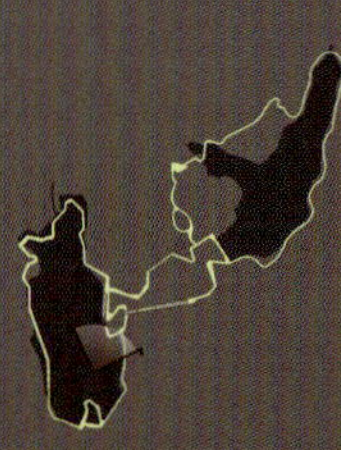

Festa Island & Boutique Hotel

Multi-purpose Field & Community center

Centrally located in the park, the multi-purpose field is the place where the neighbors and visitors enjoy sports and recreational activities and also the place for the big events. Cultural landscapes are important to set local context and this terraced water-cleansing and polishing program in the form of a traditional rice paddy landscape is a way to blend history with culture and education. In contrast to the natural first landscape scale, is the contrast of a series of manmade artificial mounds. These mounds celebrate how humans can replicate shapes found in nature but organize them to express human's ability to create patterns and language. The Community Centre is an architectural expression of how architecture and nature can come together to have a harmonious relationship.

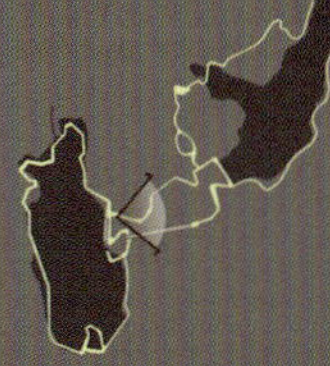

Multi-purpose Field & Community Center

Forest rooms

Outdoor rooms are spaces that with minimal intervention, can feel like natural clearings that are carved out of beautiful forest settings. These rooms will be themed to have unique characteristics that celebrate landscape paradigms that normally may not be found in the middle of a forest. The linear arrangement of a flowering orchard in the spring or a communal vegetable garden are examples of how these spaces can be created and can educate children and young campers about their landscape.

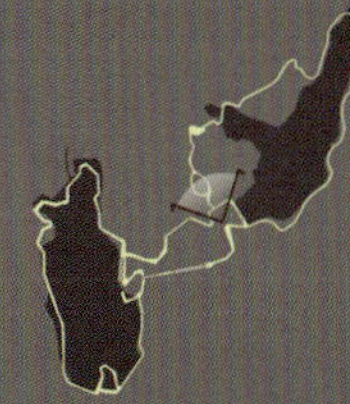

Forest Rooms

Outdoor eco-classroom

The forest needs to be replenished in places and needs new diversity and thinning. In other places, the invasive species need to be removed to create healthier growth. This needs to be interpreted and explained to young and old hikers, to get buy-in which in turn helps with future forest management practices. Similarly, although the outdoor eco-classroom will be abundant throughout the park, there will be areas for larger gatherings suited to multiple-classroom or even small school sizes.

Outdoor Eco-classroom

Weaving flux

At night, the villa complex will be seen as lighthouses which float over the sea of thousands of point lights. With the help of its light, People can navigate hills and lakes freely and pleasantly. Whenever they stroll along the lakeside, before long they are to come across crowded cafes and community hubs full of lights and enjoyment.

Against myriads of illuminating spaces which light up the whole lake park, various types of cultural events such as concert, cinema play and game, will be offered to the public, and accordingly to the residents of the villas. Therefore, the lake park will be function like a front garden to the villas, while the villas will attract and guide visitors like a beacon to the lake park.

Weaving Flux

SERENE VITALITY

GAONE Design Group (주)가원조경기술사사무소
Partners _ Dongsimwon (주)동심원조경기술사사무소 + SamWoo ENG (주)삼우엔지니어링
Choi, Jung Kwon최정권 + Yang, Woo Hyun양우현

Strategies

Communicate the City and the Lakeside Park

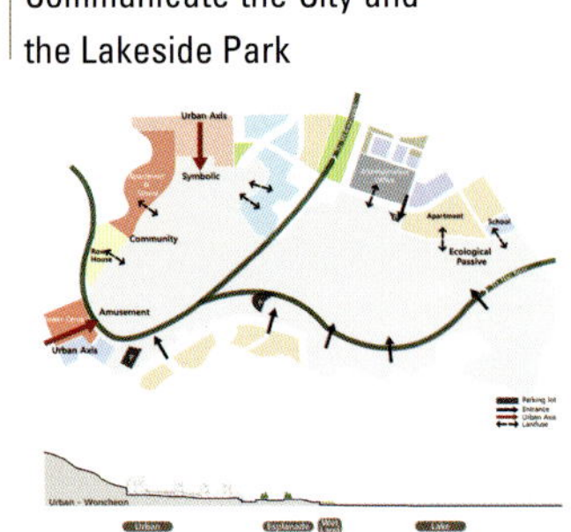

Organize a Recreation space and a Poetic space

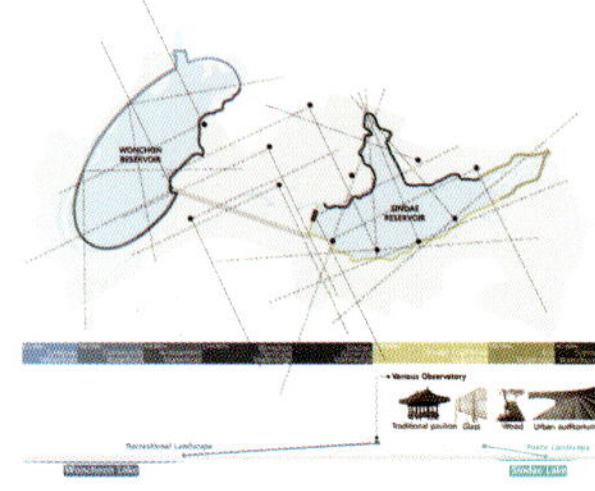

Connect the Westside lake and the Eastside lake

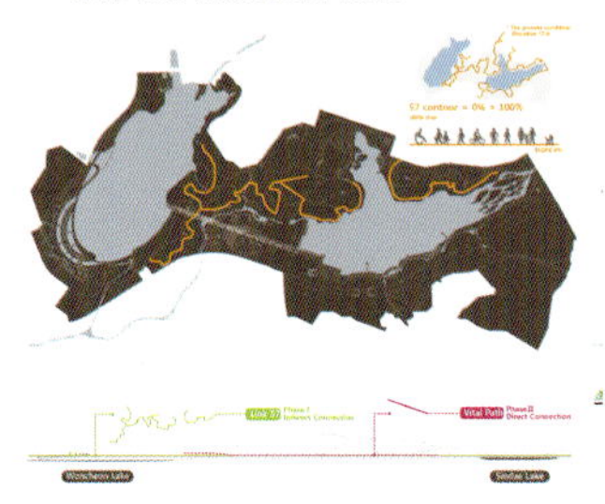

Weave the Land and the Water

Woncheon Reservoir

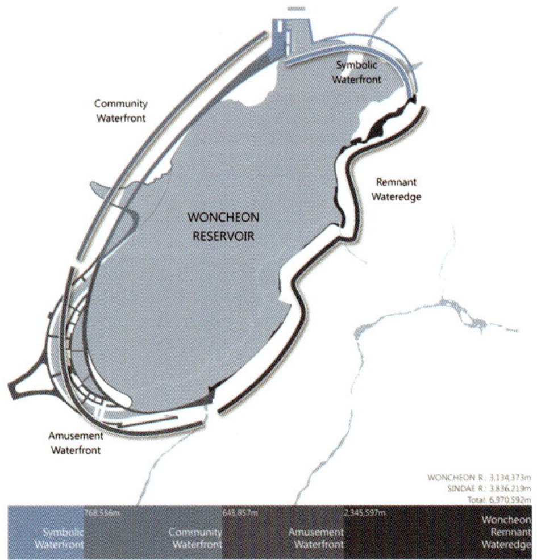

Types of Waterfront & Edge

Symbolic Waterfront
Community Waterfront
Amusement Waterfront
Remnant Water Edge

With the basis of diverse urban contacts and regional approaches to the site, Woncheon reservoir is the lake contained the various vital faces of urban. A single oval shaped esplanade takes in programs for citizens and it is characterized by 4 different waterfronts responding to urban context. The whole green system is connected by the central open space to the regional green system descending from the Mountain Gwanggyo towards the southern forest, the ecological bridges flow into the neighborhood park which designated as a neighborhood park.

원천저수지는 다양한 도시와의 접점과 광역접근체계 등을 기반으로 도심 속 워터프론트로 작용하며 다양한 도시의 활발한 표정을 담은 호수이다. 하나의 타원형 산책길은 저수지 주변의 도시적 컨텍스트와 반응하면서 네 가지 성격의 워터프론트로 구분되며, 각 워터프론트 성격에 따른 공간 프로그램을 계획하였다. 또한 광교산에서 내려오는 광역녹지체계와 연결된 대상지 중심부의 녹지를 원천저수지 아래쪽의 숲 계획과 생태육교 계획을 통하여 연립주택과 주상복합 뒤쪽의 근린공원으로 지정되어 있는 구릉의 녹지를 연결, 전체적인 녹지체계를 연계하였다.

Amusement Waterfront

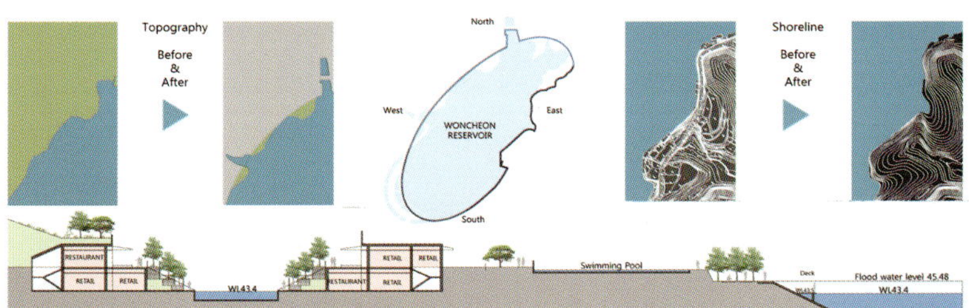

Community Waterfront & Remnant Water Edge

Community Waterfront

Remnant Water Edge

Sindae Reservoir

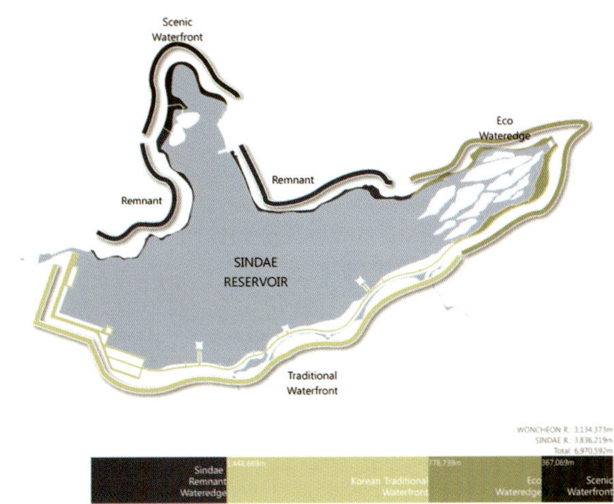

Types of Waterfront & Edge

- Scenic Waterfront
- Remnant Water Edge
- Traditional Waterfront
- Ecological Water Edge

Sindae Reservoir is planned a Korean traditional and ecological lake motivated Anapji(traditional pond). It is designed with a right angular stone embankment witch is connected with a levee, natural shore and ecological wetland at inflow area of the stream to purify naturally. Sindae Reservoir has various landscape elements, deck and traditional pavilions to observe. Separating of pedestrian, bicycles, vehicles way and each specialized area will be different strategy of Sindae reservoir. Stations to take a rest for pedestrian, bicycle and these space elements is reflected Sindea reservoir's natural context. It matches four waterfront plans. Especially Korean traditional pavilions of eight landscape points will maximize excellent experience of Sindae reservoir.

보다 한국적이며 생태적인 호수로 계획된 신대저수지는 안압지를 모티브로 하여 제방과 연계한 직각형의 석축과 자연스러운 호안을 적용하고 하천수 유입부에 생태습지를 조성하여 자연적인 정화를 통해 저수지로 유입되도록 하였으며, 다양한 경관요소와 바라보기 위한 데크, 전통정자 등이 도입되었다. 보행 · 자전거 · 차량동선의 분리와 각 구간별 특화방안은 신대저수지의 또 다른 전략이 된다. 보행자와 자전거를 위한 쉼터 개념의 정거장 계획과 더불어 이러한 공간 요소들은 신대저수지 주변의 자연적 컨텍스트를 반영한 네 가지 워터프론트 계획에 부합하며, 특히 8개 경관 포인트의 한국 전통 파빌리온은 신대저수지의 우수한 수경관 체험을 극대화할 것이다.

8 Scenic Landscape of Sindae Reservoir

1. Sindae-Jung
2. Banghwasuryu-Jung
3. Dasan-Jung
4. Tongkyung-Jung
5. Hayang-Jung
6. Wolyoung-Jung
7. Wunyeonhwa-Jung
8. Cheongpoong-Won

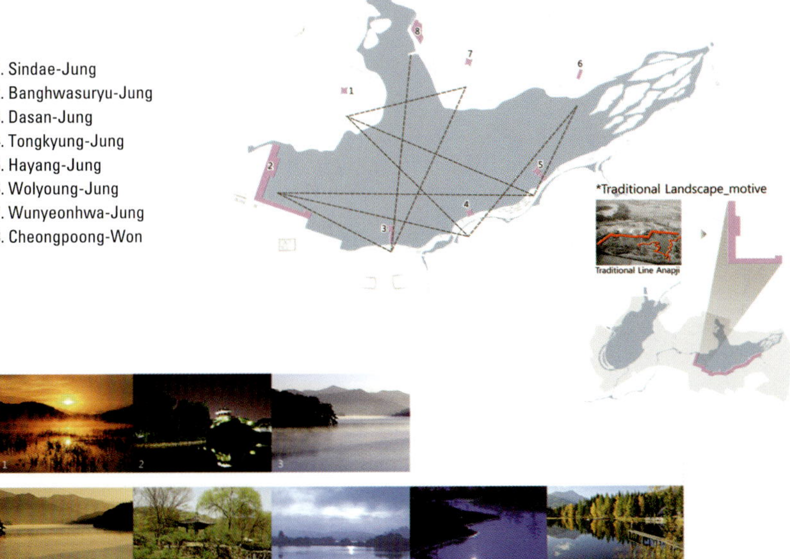

Ecological Waterfront

Forest & Field

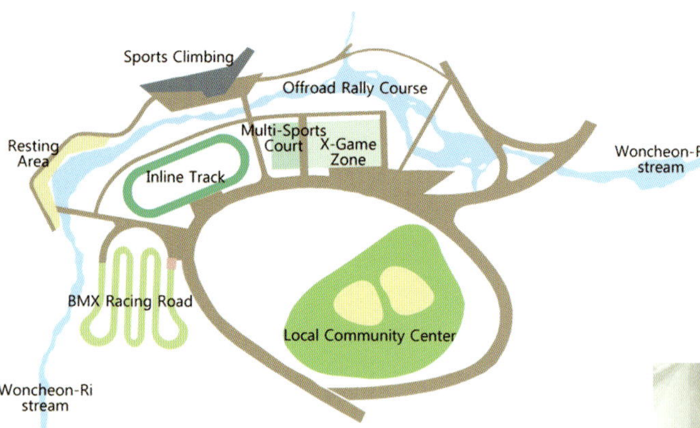

Forest and field will be used low sensitivity zone as a programed space and preserve or restore forest connected with regional green system from Mt Gwanggyo mountain. The trail which considered the existing topography does not only connect each field and extend to Gwanggyo mountain through neighborhood parks but also path plan is connected with Woncheon and Sindae reservoir.

광교산에서 내려오는 광역녹지체계와 연결된 대상지 중심부의 기존 식생을 보전 · 복원함과 동시에 일부 가용지를 프로그램 공간으로 활용한다. 이들을 연결하는 트레일은 최대한 기존 지형에 순응하면서 각 필드를 연결하며 근린공원을 통하여 광교산으로까지 이어지는 동선, 그리고 원천-신대 저수지를 직 · 간접적으로 연결하는 동선 계획 또한 포함하고 있다.

7Field Characteristics + Program

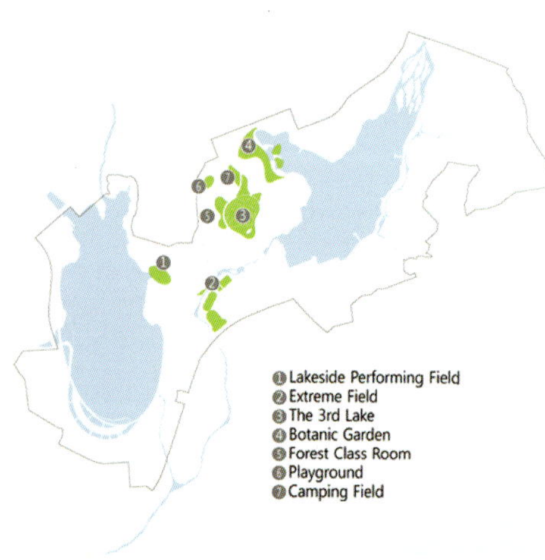

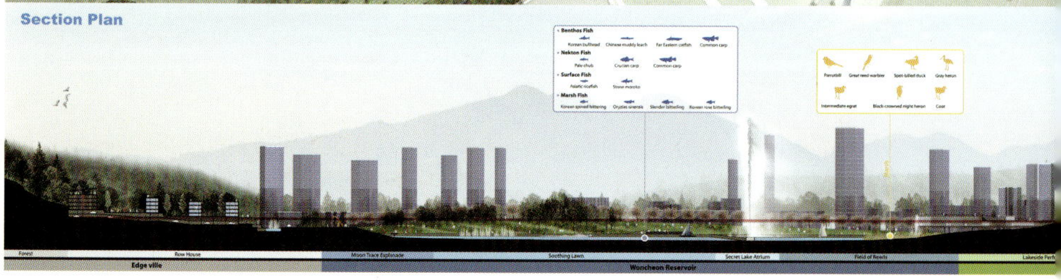

Section Plan

LINK 57

Observational Cabin

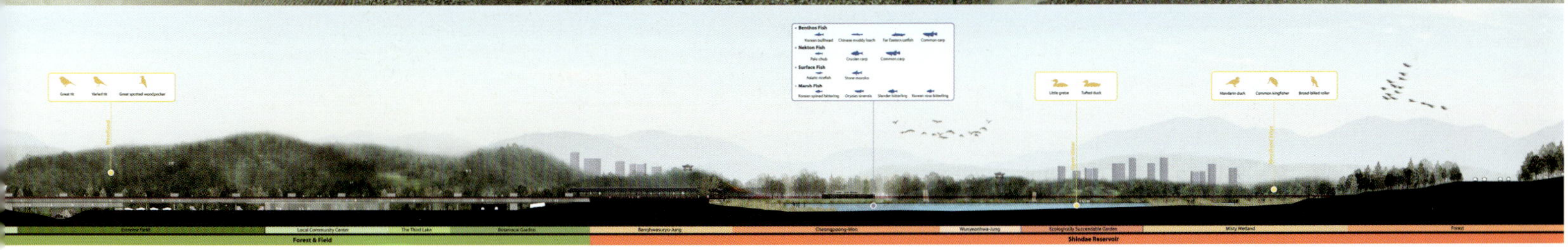

Gwanggyo Myeong San Su Won -Proposed name of park
Eventually we suggest the name of 'Gwanggyo Myeong San Su Won'(광교명산수원) instead of 'Gwanggyo Lakeside Park'(광교 호수공원)

1 Convention Center
2 Apartment & Stores
3 Neighborhood Park
4 *Ajou* University
5 Row House
6 Apartment & Stores
7 Bufer Greenery
8 Elementary School
9 Heung-duck Residential District
10 Woncheon-ri Stream
11 Taegwang Golf Course
12 Elementary School
13 Apartment
14 Law Office
15 Neighborhood Park
16 Apartment
17 Business / Office
18 Yeo Stream
19 Neighborhood Park
20 Elementary School
21 Water Supply Reservoir

1 Marina
2 Restored Existing Hills
3 Field of Reeds
4 Secret Lake Atrium
5 Soothing Lawn
6 Moon Trace Esplanade
7 *Woncheon* Waterfall
8 Water Forest
9 Retail
10 Open Market
11 The Crescent Canal
12 Lake Pool
13 Shuttle Bus Station
14 Parking Lot
15 Flying Shed

1 Lakeside Performing Field
2 Extreme Field
2-a Artificial Rock wall
2-b Inline Skating Field
2-c Basketball Court
2-d X-Game Zone
2-e BMX Road
3 Parking Lot
4 Local Community Center
5 Vital Path
6 Observational Cabin : Glass Type
7 Observational Cabin : Wood deck Type
8 Forest Classroom
9 Playground
10 The Third Lake
11 Camping Field
12 Botanocal Garden

1 Morning Calm Island
2 *Sindae-Jung*
3 *Banghwasuryu-Jung* : Water gate Pavillion
4 *Dasan-Jung* : *Dasan* is Jung Yak Yong who is a scientist of Chos
Dynasty and Built up the Suwonhwasung
5 *Tongkyung-Jung* : Waterscape Pavillion
6 *Hayang-Jung* : *Hayang* means lakeside willow
7 Park Station
8 Shuttle Bus Station
9 Misty Wetland
10 Open School
11 *Wolyoung-Jung* : *Wolyoung* means lunar shadow
12 Ecologically Succeedable Garden
13 *Wunyeonhwa-Jung* : *Wunyeonhwa* means waterlily on the clea
14 *Cheongpoong-Won* : *cheongpoong* means refreshing breeze
15 Parking Lot
16 Triple Esplanade

* -Jung / -Won : Korean Traditional Pavillion

Woncheon Reservoir : The Crescent Canal

Forest & Field : The 3rd Lake

Sindae Reservoir : Misty Wetland

TWO LAKES

Balmori Associates Inc

Partners _HeeRim Architecture & Planners 희림종합건축사사무소 + ARUP + N.E.E.D

Not long ago, the South Korean government began decentralizing its power to control urban planning to provincial governments. Subsequently, an onslaught of unregulated urban development plans emerged. Some were successful, some failed, and others never made it off the drawing board. Clearly, the market and desire for new urban development in Korea is robust, and relatively young in Korea, making its own rules on the way. However, it is still an unstable battle field full of untamed urban competitions.

Among this development market is the city of Gwanggyo. In order to thrive, Gwanggyo and its public amenities must be latently resilient against its urban competitors. They must be equipped to survive unforeseeable urban challenges and catastrophes such as a saturation of new development, environmental problems, and domestic or global financial crises. We envision Two Lakes Park as an adaptable organism that can respond to a variety of cultural, societal, programmatic, economic and political circumstances.

Urban Competition

Area Identification

dev. typ. A
dev. typ. B
dev. typ. C

얼마 전부터 대한민국 정부는 지방자치단체들에게 도시 계획에 대한 권한을 분산하기 시작했다. 그 결과로, 규제되지 않은 무분별한 도시개발계획들이 남발하게 되었고, 몇몇 계획들은 성공적으로 이루어졌지만, 그에 못지않게 많은 계획들이 실패로 끝나거나 그다지 좋은 반응을 얻지 못하였다. 이유야 어떻든, 한국에서의 도시 개발에 대한 갈망과 시장성은 무척 활발하고, 그를 위한 원칙은 만들어져가는 과정에 있다. 신도시끼리 서로 경쟁하는 불안정한 상황이 계속되고 있는 것이다. 이러한 상황에서 나온 개발계획 중의 하나가 바로 광교신도시이다. 이러한 상황에서 광교신도시와 공공 어메니티가 성공하기 위해서는, 새로운 개발 욕망과 환경 문제, 국내외 금융위기와 같은 예측할 수 없는 도시의 도전과 위기에 대항해서 잠재적으로 탄력적인 대응을 추구해야 한다. 우리는 다양한 문화적, 사회적, 계획적, 경제적 그리고 정치적 상황에 대응할 수 있는 유기적인 조직체로서 두 개의 호수공원을 그려본다.

Our design concept transcends the notion of designing for a fixed state, and embraces the future as a responsive pattern of cultivation that advances the evolution of the park's landscape. In order to plan for Lakeside Park's long term survival, 'Insertions' that maintain the park in its basic sustainable state are constructed first; such as circulation paths, retaining walls, and any water/energy related system, which will be a permanent backbone of the park.

우리의 설계안은 고정된 상태의 공원을 계획해야 한다는 생각을 초월해, 공원 설계의 진화를 이끌 수 있는 유연한 패턴을 가꾸어나감으로써 미래의 변화를 수용코자 한다. 호수공원의 장기적인 비전을 위해, 우리는 기본적인 지속가능한 상태에서 공원이 유기적으로 작동될 수 있도록 핵심 인프라스트럭처를 '삽입' 하고, 그를 바탕으로 공원 시나리오를 제안한다.

Resilient Park

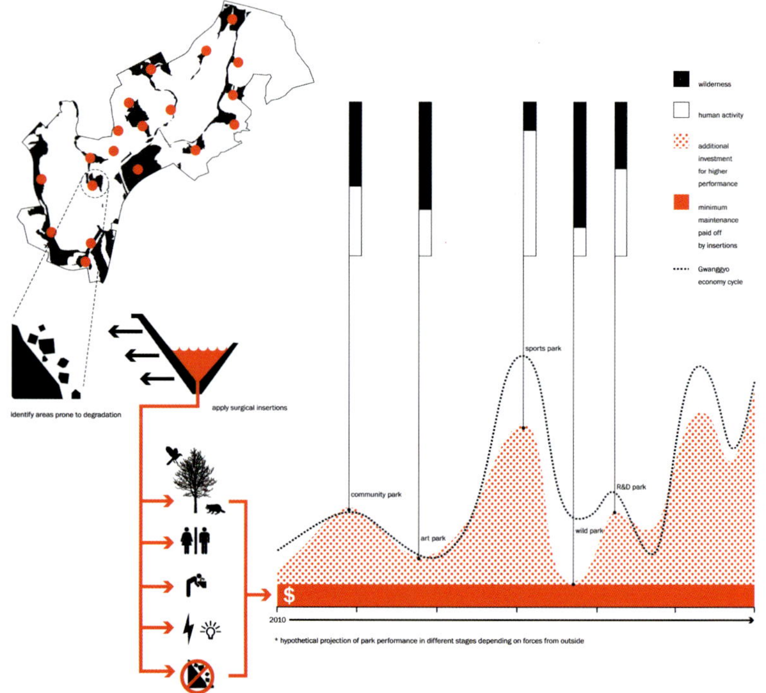

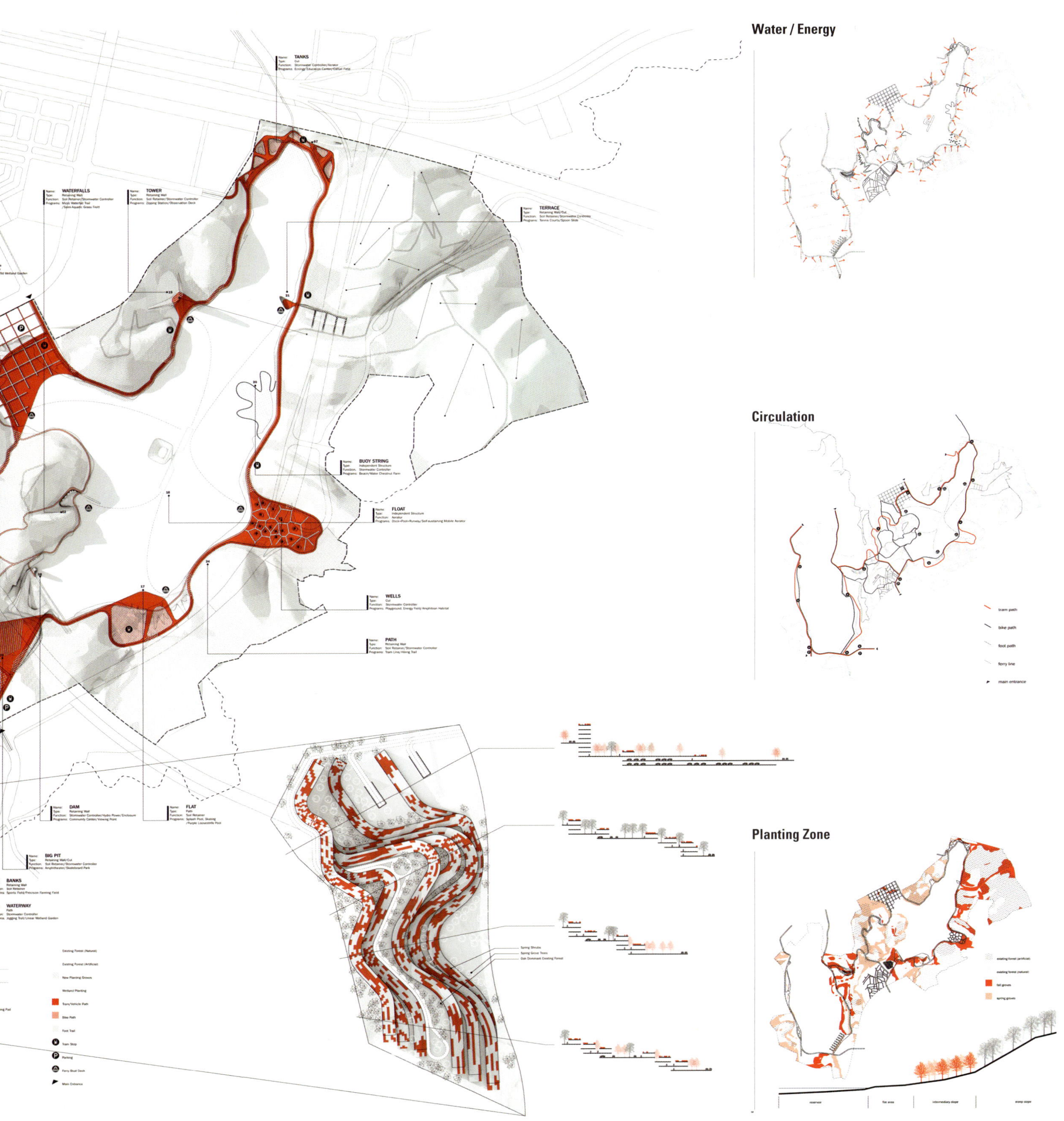
Water / Energy
Circulation
Planting Zone
TANKS
WATERFALLS
TOWER
TERRACE
BUOY STRING
FLOAT
WELLS
PATH
DAM
FLAT
BIG PIT
BANKS
WATERWAY

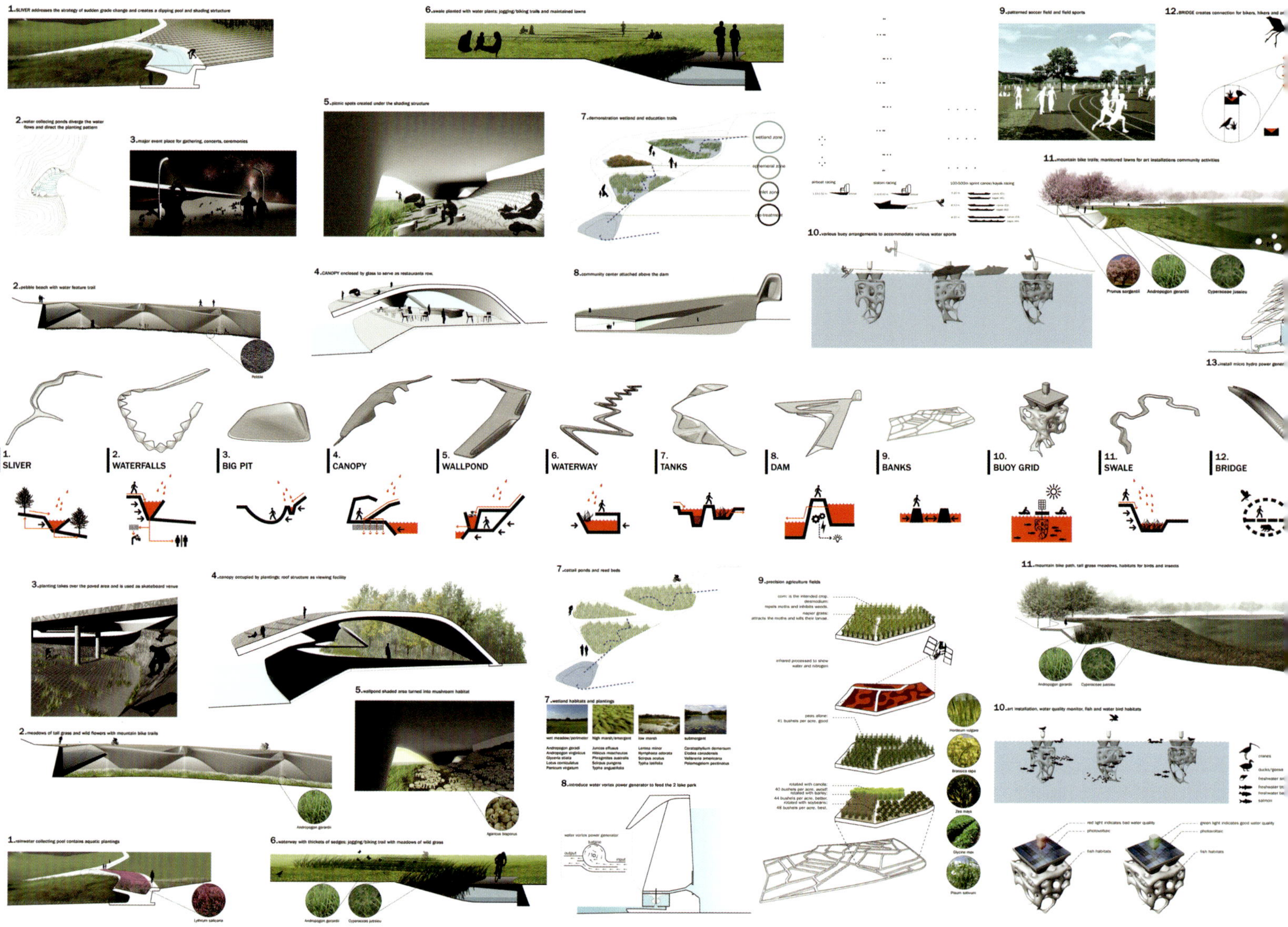

1. SLIVER addresses the strategy of sudden grade change and creates a dipping pool and shading strtucture

2. water collecing ponds diverge the water flows and direct the planting pattern

3. major event place for gathering, concerts, ceremonies

5. picnic spots created under the shading structure

6. swale planted with water plants; jogging/biking trails and maintained lawns

7. demonstration wetland and education trails

wetland zone
ephemeral zone
inlet zone
pre-treatment

9. patterned soccer field and field sports

12. BRIDGE creates connection for bikers, hikers and an

11. mountain bike trails; manicured lawns for art installations community activities

airboat racing
slalom racing
100-500m sprint canoe/kayak racing

10. various buoy arrangements to accommodate various water sports

Prunus sargentii
Andropogon gerardii
Cyperaceae jussieu

2. pebble beach with water feature trail

4. CANOPY enclosed by glass to serve as restaurants row.

8. community center attached above the dam

13. install micro hydro power gener

1. SLIVER
2. WATERFALLS
3. BIG PIT
4. CANOPY
5. WALLPOND
6. WATERWAY
7. TANKS
8. DAM
9. BANKS
10. BUOY GRID
11. SWALE
12. BRIDGE

3. planting takes over the paved area and is used as skateboard venue

4. canopy occupied by plantings; roof structure as viewing facility

7. cattail ponds and reed beds

9. precision agriculture fields

11. mountain bike path, tall grass meadows, habitats for birds and insects

5. wallpond shaded area turned into mushroom habitat

7. wetland habitats and plantings

wet meadow/perimeter	high marsh/emergent	low marsh	submergent
Andropogon geradi Andropogon virginicus Glyceria stiata Lotus corniculatus Panicum virgatum	Juncus effusus Hibicus moscheutos Phragmites australis Scirpus pungens Typha angustifolia	Lemna minor Nymphaea odorata Scirpus acutus Typha latifolia	Ceratophyllum demersum Elodea canadensis Valisneria americana Potomogetom pectinatus

2. meadows of tall grass and wild flowers with mountain bike trails

Andropogon gerardii

Agaricus bisporus

8. introduce water vortex power generator to feed the 2 lake park

water vortex power generator
output
turbine
input

peas alone:
41 bushels per acre, good

rotated with canola:
40 bushels per acre, avoid!
rotated with barley:
44 bushels per acre, better.
rotated with soybeans:
48 bushels per acre, best.

Hordeum vulgare
Brassica rapa
Zea mays
Glycine max
Pisum sativum

10. art installation, water quality monitor, fish and water bird habitats

cranes
ducks/geese
salmon

red light indicates bad water quality
photovoltaic
fish habitats
green light indicates good water quality
photovoltaic
fish habitats

1. rainwater collecting pool contains aquatic plantings

Lythrum salicaria

6. waterway with thickets of sedges; jogging/biking trail with meadows of wild grass

Andropogon gerardii
Cyperaceae jussieu

Basin : Swimming pool turned into wetland skateboarding park

Walk : Municipal library an wetland

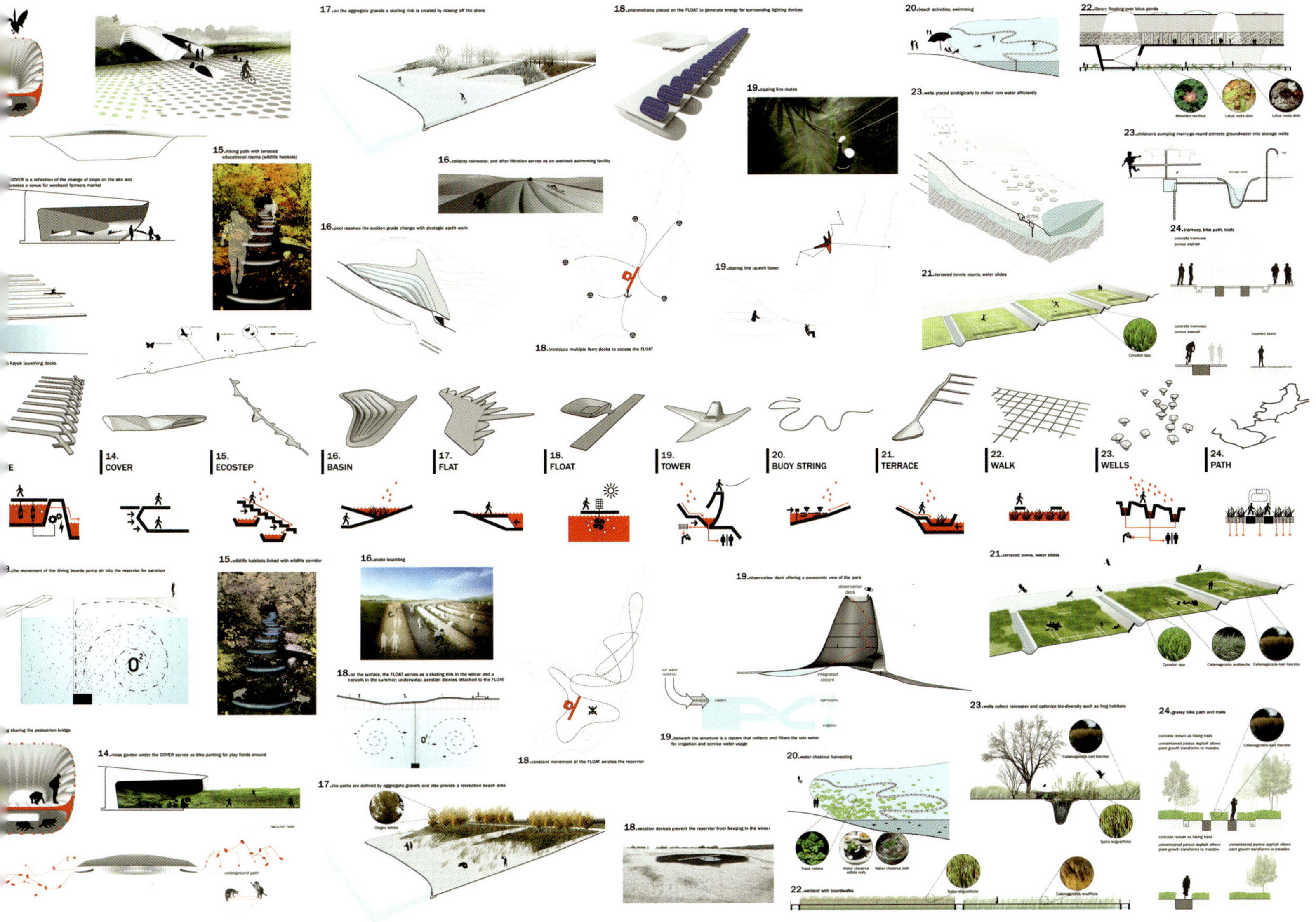

14. COVER
15. ECOSTEP
16. BASIN
17. FLAT
18. FLOAT
19. TOWER
20. BUOY STRING
21. TERRACE
22. WALK
23. WELLS
24. PATH

Bouy Grid : Floating lighting field

Swale : Wild flower field for art installation

GWANGGYO LAKESIDE PARK

GROSS.MAX.(Eelco Hoofman, Bridget Baines, Nigel Sampey)

The two reservoirs create a unique experience of contrasting atmospheres. Wocheon reservoir is extrovert ; here the lake provides a stage for activities. Sindae reservoir is introvert; a place for solitude and the spiritual experience of nature. We also extended the Gwanggyo lakeside park by including a third area in between the two reservoirs and alongside the waterways consisting of a patchwork of cultivated fields. Interestingly during the renaissance period in the West there was a characterisation of nature as respectively wilderness, cultivated landscape and designed garden. In Gwanggyo Lakeside Park we found all three categories. There will be a series of public space categories ranging from waterfront plaza, lakeside promenade, expo park, water resort, woodland trail, neighbourhood park, wetland gardens as well cultivated fields. The contrast between natural and artificial conditions will create a dramatic setting for contemporary urban living.

두 개의 저수지는 각기 다른 분위기를 갖고 있어서, 색다른 경험을 체험할 수 있다. 원천저수지는 외향적이어서, 활동을 위한 무대를 제공해주는 반면, 신대저수지는 내향적이어서, 고요히 자연을 느끼고 사색할 수 있는 공간이다. 우리는 두 개의 저수지와 수변을 따라 늘어선 경작지 사이에 있는 공간들을 통해서, 광교호수공원을 확장시키코자 한다.

흥미로운 점은 서구의 르네상스 시기에 야생의 거친 황야와 양식적인 풍경, 잘 가꾸어진 정원이, 각각 자연의 특징으로 존재했듯이, 광교호수공원에서도 마찬가지로 세 가지 특징이 존재한다는 점이다. 이런 특징들은 워터프론트 플라자, 수변 프로미나드, 엑스포 파크, 워터 리조트, 숲속 오솔길, 근린공원, 습지원과 같은 다양한 공공영역의 범주에 걸쳐있다. 이런 다양함 속에 담겨 있는 자연스러움과 인공적인 것의 대조는 현대 도시 생활을 다이나믹하게 만들어줄 것이다.

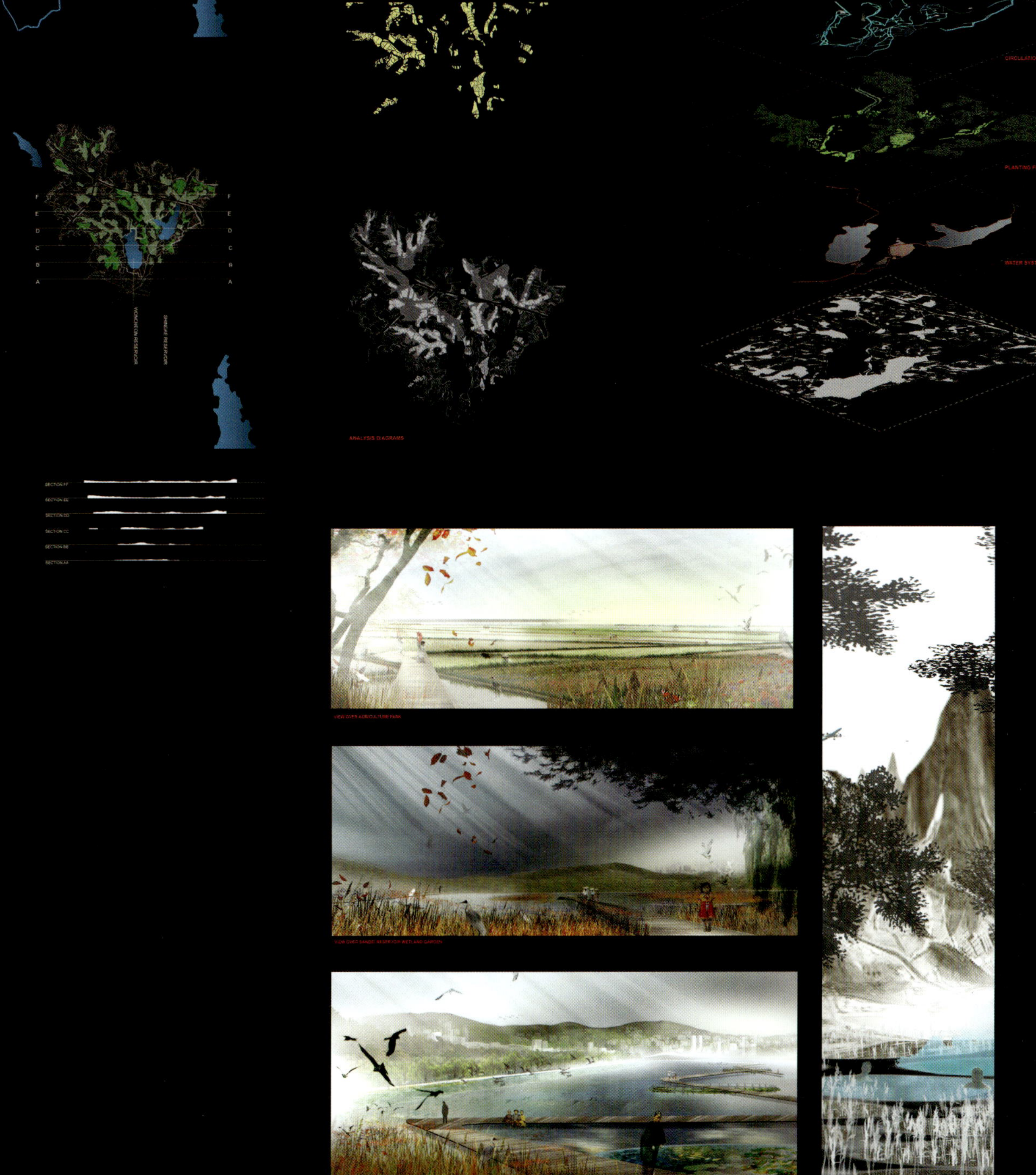

VIEW OVER AGRICULTURE PARK

VIEW OVER SANDEI RESERVOIR WETLAND GARDEN

DRAGON WATER JETS
WATER CLOUD ROCK RESORT
DRAGON JET SEQUENCE
Spring
Summer
Autumn
Winter

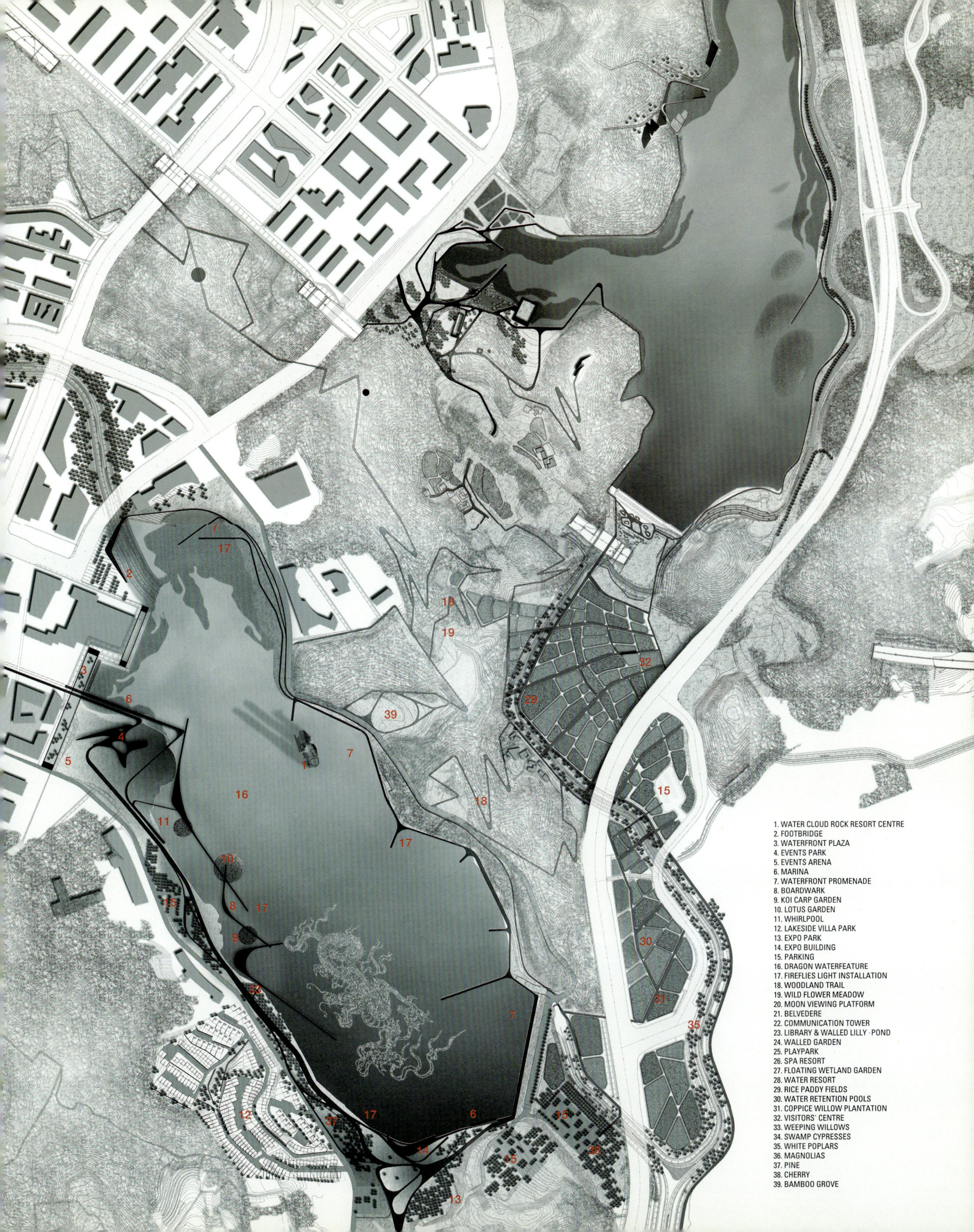
1. WATER CLOUD ROCK RESORT CENTRE
2. FOOTBRIDGE
3. WATERFRONT PLAZA
4. EVENTS PARK
5. EVENTS ARENA
6. MARINA
7. WATERFRONT PROMENADE
8. BOARDWARK
9. KOI CARP GARDEN
10. LOTUS GARDEN
11. WHIRLPOOL
12. LAKESIDE VILLA PARK
13. EXPO PARK
14. EXPO BUILDING
15. PARKING
16. DRAGON WATERFEATURE
17. FIREFLIES LIGHT INSTALLATION
18. WOODLAND TRAIL
19. WILD FLOWER MEADOW
20. MOON VIEWING PLATFORM
21. BELVEDERE
22. COMMUNICATION TOWER
23. LIBRARY & WALLED LILLY · POND
24. WALLED GARDEN
25. PLAYPARK
26. SPA RESORT
27. FLOATING WETLAND GARDEN
28. WATER RESORT
29. RICE PADDY FIELDS
30. WATER RETENTION POOLS
31. COPPICE WILLOW PLANTATION
32. VISITORS' CENTRE
33. WEEPING WILLOWS
34. SWAMP CYPRESSES
35. WHITE POPLARS
36. MAGNOLIAS
37. PINE
38. CHERRY
39. BAMBOO GROVE

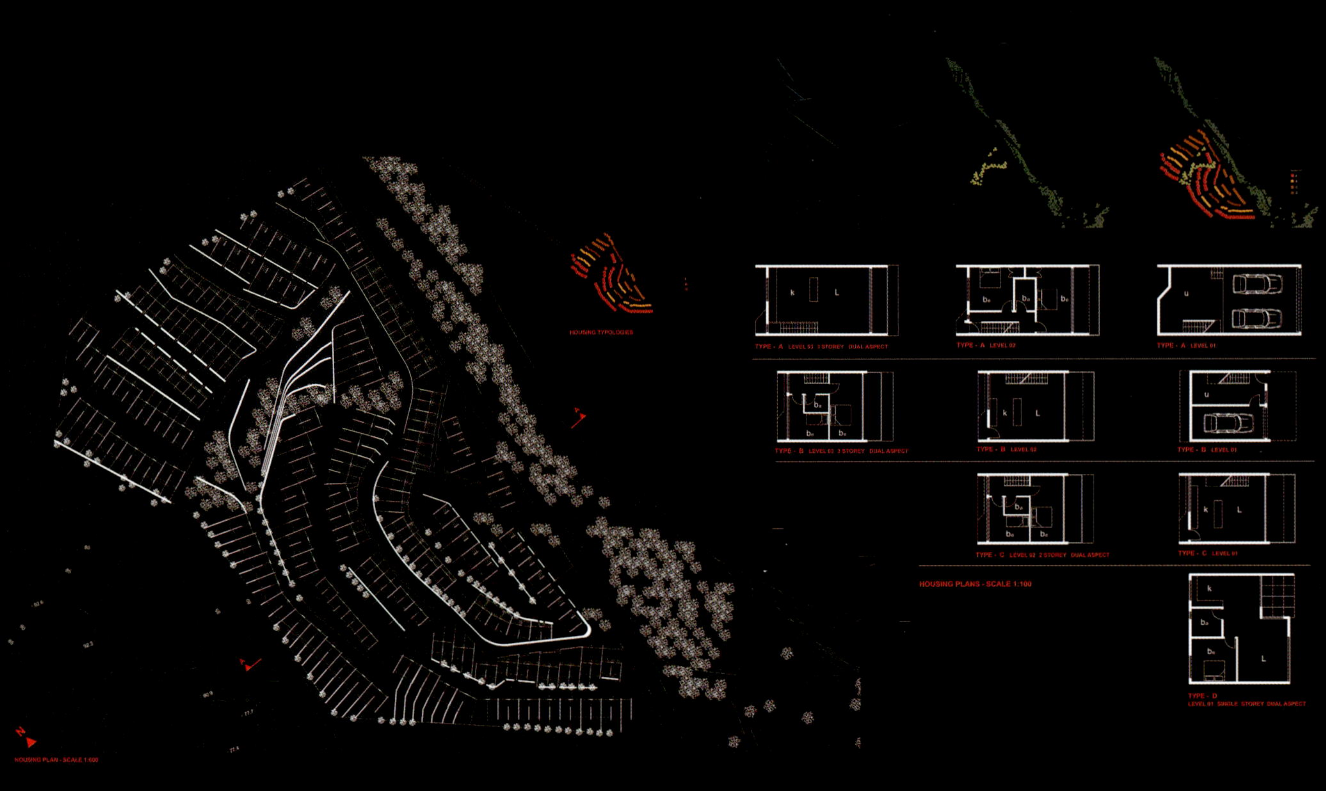

EVENTS ARENA SECTION - BB

NEIGHBOURHOOD PARK SECTION - BB

GWANGGYO PIER LAKESIDE PARK

Stan Allen Architect

Partners _Omnivorous + Terence Gower + Tillotson design + Coenpartners + Bioengineering + ARUP

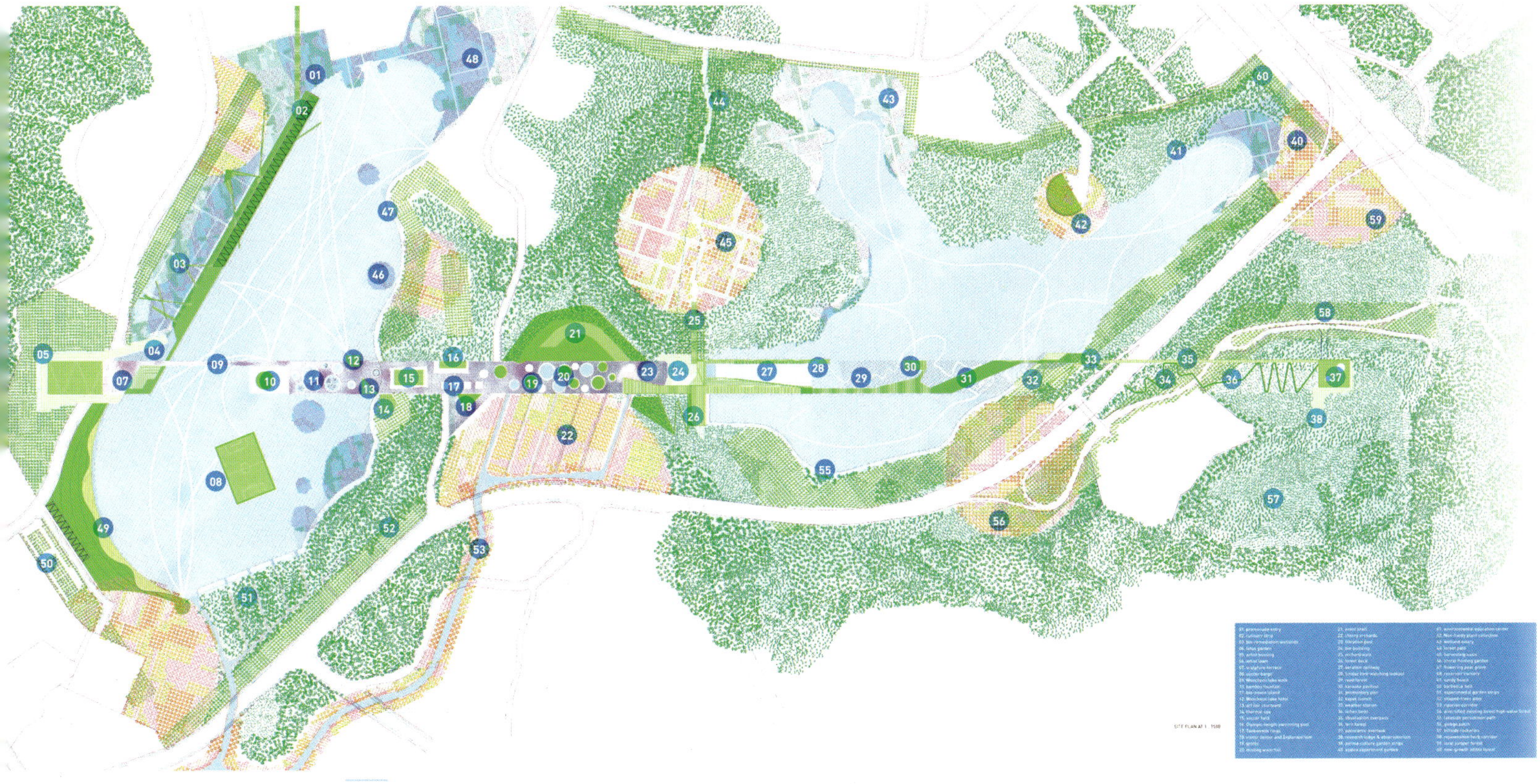

WHERE IS THE PARK? The site currently lacks definition:

- an already extensive green network exists
- city fabric risks disrupting ecological continuity
- future development is unpredictable

- 이미 형성된 풍부한 그린 네트워크의 존재
- 생태적인 지속성을 붕괴시킬 수 있는 도시 구조의 위험성
- 미래 개발에 대한 불확실성

From aeration spillway looking at sindae lake

Gwanggyo pier looking at woncheon lake

MAKE A 21ST CENTURY PARK AT GWANGGYO LAKESIDE
: FROM A VOID IN THE CITY TO A LANDSCAPE INFRASTRUCTURE

Instead of conceiving the park as a Void in the City - a green oasis surrounded by built fabric - we will construct the new park as a Working Landscape Infrastructure: A new concept that integrates an expanded ecology with a new cultural infrastructure.

도시 구조에 둘러쌓인 녹색 오아시스처럼 도시의 여백으로 계획된 공원이 아니라, 랜드스케이프 어바니즘으로 회자되고 있는, 작동하는 랜드스케이프 인프라로서의 새로운 공원을 제안한다. 이 새로운 컨셉은 새로운 문화적 인프라와 확장된 생태계를 통합시킬 것이다.

EXPAND THE FIELDS

- DISSOLVE the boundary between culture and nature
- EXTEND the park out to the city
- PROTECT the local ecologies from disruption

- 문화와 자연 사이의 경계선을 분리
- 공원에서 도시로 확장
- 지역 생태계를 보호

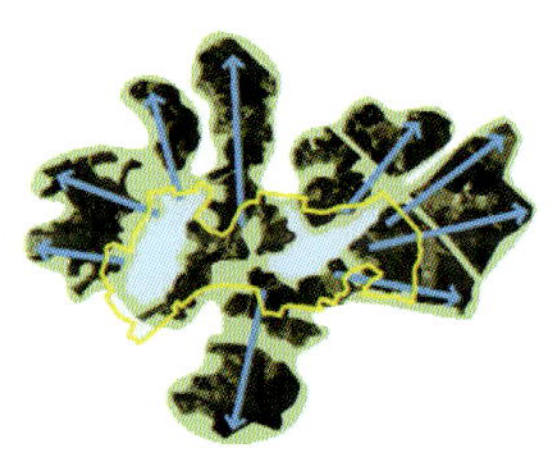

FIVE STEPS TO MAKE A PARK

CREATE A CULTURAL INFRASTRUCTURE

- CONNECT the site end-to-end with the 5K pier
- CONSOLIDATE all active programs on the pier
- INTEGRATE environmental education and research

- 5km에 이르는 교각의 끝과 끝을 이어주는 접속
- 교각에서 이루어지는 활동적인 프로그램의 통합
- 환경에 관한 교육과 연구방법에 대한 총괄

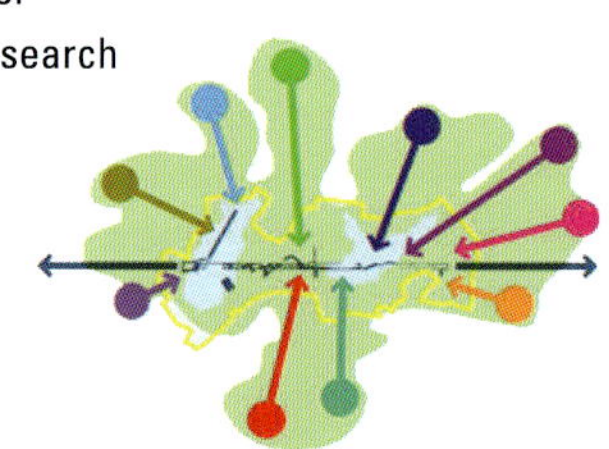

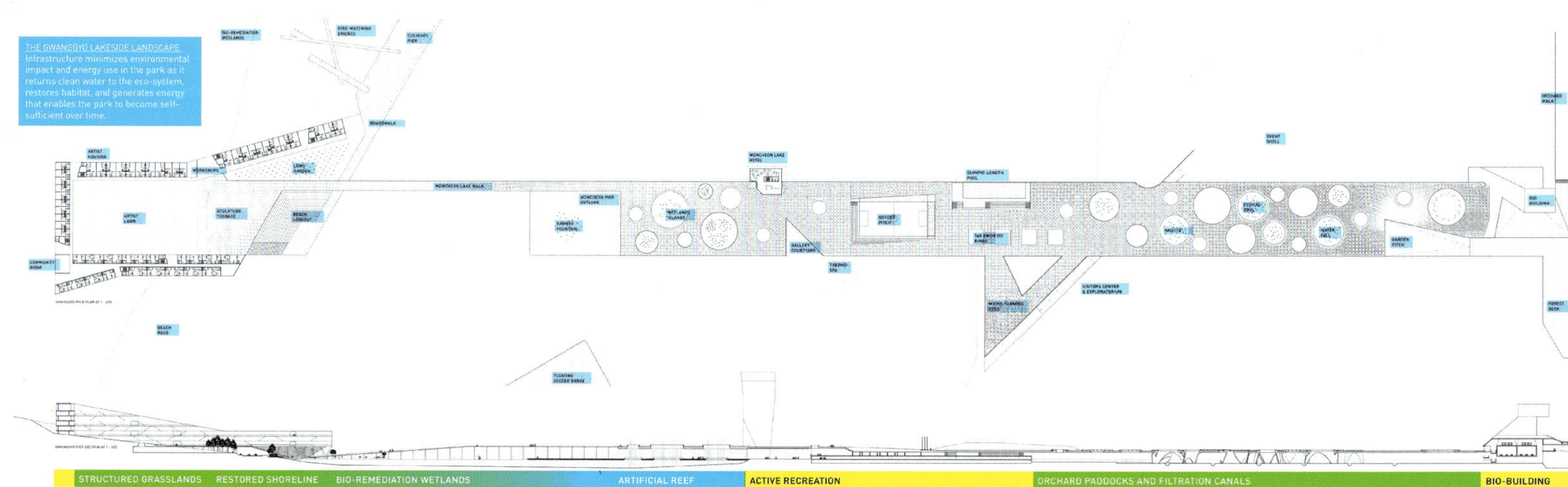

Foam bridge at orchard paddocks

From wetlands looking at culinary strip and lakeside hotel

CURATE THE PROGRAM/EMBRACE CHANGE

- LINK UP all the active programs into a central spine
- CREATE a single cultural route through the site
- EXPAND the activities and programs in the park

- 핵심부에 필요한 활동적인 프로그램 연계
- 부지를 가로지르는 독창적 문화루트 창조
- 공원에 있는 활동들과 프로그램들의 확대

MANUFACTURE A NEW ECOLOGY ON SITE

- INTERGRATE architecture and landscape
- REDEFINE architecture as biological machine
- CULTIVATE productive landscapes

- 건축과 조경의 통합
- 생물학적 기기로서 건축의 재정의
- 생산적인 조경의 경작

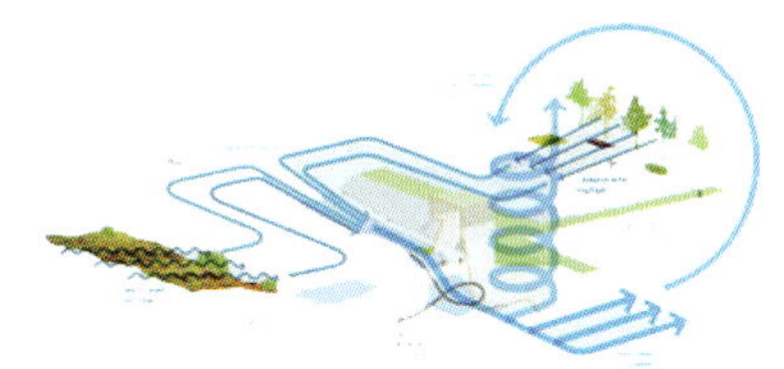

RE-ACTIVATE THE FIELDS/FOSTER DIVERSITY

- RESTORE AND RENEW existing forest habitats
- DISTRIBUTE learning and quiet recreation within the site
- DIVERSIFY the biomass of the site with new plantingsprograms in the park

- 기존의 산림 서식지들의 쇄신과 부흥
- 부지 내에 조용하고 교육적인 레크레이션에 기여
- 새로운 식재로 인한 생물종의 다양화

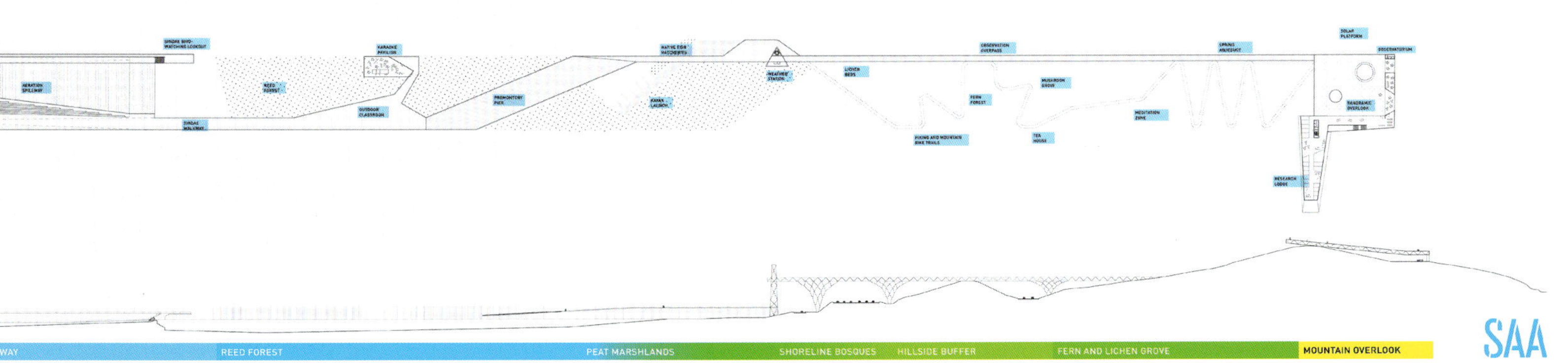

SAA

CUT-AWAY AXONOMETRIC SHOWING WATER FOAM BRIDGE AND FILTRATION BEDS

ECOLOGY
- consolidation and extension of high-impact forest fields which link up to local greenways creating a region green network
- restoration of riparian and water ecologies improving the environment and creating opportunities for active use and passive recreation
- pier contains active programs in a defined site and functions as a biological machine to clean and filter water and air

CULTURE
- major new cultural venue for artists-in-residence and international exhibitions
- research stations and environmental education as programmatic theme
- art, music and culture that engages landscape, ecology, and the environment

MOVEMENT
Three overlapping systems:
- dispersed network of low-impact pedestrian and bicycle paths
- linear concentration of high-impact circulation along the route of the pier
- water taxis and active recreation at Woncheon Lake; canoes and kayaks at Sindae Lake

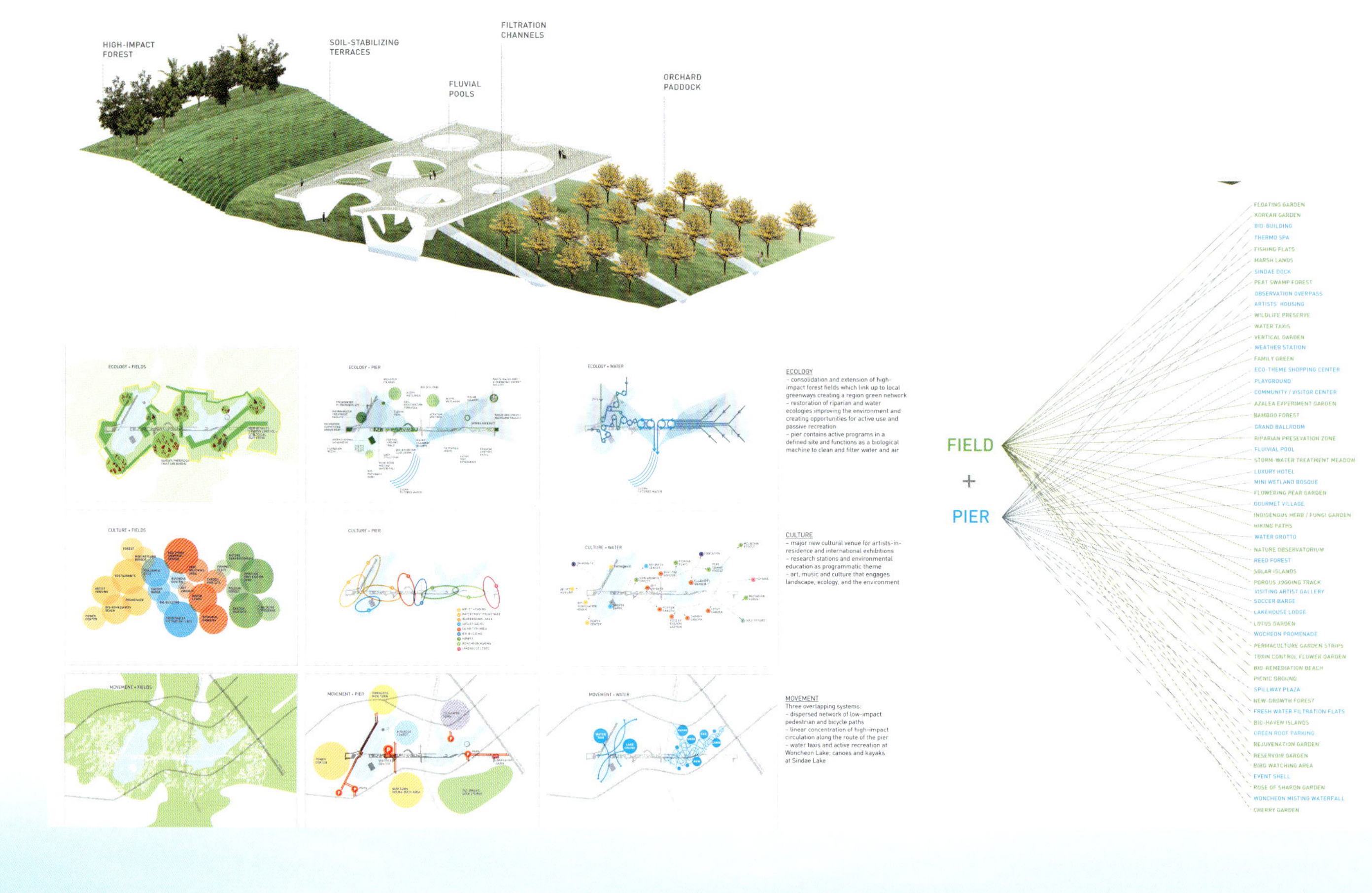

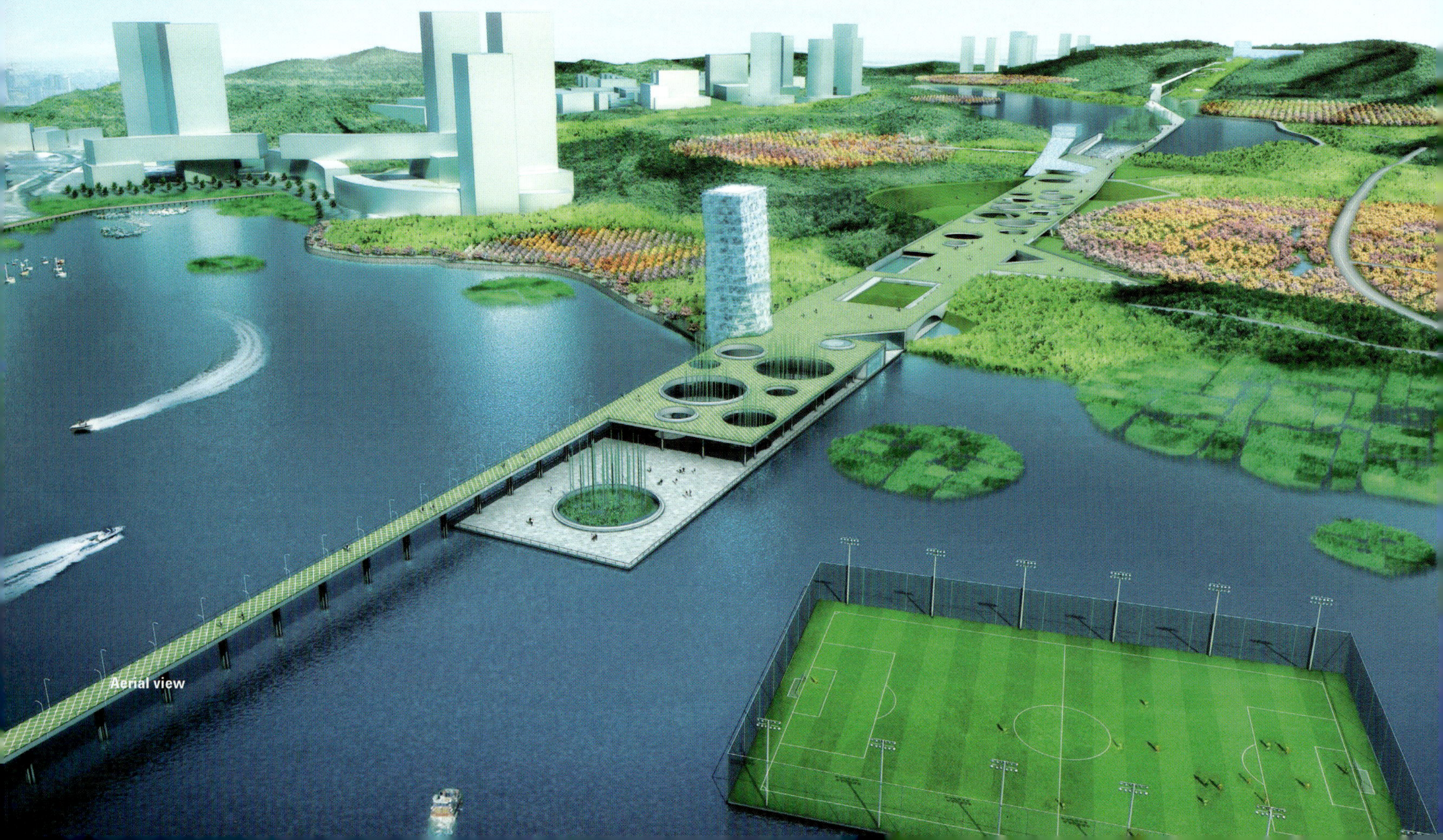

Aerial view

CUT-AWAY AXONOMETRIC SHOWING AERATION SPILLWAY AT SINDAE LAKESHORE

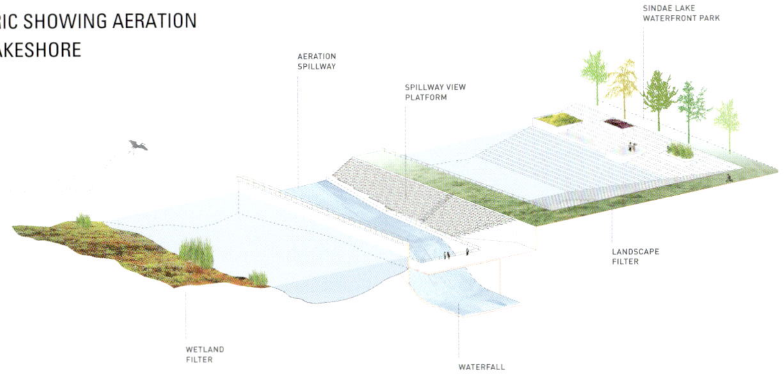

SINDAE LAKE WATERFRONT PARK

AERATION SPILLWAY

SPILLWAY VIEW PLATFORM

LANDSCAPE FILTER

WETLAND FILTER

WATERFALL

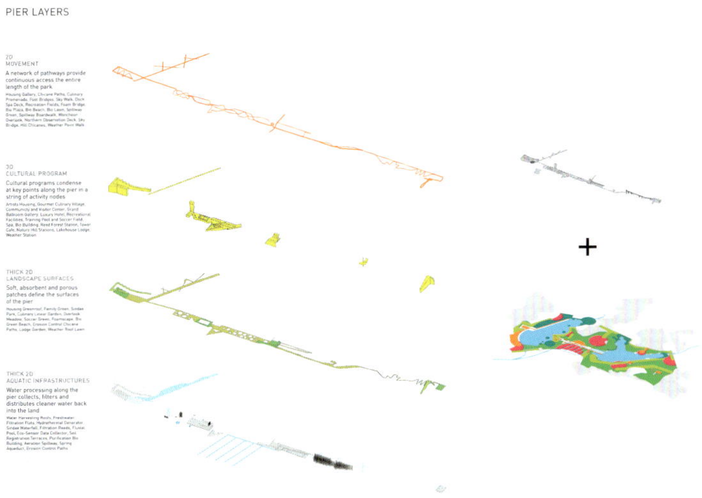

PIER LAYERS

2D
MOVEMENT

A network of pathways provide continuous access the entire length of the park

Housing Gallery, Chicane Paths, Culinary Promenade, Foot Bridges, Sky Walk, Dock Spa Deck, Recreation Fields, Foam Bridge, Bio Plaza, Bio Beach, Bio Lawn, Spillway Green, Spillway Boardwalk, Woncheon Overlook, Northern Observation Deck, Sky Bridge, Hill Chicanes, Weather Point Walk

3D
CULTURAL PROGRAM

Cultural programs condense at key points along the pier in a string of activity nodes

Artists Housing, Gourmet Culinary Village, Community and Visitor Center, Grand Ballroom Gallery, Luxury Hotel, Recreational Facilities, Training Pool and Soccer Field, Spa, Bio Building, Reed Forest Station, Tower Cafe, Nature Hill Stations, Lakehouse Lodge, Weather Station

THICK 2D
LANDSCAPE SURFACES

Soft, absorbent and porous patches define the surfaces of the pier

Housing Greenroof, Family Green, Sindae Park, Culinary Linear Garden, Overlook Meadow, Soccer Green, Foamscape, Bio Green Beach, Erosion Control Chicane Paths, Lodge Garden, Weather Roof Lawn

THICK 2D
AQUATIC INFRASTRUCTURES

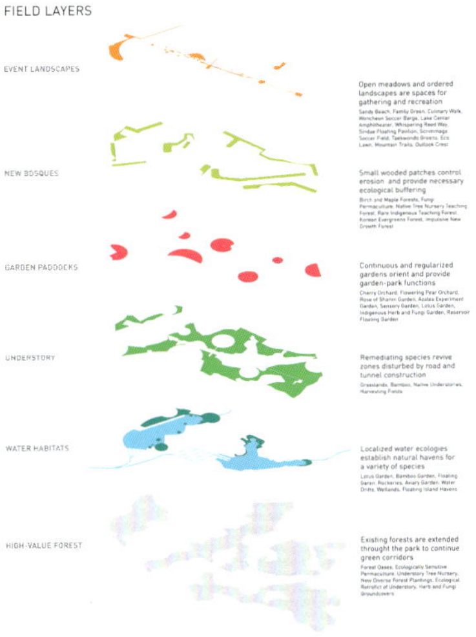

Water processing along the pier collects, filters and distributes cleaner water back into the land

Water Harvesting Roofs, Freshwater Filtration Flats, Hydrothermal Generator, Sindae Waterfall, Filtration Reeds, Fluvial Pool, Eco-Sensor Data Collector, Soil Registration Terraces, Purification Bio Building, Aeration Spillway, Spring Aqueduct, Erosion Control Paths

+

FIELD LAYERS

EVENT LANDSCAPES

Open meadows and ordered landscapes are spaces for gathering and recreation

Sandy Beach, Family Green, Culinary Walk, Woncheon Soccer Barge, Lake Center Amphitheater, Whispering Reed Way, Sindae Floating Pavilion, Scrimmage Soccer Field, Taekwondo Greens, Eco Lawn, Mountain Trails, Outlook Crest

NEW BOSQUES

Small wooded patches control erosion and provide necessary ecological buffering

Birch and Maple Forests, Fungi Permaculture, Native Tree Nursery Teaching Forest, Rare Indigenous Teaching Forest, Korean Evergreens Forest, Impulsive New Growth Forest

GARDEN PADDOCKS

Continuous and regularized gardens orient and provide garden-park functions

Cherry Orchard, Flowering Pear Orchard, Rose of Sharon Garden, Azalea Experiment Garden, Sensory Garden, Lotus Garden, Indigenous Herb and Fungi Garden, Reservoir Floating Garden

UNDERSTORY

Remediating species revive zones disturbed by road and tunnel construction

Grasslands, Bamboo, Native Understories, Harvesting Fields

WATER HABITATS

Localized water ecologies establish natural havens for a variety of species

Lotus Garden, Bamboo Garden, Floating Garen, Rockeries, Aviary Garden, Water Drifts, Wetlands, Floating Island Havens

HIGH-VALUE FOREST

Existing forests are extended throught the park to continue green corridors

Forest Oases, Ecologically Sensitive Permaculture, Understory Tree Nursery, New Diverse Forest Plantings, Ecological Retrofit of Understory, Herb and Fungi Groundcovers

FIELD PLANTINGS

From aeration spillway deck looking at sindea lake

JOURNEY OF WATER

WRT

Partners _ DANU 단우건축사사무소 + Han, Kwang Ya 한광야 + Keih Bowers + Hwang, Bo Chul 황보철

Frame Work

New Park Paradigm

Gwanggyo Lakeside Park lies at the confluence of two regional watersheds - a receiving basin for the Mt. Gwanggyo water system. Over time, this water system has been modified to support human activity. A hydrological landscape once altered to create dammed reservoirs for agricultural fields now awaits transformation into a new type of urban park. Our proposal envisions Gwanggyo Lakeside Park as an integrated urban water infrastructure that works symbiotically with the city to cultivate community, economic development, well-being, and complete and connected urban lives.

As water journeys through Gwanggyo Lakeside Park, it connects the city to nature and completes the cycles of the city. This multifunctional symbiotic role of Gwanggyo Lakeside Park is the new park paradigm. Yesterday's idea of an urban park as a bounded green box within a city or at the edge of a city no longer suits the needs of tomorrow's city.

Viewing the Woncheon waterfalls at the library / Culture center plaza

광교호수공원은 광교산으로부터 흘러드는 두 개의 물줄기가 모인 합류부에 위치해 있다. 이러한 수계는 다양한 인간 활동에 도움이 되는 방향으로 조금씩 변해왔다. 예전에는 농사를 위한 저수지로 그 용도가 바뀌었다가, 이제는 새로운 유형의 도시 공원으로 변신할 채비를 차리고 있다. 우리가 제안하는 광교호수공원은 커뮤니티, 경제발전, 웰빙에 기여하고 도시생활과 밀접하게 연계된 통합적인 도시 수변 인프라이다. 광교호수공원에서 우리가 제안하는 '물의 여정' 은 자연과 도시를 연결하고, 도시의 순환체계를 완성시킬 것이다. 광교호수공원의 복합적인 공생 기능은 새로운 공원 패러다임이다. 도시 속에서 동떨어진, 혹은 도시 가장자리의 녹색 상자와 같은 과거의 공원은 더 이상 미래 도시에 부합하지 않는다.

Park form above

Responsive Landscape

The context of the region informs the design of the site. Three disconnected green ribbons reach out from Gwanggyo Mountain, enveloping and penetrating the site. Another distinguishing characteristic of the site and its context is its varying topography of landforms that sculpt two basins of water from two separate watersheds. These lakes are almost tangent to each other yet completely separated by a ridge line - this juxtaposition of proximity and separation makes this landscape truly unique. Gwanggyo Lakeside Park is strategically nestled at the focal point of these three disconnected greenways and two noncontiguous watersheds. At this important junction, the park plays a key role in connection - the park connects the three greenways and links the two watersheds.

사이트 디자인은 지역의 컨텍스트가 방향을 일러주는 법이다. 대상지의 경우, 광교산에서부터 뻗어나온 세 개의 분리된 그린웨이가 부지를 감싸고 지나고 있다. 부지의 눈에 띄는 또 다른 특징과 맥락은 두 갈래의 물줄기로부터 형성된 두 곳의 수원이 형성하는 다양한 지형에 있다. 산등성이에서 완전히 떨어져 나온 이 두 개의 호수는 아주 근거리에 위치해 있는데, 경관적으로 매우 독특한 형태를 띄고 있다. 광교호수공원은 이 세 개의 분리된 그린웨이와 두 개의 인접한 호수가 형성하는 지형에서 전략적으로 중요한 위치에 자리잡고 있다. 때문에 앞으로 호수공원은 세 개의 그린웨이와 두 개의 수계를 연결하는 중요한 역할을 담당해야 한다.

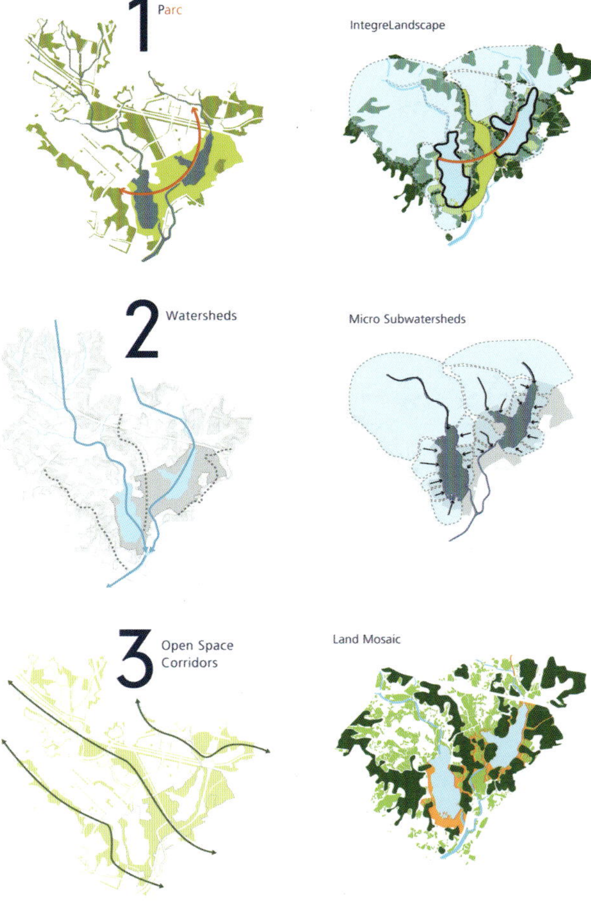

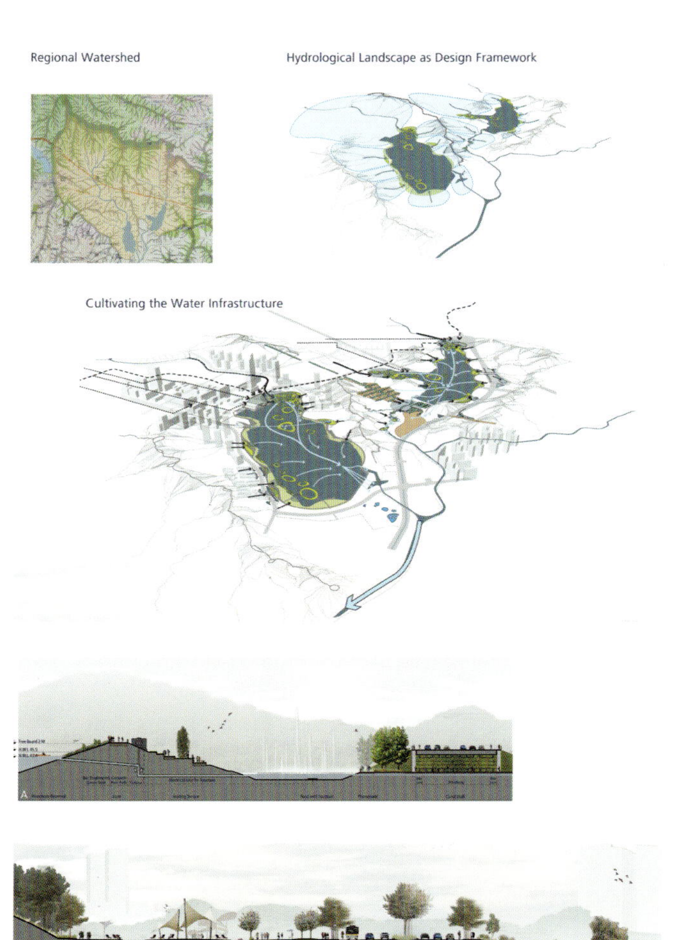

Role of Water

The vision of Gwanggyo Lakeside Park is not complete without water. The role of water is regional and relational. The quantity and quality of water in Gwanggyo Lakeside Park is affected by water upstream in the respective local watersheds and the larger regional watershed. An increasing amount of stormwater runoff and a decreasing quality of this runoff water is an inevitable result of growing urbanization. Gwanggyo Lakeside Park is a resting place for water in its journey that begins at the watershed and continues into Seoh Hae (West Sea) of Korea. The hydrological pattern on the site sets the framework for the design of the park. The flow of water informs the location of programmed spaces on the water surface - floating wetlands in the path of water flow slows down the speed of water and improves the quality of water.

광교호수공원의 미래는 '물' 이 완성하는 것이라고 해도 과언이 아니다. 여기에서 물의 역할은 지역적이며 상관적이다. 광교호수공원에서 물의 양과 질은, 각각 근거리 지역 수계와 광역 지역 수계 내에서 상류 지역의 물의 흐름에 영향을 받는다. 호우로 인한 유거수 양의 증가와 수질의 저하는 도시화 과정에서 피할 수 없는 부분이다. 수계로부터 시작된 물이 서해 바다로 흘러가는 물의 여정에서, 광교호수공원은 거점의 역할을 하게 될 것이다. 대상지에서 수문학적인 패턴은 공원 설계를 위한 기본적인 뼈대라 할 수 있다. 물의 흐름은 수면에서, 수질을 정화하고 유속을 조절하는 프로그램화된 공간의 위치를 알려준다.

Park & Program

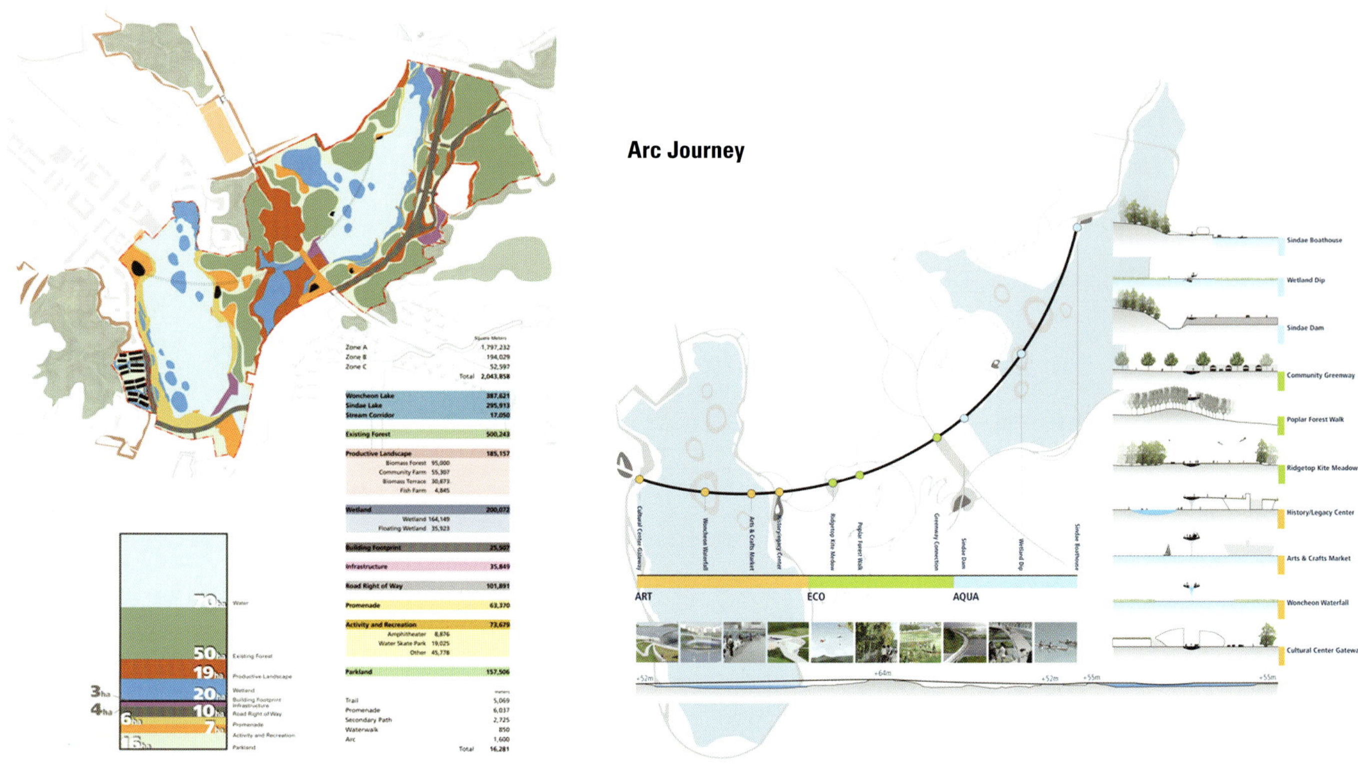

	Square Meters
Zone A	1,797,232
Zone B	194,029
Zone C	52,597
Total	2,043,858
Woncheon Lake	387,621
Sindae Lake	295,913
Stream Corridor	17,050
Existing Forest	500,243
Productive Landscape	185,157
Biomass Forest	95,000
Community Farm	55,307
Biomass Terrace	30,673
Fish Farm	4,845
Wetland	200,072
Wetland	164,149
Floating Wetland	35,923
Building Footprint	25,507
Infrastructure	35,849
Road Right of Way	101,891
Promenade	63,370
Activity and Recreation	73,679
Amphitheater	8,876
Water Skate Park	19,025
Other	45,776
Parkland	157,506

	meters
Trail	5,069
Promenade	6,037
Secondary Path	2,725
Waterwalk	850
Arc	1,600
Total	16,281

Connection

History/Legacy Center The Arc Community/Eco Center Greenway Corridor Water Pavilion Boathouse

Cultural Center/Library

Concept

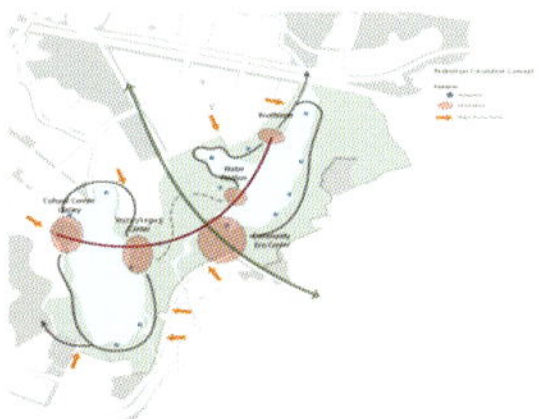

Regional Connections

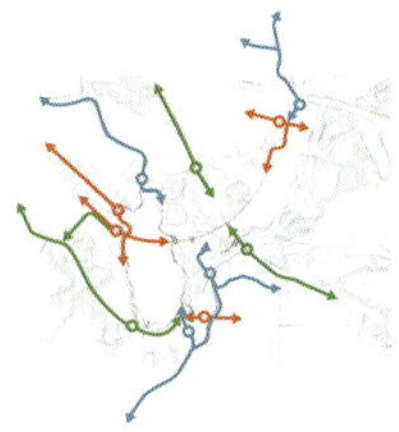

Sustainable Modes of Transportation

Vehicular

Pedestrian

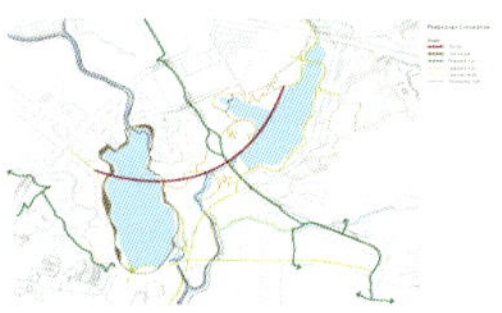

Completion

The Journey of Water completes the cycles between the city and park. Gwanggyo Lakeside Park cleans stormwater, recycles refuse, generates energy, promotes good health and well being, restores the native ecosystem, produces food, improves air quality - in essence, the park is a green living organism that connects people to their environment and completes the processes of city life.

물의 여정은 도시와 공원의 사이클을 완성시킨다. 광교 수변공원은 유거수 정화, 폐기물의 재생, 에너지 생산, 건강과 복지의 증진, 자생 생태계의 복원, 식량 생산, 삶의 질의 향상 등에 기여하고, 사람과 환경을 연계시키고, 도시 생활의 과정을 완성하는 살아있는 녹색 유기체로 기능할 것이다.

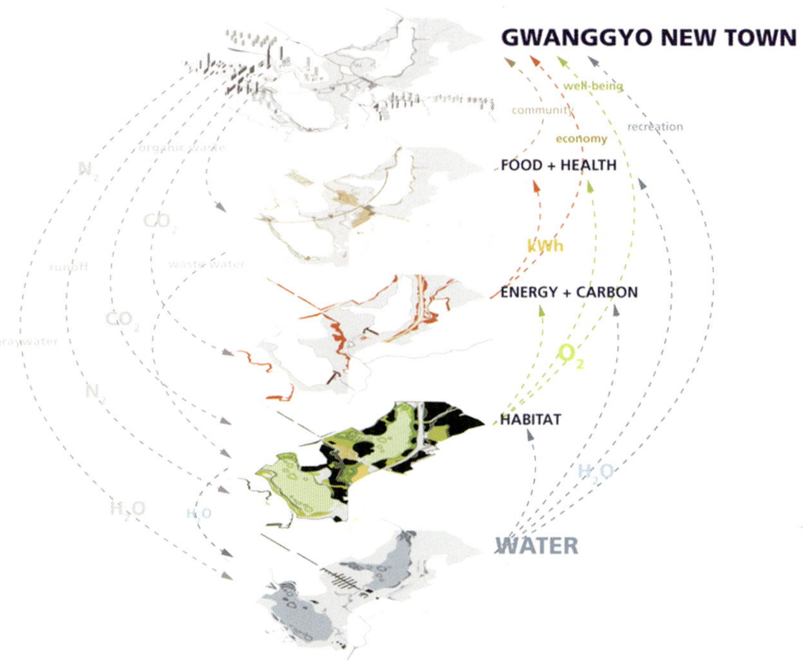

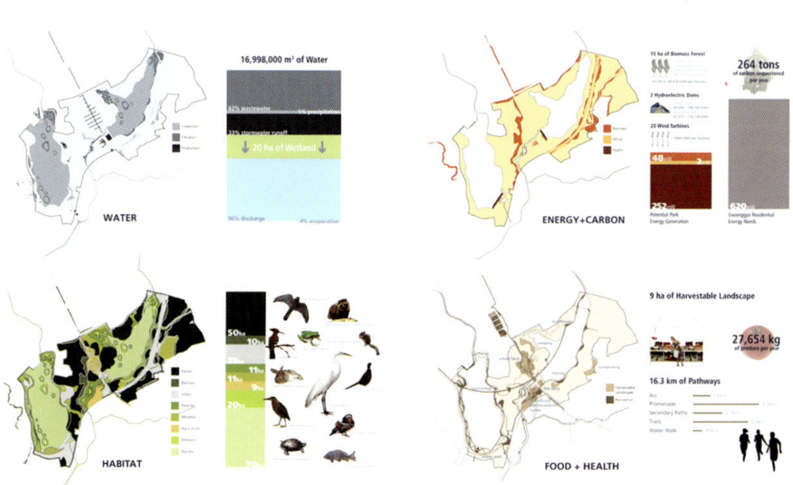

Bird watching at poplar forest

Fishing from the Sindae Piers

Observing wetland habitats from above and below

Floating wetlands on Sindae Lake

Terraced roof gardens along the hillside

Floating Wetland
Fish Pond
Turbine
Algae Pipes
Hatchery House
Marketplace
Arc Path
Riparian Wetland Edge w/ Stormwater Mitigation & Infiltration
Stream
Constructed Wetland
Algae Pool
Biofuel Crop Terrace
Park Path w/ Permeable Paving
Hydroponic Garden
Cistern
Community Wellness Center
Underground Parking
Rooftop Garden
Harvest Landscape
Road

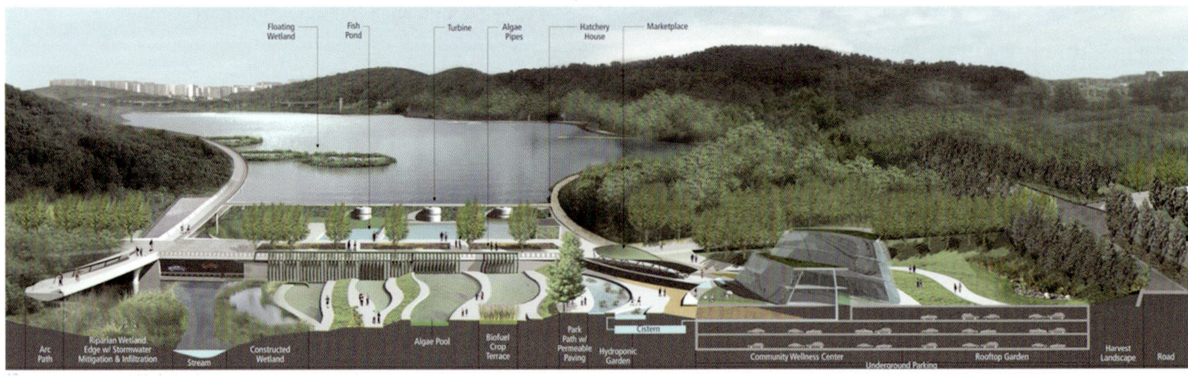

Section BB: Harnessing water and CO_2 for fish farming, algae cultivation, and community agriculture

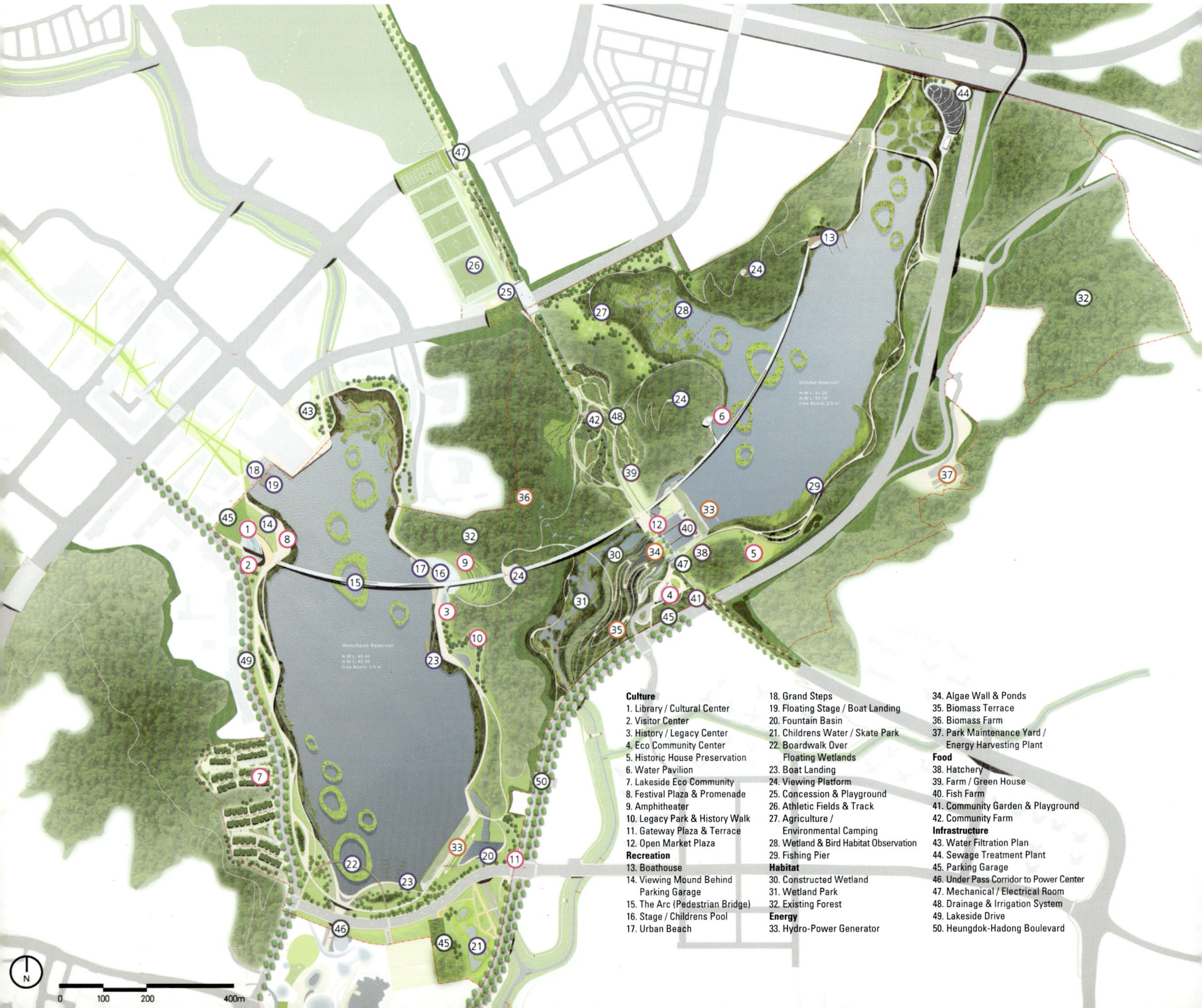

Culture
1. Library / Cultural Center
2. Visitor Center
3. History / Legacy Center
4. Eco Community Center
5. Historic House Preservation
6. Water Pavilion
7. Lakeside Eco Community
8. Festival Plaza & Promenade
9. Amphitheater
10. Legacy Park & History Walk
11. Gateway Plaza & Terrace
12. Open Market Plaza
Recreation
13. Boathouse
14. Viewing Mound Behind Parking Garage
15. The Arc (Pedestrian Bridge)
16. Stage / Childrens Pool
17. Urban Beach
18. Grand Steps
19. Floating Stage / Boat Landing
20. Fountain Basin
21. Childrens Water / Skate Park
22. Boardwalk Over Floating Wetlands
23. Boat Landing
24. Viewing Platform
25. Concession & Playground
26. Athletic Fields & Track
27. Agriculture / Environmental Camping
28. Wetland & Bird Habitat Observation
29. Fishing Pier
Habitat
30. Constructed Wetland
31. Wetland Park
32. Existing Forest
Energy
33. Hydro-Power Generator
34. Algae Wall & Ponds
35. Biomass Terrace
36. Biomass Farm
37. Park Maintenance Yard / Energy Harvesting Plant
Food
38. Hatchery
39. Farm / Green House
40. Fish Farm
41. Community Garden & Playground
42. Community Farm
Infrastructure
43. Water Filtration Plan
44. Sewage Treatment Plant
45. Parking Garage
46. Under Pass Corridor to Power Center
47. Mechanical / Electrical Room
48. Drainage & Irrigation System
49. Lakeside Drive
50. Heungdok-Hadong Boulevard
N
0
100
200
400m

Jury Comments
심사평

The Urban Reservoir Project prepared by Synwha Consulting for the Gwanggyo Lakeside Park was selected as the winner based on a strong and poetic design vision. This design vision was grounded in the reinvention of a forgotten landscape place- the urban reservoir- that holds deep cultural and social meanings for the local inhabitants charged with a new and fresh approach to ecology, nature and the sustainable future of Gwanggyo New City. In doing so, it provided a dispersed scheme of landscape and cultural programs and spaces.

The Committee in particular praised a number of the broad ideas in the submission- "the strong narrative" linking the tradition of reservoir levees and dumbungs(ponds that never go dry) as strong and subtle outdoor public spaces configured throughout the park layout. Discussion among the Committee members on the treatment of the two lakes resulted in energetic debate and acknowledgement that the design proposals differentiation of the treatment of each lake—one quiet, serene and verdant, the other more urbane, public and busy in character was a creative and sensitive response to the local conditions of the site. A controversial design idea that generated considerable discussion among the jurors was the expression of the western edge of Woncheon Lake, with multiple levels of parallel pathways, intersecting and weaving a distinctive and complex braided circulation pattern. While the Committee recognized that editing of these elements would produce a more coherent planning approach the existing sectional relationship to the lake was considered by some jurors as producing a rich and imaginative experience by the water edge.

In summary, the Committee praised in this scheme the thorough development of the ideas and natural and cultural processes expressed on the site as an example of the new Korean urban park that would provide local, national and international interest, engagement and pleasure.

Text _ Niall Kirkwood

광교신도시 호수공원 국제설계공모에 (주)신화컨설팅이 제출한 "Urban Reservoir"가 당선작으로 선정되었다. 이 작품의 강하면서도 시적인 디자인 비전은 우리가 잃어버린 경관과 장소—도시적 저수지—의 가치를 재발견하는데 토대를 두고 있다. 이 작품은 광교신도시의 생태, 자연, 지속가능한 미래에 대해 새롭고 신선한 접근을 보이고 있으며, 이러한 과정을 통해 전반적인 경관의 틀과 문화적 프로그램 및 공간을 제시했다. 이 작품은 광교 주민들에게 매우 깊은 문화적·사회적 의미를 부여해 줄 것이다.

심사위원회는 특히 이 작품의 폭넓고 다양한 아이디어—저수지 제방과 둠벙(마르지 않는 연못)의 전통을 잇는 "강한 내러티브"와 공간 전반에 편재한 섬세한 공공공간—를 높이 샀다. 두 호수의 처리에 대한 심사위원들간의 토론은 열정적인 논쟁을 낳았는데, 각 호수를 차별적으로 다룬 점—하나는 조용하고 고요한 반면, 다른 하나는 보다 도시적이고 공적이며 복잡한—은 대상지의 지역적 조건에 매우 창조적이고 민감하게 대응한 특징이라고 평가받았다. 원천저수지 서측 호안에 복잡하게 휘어지며 나란히 가는 다층적인 동선 패턴을 특징적으로 설계하였는데, 이는 여러 심사위원들로부터 논쟁적인 설계 아이디어라는 평가를 받았다. 심사위원회는 이러한 설계 요소가 일부 조정된다면 더 적절한 계획안을 생산할 수 있을 것이라고 보았다. 하지만 일부 심사위원들은 이러한 동선이 보여주고 있는 호수와의 입체적 관계가 수변에 매우 풍요롭고 생산력 넘치는 경험을 제공할 것이라고 보기도 하였다.

요약하자면, 심사위원회는 대상지의 자연적·문화적 과정을 존중한 이 작품의 설계 아이디어를 매우 높게 평가했다. 이 작품이 보여주는 공원의 모습은 광교신도시와 한국은 물론 국제적인 관심과 즐거움을 제공할 수 있는 새로운 한국적 도시공원의 패러다임을 제시할 것이다. 글 _ 니얼 커크우드

심사위원장

Niall Kirkwood · 하버드대학교 조경학과 교수

심사위원

Kim, Kee Ho 김기호 · 서울시립대학교 도시공학과 교수
Kim, Young Dae 김영대 · 대구시 도시디자인총괄본부장 / 영남대 교수
Seung, Hyo Sang 승효상 · 이로재 대표
Zoh, Kyong Jin 조경진 · 서울대학교 환경대학원 환경조경학과 교수
Dennis Pieprz · Sasaki Associates 대표
Ken Smith · Ken Smith Landscape Architects 대표
Peter Latz · Latz+Partners 대표

전문위원

Im, Seung Bin 임승빈 · 서울대학교 조경 / 지역시스템공학부 교수

Magok Waterfront

마곡 워터프론트 국제현상공모

Location	**위치**
Neighborhood of Magok-dong, Gangseo-gu, Seoul, Korea	서울특별시 강서구 마곡동 일대
1,170,780㎡ **Area**	**면적** 1,170,780㎡
Promoters	**공모주최**
Governments of Seoul Metrocity, SH Corporation	서울특별시, SH공사
Schedule	**공모일정**
Announcement of Design Guideline 2007. 11. 15	설계지침제공 2007. 11. 15
Registration 2008. 1. 31	참가등록 2008. 1. 31
Final Submission 2008. 6. 5	작품제출 2008. 6. 5
Announcement of Winner 2008. 6. 24	당선작발표 2008. 6. 24
See the results on-line	**공식홈페이지**
www.magokwaterfront.org	www.magokwaterfront.org

심사위원

조경 _ Grant Jones · 심사위원장, Jones and Jones 조경설계사무소장(미국)
Kim, Sin Won 김신원 · 경희대학교 교수(한국)

건축 _ Choi, Doo Nam 최두남 · 서울대학교 교수(한국)
Elizabeth Mortamais · Paris Val de Seine Univ. 교수(프랑스)

도시 _ Ahn, Geon Hyeck 안건혁 · 서울대학교 교수(한국)
Rodolfo Machado · Harvard Univ. 교수(미국)
Gregory Yager · RTKL부사장(미국)

수자원 _ Song, Jae Woo 송재우 · 홍익대학교 교수(한국)
Ishii Yumio · 일본토목학회 회장(일본)

전문위원

Kim, Chang Seok 김창석 · 대한국토 · 도시계획학회 회장, 서울시립대학교 교수(한국)

The point of Magok prize competition was to utilize the purpose of specialized space formation that enables cultural dynamic power and energy of the existing Magok district, maximize the sources of Han River and the Lake Park and harmonize all the factors for the project progress. The 1st prize winning 'Heart of Magok is Nature of Living Water', which is based on continuously flowing water suggests not just one perfectly completely completed park but a park with the possibility of further development and expansion under the goal of making park where nature and human beings show active life together in harmony. This forms an urban center to support water, nature, culture and community and enables Magok district to work as the gateway of Han River. This design was evaluated to consider environmental parts in a wide range by suggesting the way to utilize water and wind as important energy sources in connection with energy problems.

서울시와 SH공사는 한강 르네상스 사업 중 한강 권역별 특화사업의 일환으로 추진 중인 마곡 워터프론트를 세계적인 수변도시로 만들기 위해 국내 · 외 전문가의 폭 넓은 아이디어를 얻고 설계안을 수립하고자 국제현상공모를 실시했다. 총 41개국 439개팀(국내 235개, 국외 204개)이 참가 등록한 가운데, 최종적으로 105개 작품(국내 45개, 국외 15개국 60개)이 접수되었으며, 총 7차례의 심사를 거쳐 당선작을 선정했다.

1등에는 'Heart of Magok is Nature of Living Water' 라는 작품을 제출한 김관중이 차지했으며, 2등은 은동신, 3등은 David Leven과 조주환의 2개 작품이, 가작에는 Coxall Philip, Valentin Oleynik 및 Lee Jeff의 3개 작품이 수상했다.

1등 당선작은 자연과 커뮤니티를 잘 엮어 조화시킨 작품으로 물과 자연, 문화와 커뮤니티를 지원하는 독특한 도시적 중심을 조성함으로써, 마곡지구를 한강의 관문으로서 기능하게 하며, 특히 동서로 분리된 두 호수 사이에 길게 뻗은 녹색의 제방을 적절하게 위치시킴으로써 수면에 접하는 길이를 최대화함은 물론 '수변과 한강접근에 대한 시민권리의 회복' 을 확실하게 구현한 작품이라는 평가를 받았다.

1st Prize

Heart of Magok is Nature of Living Water

Kim, Kwan Joong 김관중 · Architect, Korea | **Partners** _ Yu, Deok Hee 유덕희

M-Park's heart is at the core of a 'state-of-the-art' hightech research development (the Magok Development District) located on the western fringes of Seoul with direct linkages to the Han River. M-Park, 'the LIVING WATER', proposes a unique vision for this 21st-century eco-urbanscape posited to redefine contemporary conditions between urban/land/water/people. With Re-Generation, Re-Cycle and Re-Connection as overarching principles, the park operates as a new sustainable fabric stimulating creative growth within the park, around the park as well as the Han River.

On 1,170,780㎡ of open farmland directly adjacent to the Han River, M-Park will be the largest connected 'waterpark' in Seoul and the first in a series of Han River. Renaissance Plan projects to restore this important feature in the life of this city. Our proposal activates multiple conditions of water as a catalyst for a natural eco-system, not simply a restoration of nature, but rather a complex union between advanced technology, nature and man originating in an innovative concept/idea of an eco-friendly community.

By consciously experimenting with different procedures of aqua-management, the land design prioritizes unique operational strategies such as water filtration, land sculpting and shaping, water retention, vegetation density, growth patterns and water circulation to create a diverse ensemble of natural spaces. These diverse ecologies each provide discrete environments and operational modes while naturally finding their balance and equilibrium as a whole.

These areas are as follows : the Lake Park, the Green Levee, the Marina / Ferry Terminal, the Commercial & Convention Center, the Reservoir Park, the Eco-Park, the Flood Gate. Architecture is the three-dimensional transformative layer creating synthesis between the multiple layers of the earth, water and sky. As water is the central thesis for the park, an extensive water management system has been developed where filtration and revitalization is vital to the life and health of the park. An extensive water filtration system provides for the reuse of grey water and storm water runoff for site irrigation as well as healthy recreation water system.

The use of renewable and recycled materials will provide tangible evidence of responsive stewardship of the natural resources and further perpetuates the sustainable goals of the development. Carbon neutrality through smart development, forested growth and sustainable natural environments is of high priority within the district. Reconnecting the organism of life, nature and society is the first step towards sustainable longevity. On-site power creation provided by clean renewable energy sources allows M-Park to thrive completely outside the regulations of the municipal power grid. Solar and wind energy both reinforce the independent spirit of the development.

M-Park의 핵심이자 하이테크 연구개발단지인 마곡지구는 한강과 직접 연결된 서울의 서쪽 변두리에 위치한 곳이다. '살아있는 물' 을 중심으로 한 M-Park는 도시, 토지, 물, 사람 간에 있어서 현대적 조건을 재정의한다는 의미로 21세기형 eco-urban scape의 독특한 비전을 제시하고 있다.

한강과 인접한 1,170,780㎡ 규모의 개방된 농경지에 자리하는 M-Park는 서울시 최대의 연결형 워터파크가 될 것이며, 한강 르네상스 플랜은 이 도시의 중요한 특징인 한강을 복원하는데 있다. 설계안은 자연친화적 에코시스템을 위한 촉매제로서 물이 가지는 다각적 조건을 활성화하는 것으로 단순히 자연적인 외형을 복원하기 위한 것이 아니라 보다 발전된 기술로 인간과 자연의 파생된 환경 친화적인 커뮤니티의 혁신적인 컨셉을 구축하는데 있다.

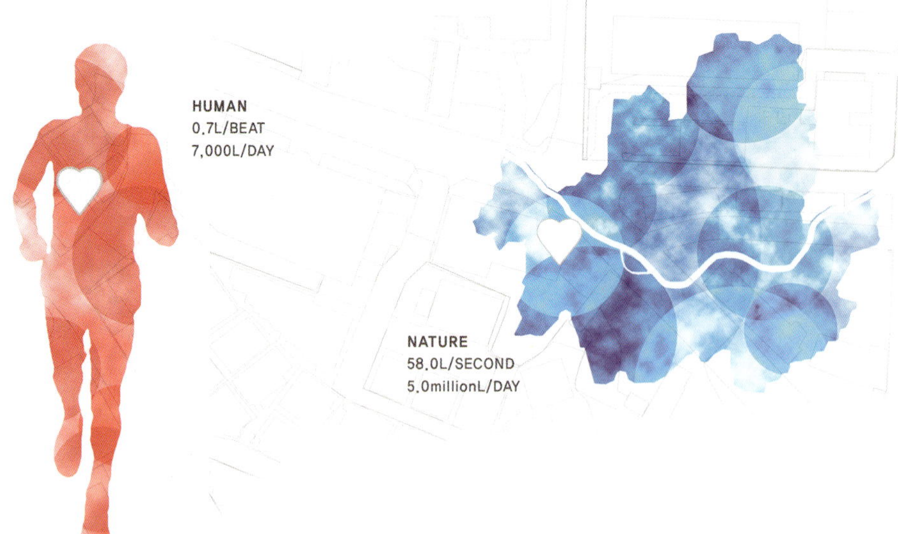

LANDUSE

NAME	AREA	%
Convention/hotel	70,000	5.98
Retail/waterfront	66,776	5.70
Boat facilities	37,623	3.21
Community	21,466	1.83
Pump	10,239	0.87
Intrastructure	49,994	4.27
Flood gate	6,776	0.58
Aqua	26,976	2.30
River front park	37,698	3.22
Green levee	68,592	5.86
Eco park	242,955	20.75
Reservoir	114,900	9.81
Lake park	63,936	5.46
Ferry terminal	7,408	0.63
Marina	27,111	2.32
Water way	210,977	18.02
Lake	84,173	7.19
Road / rail road	23,180	1.98
TOTAL	1,170,780	100.00

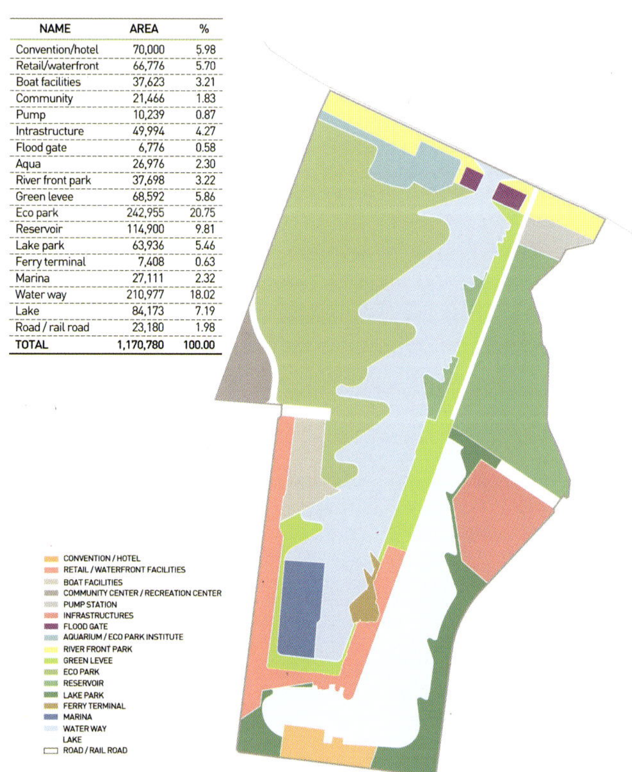

토지설계를 위해 수자원관리의 여러과정을 의도적으로 시도함으로써 정수작용, 부지형성 및 구축, 관리, 식생의 밀도 및 패턴 등과 같은 운영 전략에 대해 우선 순위를 두었으며, 이는 자연공간의 다양한 조화를 구축하는데 있다. 이렇게 복합적인 환경보전정책들은 자연과 인간과의 자연스런 평형관계 및 균형을 모색함과 동시에 개별적인 환경과 운영적인 방식을 제공한다.

이러한 정책들이 포함된 지역은 호수공원, 녹화된 제방, 강변산책로/페리선착장, 상업 및 컨벤션센터, 저수지공원, 생태공원, 수문 등이다. 건축은 지면, 강, 하늘로 구성된 복합적 레이어들을 포괄하여 3차원으로 변형된 레이어로 제작된다. 물은 공원에서의 핵심적인 요소로, 물의 여과작용 및 활성화는 공원의 보존상태 및 생활에 필요한 요소를 내포한 광범위한 수처리 시스템을 발전시켜왔다. 물의 여과시스템은 중수도 용수 및 우수를 재활용함으로써 관개용수로 제공하는데 있다.

사용할 수 있는 자재들을 재활용하는 것은 천연자원을 보존한다는 책임감에 대한 실질적인 증거이며 자원개발을 지속시키고자 하는 목적을 영속시킬 수 있음을 보여준다. 현명한 개발정책을 통하여 이산화탄소를 중성화하는 것이나 자연환경을 지속하고 성장시키고자 하는 것이야 말로 가장 우선순위이며, 인간의 생활과 자연 및 사회의 유기조직을 재결합하는 것은 생명을 보다 지탱할 수 있는 것에 한걸음 앞서감을 의미한다.

친환경적인 자원의 재활용으로 현장에서 가용할 수 있는 에너지를 창출해 낸다는 것은 M-Park가 시 자체에서 소비되는 전력상황에 대한 규제로 부터 벗어날 수 있음을 말한다.

CONNECTIONS

HOMO

- WATER TO WATER
- PEDESTRIAN TO PEDESTRIAN
- PARK TO PARK
- URBAN TO URBAN
- BLOCK TO BLOCK
- INFRASTRUCTURE TO INFRASTRUCTURE
- ECOSYSTEM TO ECOSYSTEM

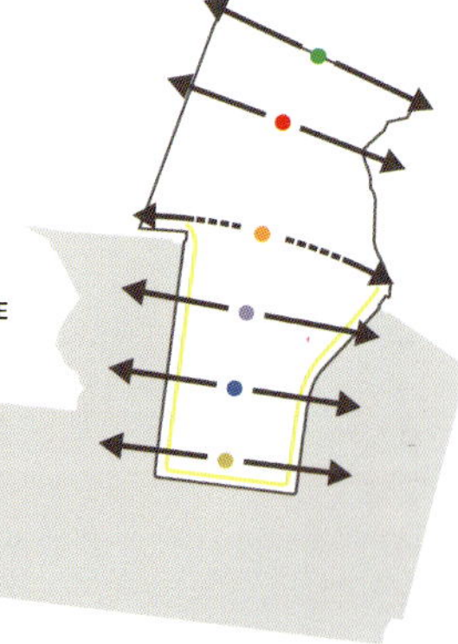

HETERO

- OLD TO NEW
- DIRTY WATER TO CLEAN WATER
- GROUND TO UNDERGROUND
- GROUND TO ABOVEGROUND
- WATER TO LAND
- OFFICE TO R&D
- NEIGHBORHOOD TO CITY
- COMMUNITY TO COMMUNITY

URBAN AXIS

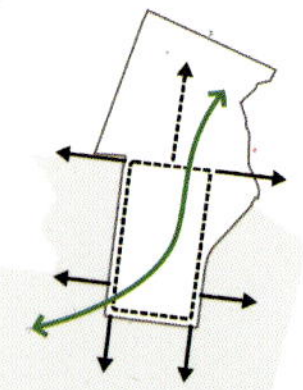

GREEN AXIS

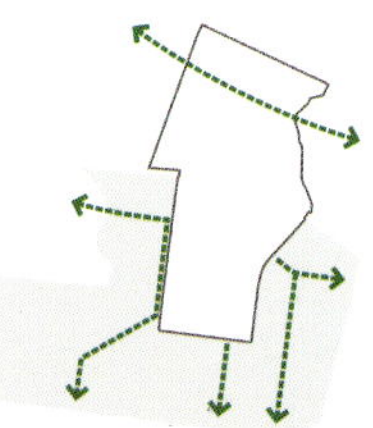

BLUE AXIS

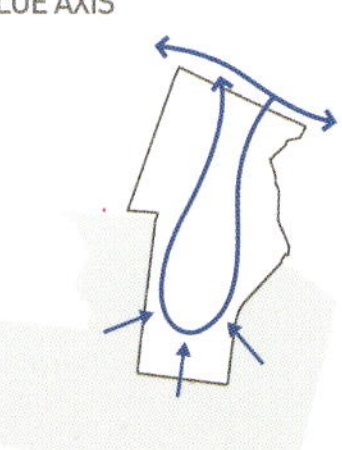

NATURE / ECO

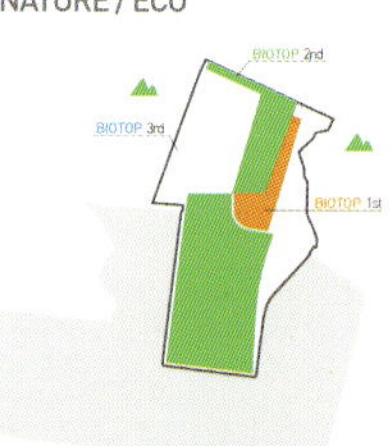

ACTIVITY ACCESS

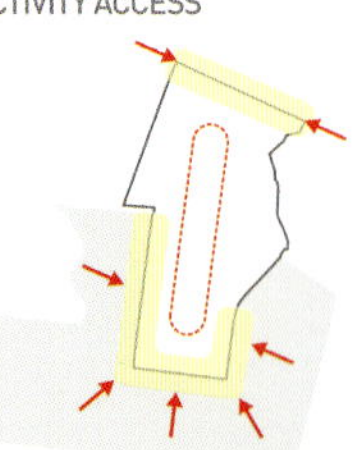

WATER ACCESS

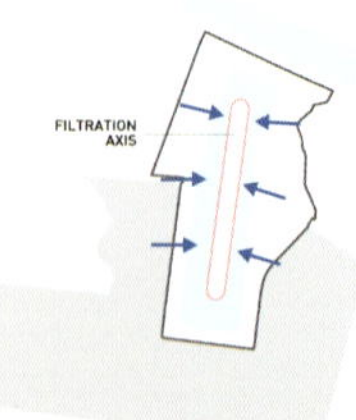

PROGRAM

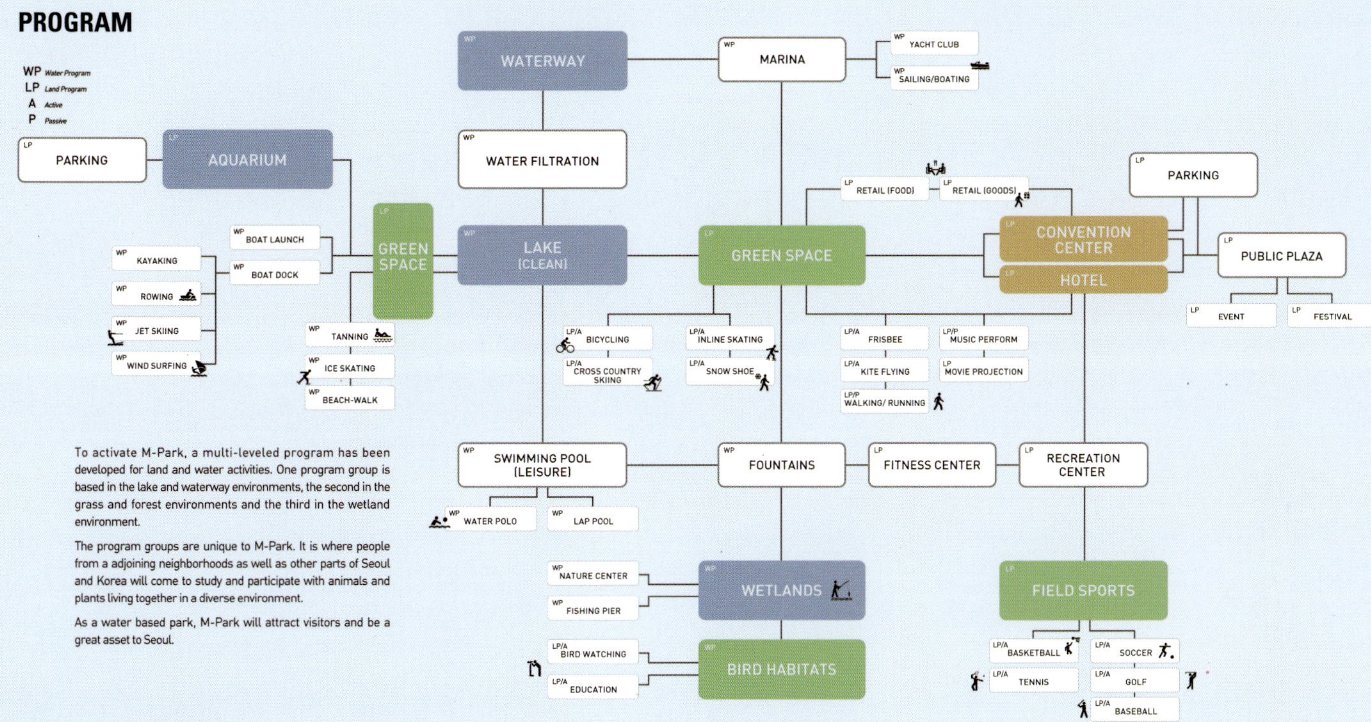

To activate M-Park, a multi-leveled program has been developed for land and water activities. One program group is based in the lake and waterway environments, the second in the grass and forest environments and the third in the wetland environment.

The program groups are unique to M-Park. It is where people from a adjoining neighborhoods as well as other parts of Seoul and Korea will come to study and participate with animals and plants living together in a diverse environment.

As a water based park, M-Park will attract visitors and be a great asset to Seoul.

METRIC SEQUENCE

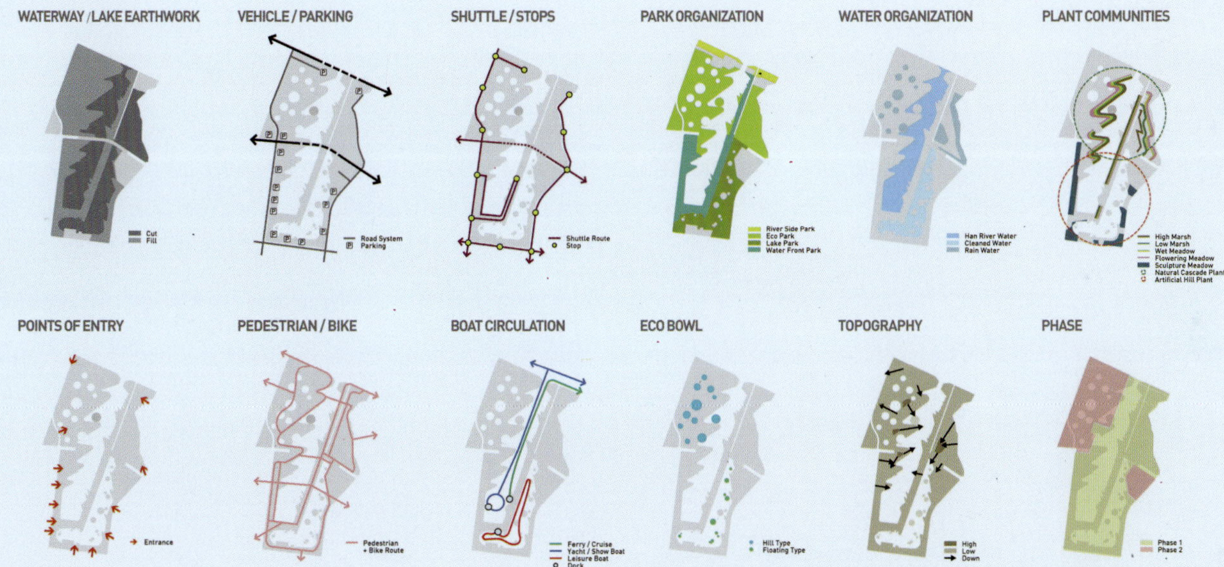

Eco Park

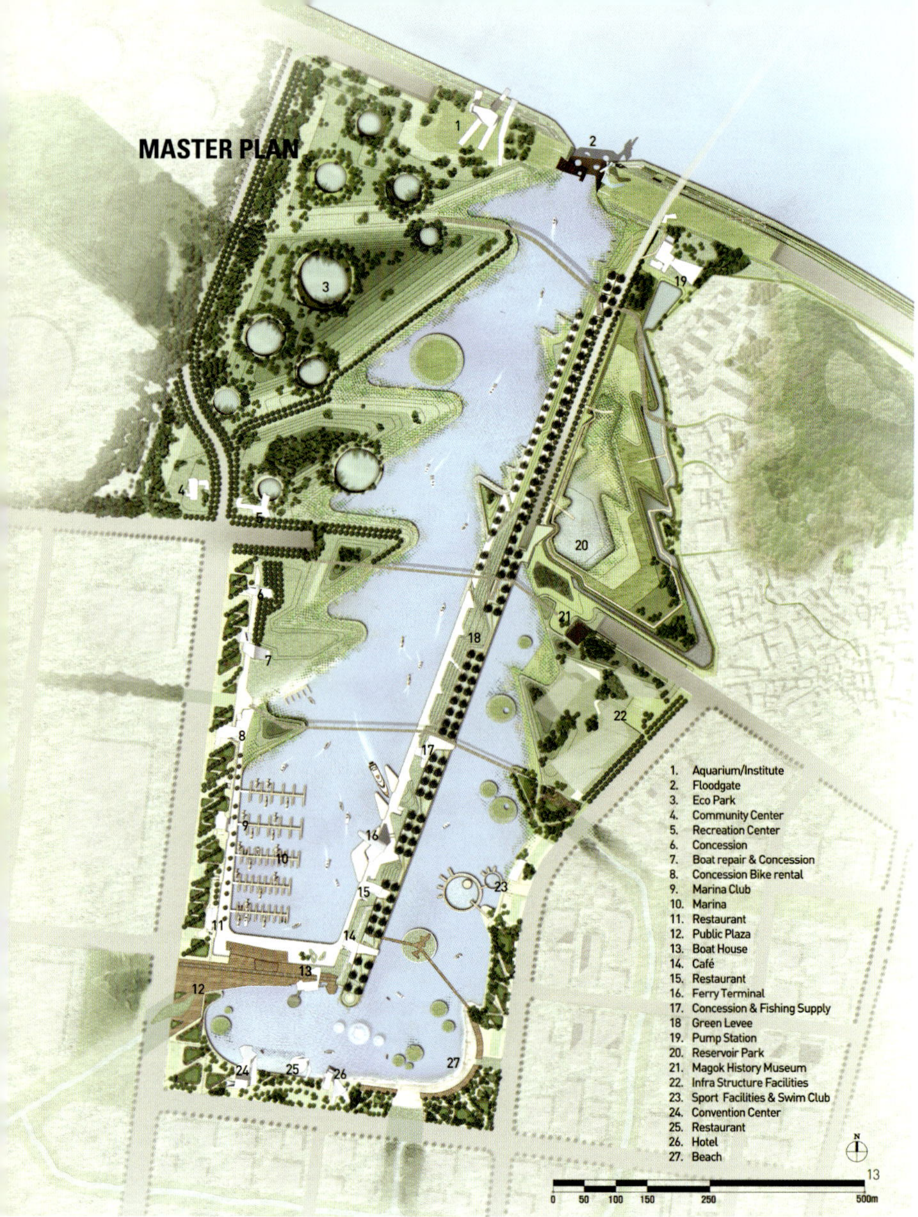
MASTER PLAN
1. Aquarium/Institute
2. Floodgate
3. Eco Park
4. Community Center
5. Recreation Center
6. Concession
7. Boat repair & Concession
8. Concession Bike rental
9. Marina Club
10. Marina
11. Restaurant
12. Public Plaza
13. Boat House
14. Café
15. Restaurant
16. Ferry Terminal
17. Concession & Fishing Supply
18 Green Levee
19. Pump Station
20. Reservoir Park
21. Magok History Museum
22. Infra Structure Facilities
23. Sport Facilities & Swim Club
24. Convention Center
25. Restaurant
26. Hotel
27. Beach
N
0 50 100 150 250 500m
13

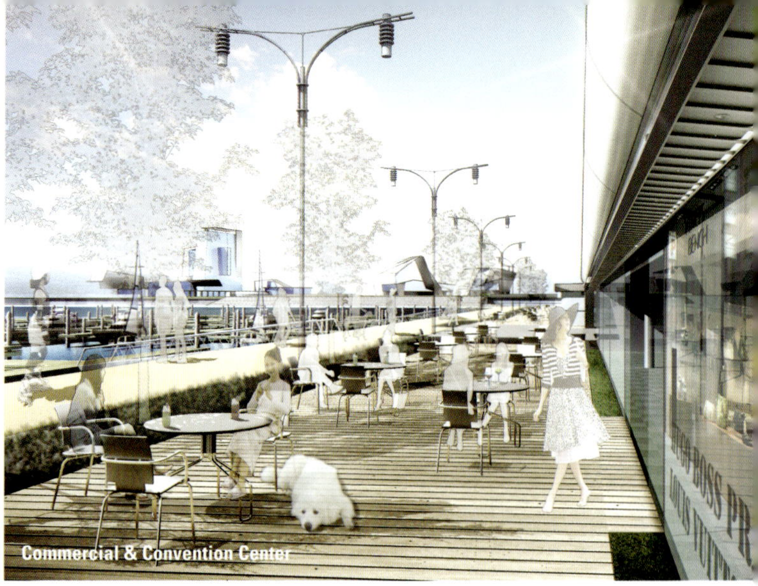

Commercial & Convention Center

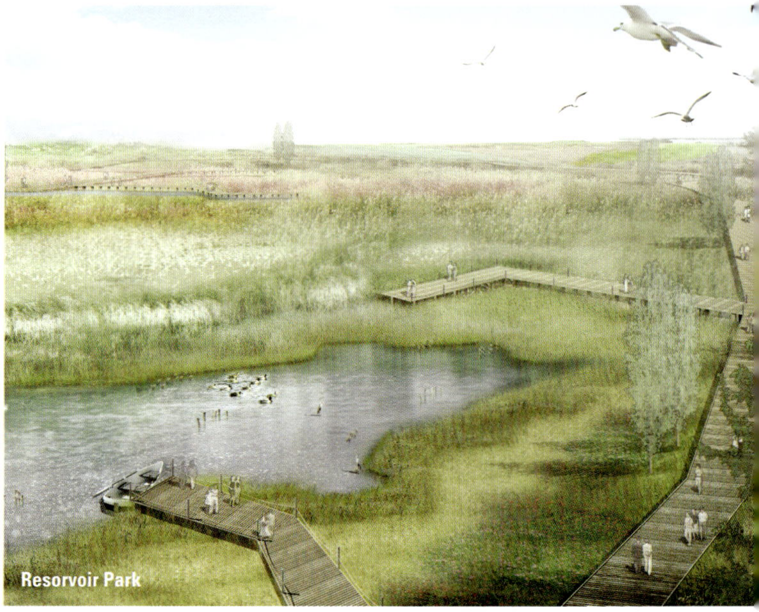

Resorvoir Park

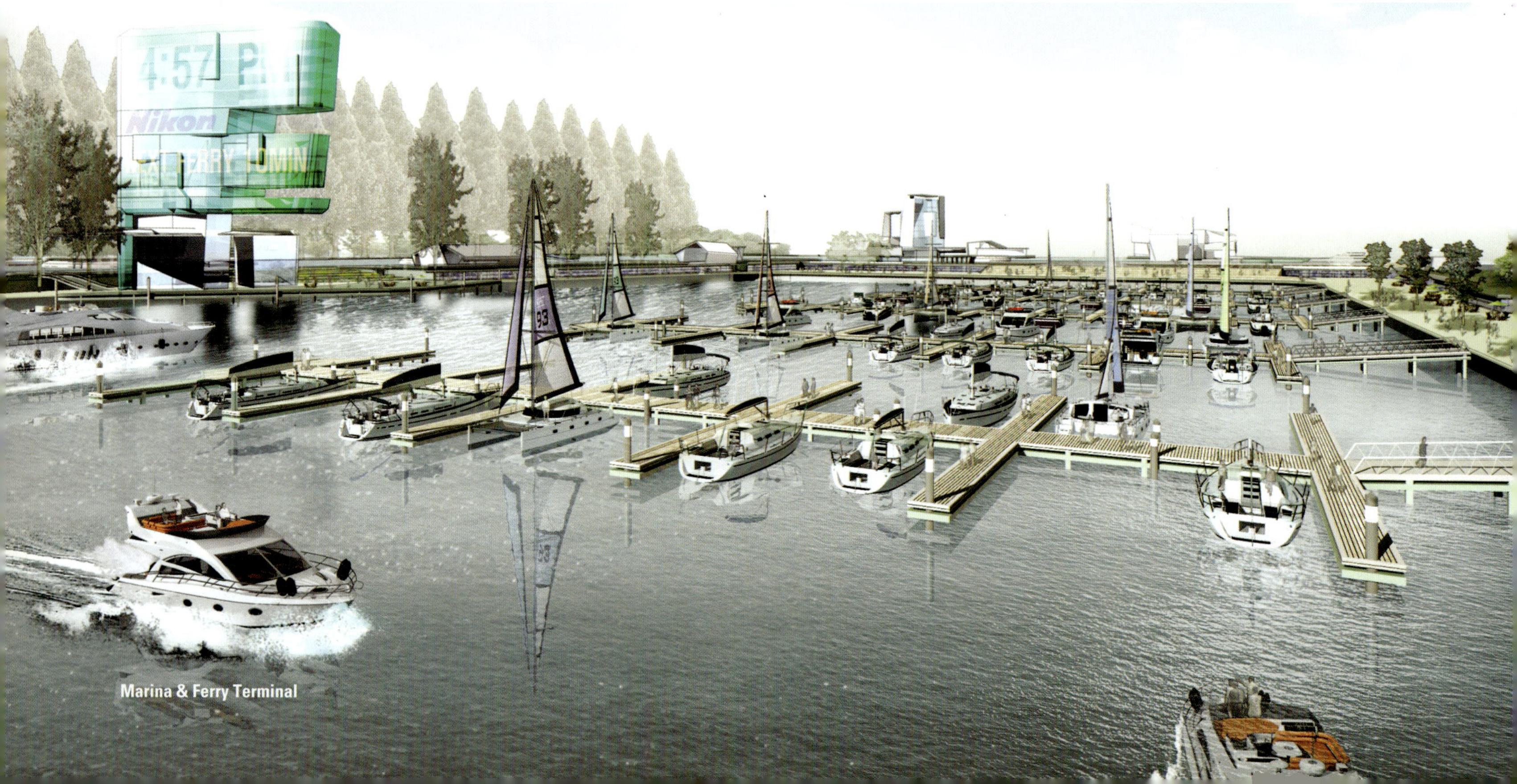
4:57 P
Nikon
FERRY TOMIN
93

Marina & Ferry Terminal

LANDSCAPE

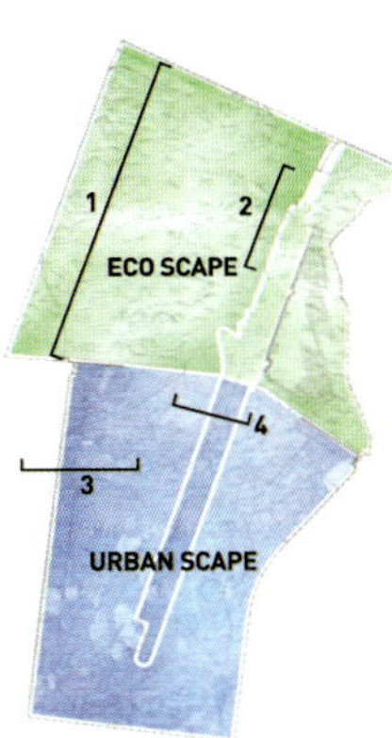

ECO SCAPE

Eco-scape has an energy saving effect, reduces air-pollution and protects the environment and life. The north-south axis of M-Park is part of migratory bird's pathway and connects to Gangseo greenbelt. The east-west axis of M-Park is a river mouth coast wetland. The eco bowl and vegetation filter system makes M-Park to natural purifying system.

· Eco bowl

: Wet land utilization, stable natural ecological system, provides aquatic and animal habitat

· Vegetation Filter System

: Cascading landscape, eco-friendly and energy saving water, purification system using natural organic vegetation.

URBAN SCAPE

Urban-scape incorporates and modifies elements of urban infrastructure. The linear shaped 'Green Levee' has the new airport railroad and subway ventilation towers which are reluctant to be exposed. These elements can be modified into architectural elements. Left to site, the sewage treatment facility is moved underground and incorporated into the green park.

Mussel-shaped landscape units (Rhythm, Flat) are spread out all over M-Park site at 9m to become significant unique elements of M-Park's urban-scape.

· Rhythm : a unit combined small and big trees makes more natural undulating landscape at east side.

· Flat : a unit makes digitized pattern at west side.

Lake Park

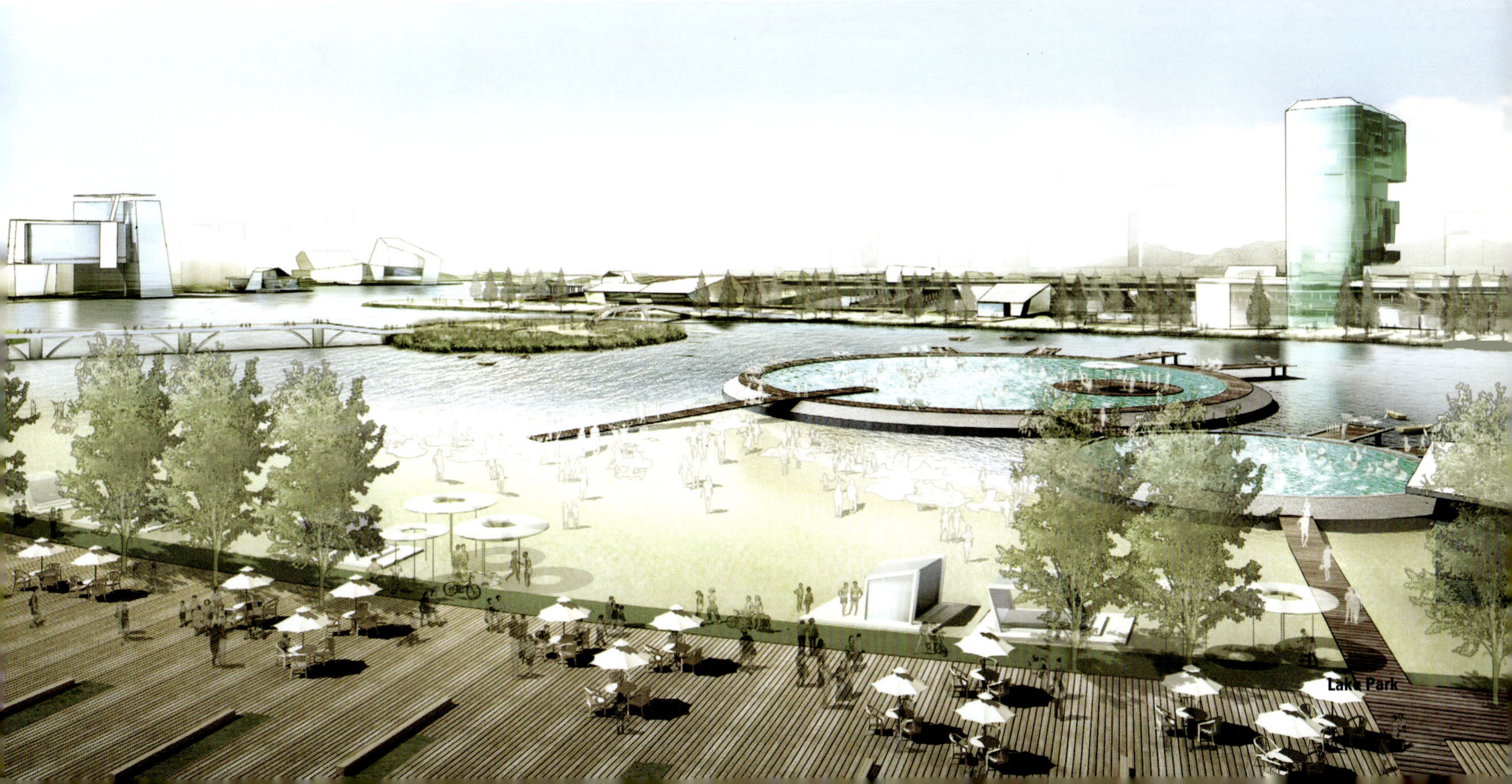

ECO SCAPE

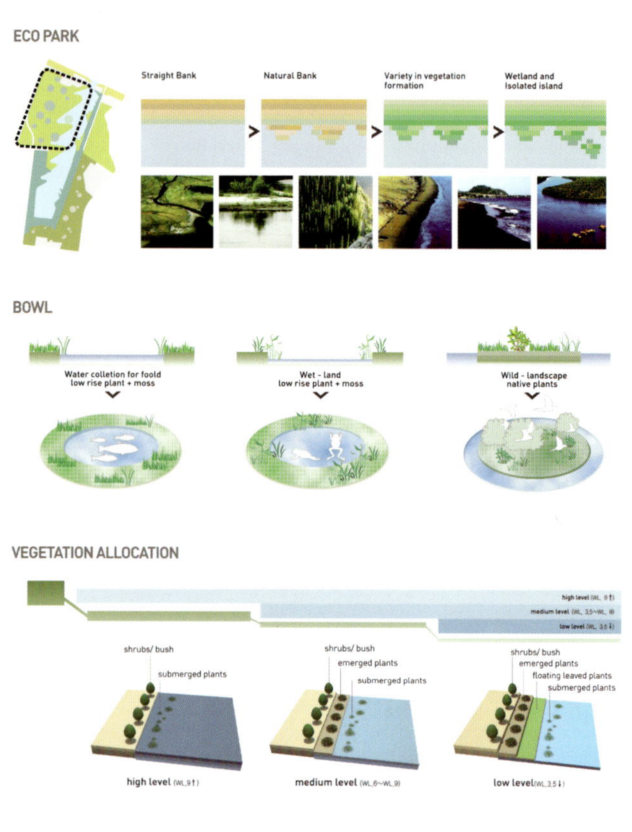

RESERVOIR PARK

Flitration from multiple vegetation

- Support spawning and nesting ground
- Feathered tribe: mallard, teal, wild goose, swan and heron
- Insect species: yellow butterfly, dragonfly, bug
- Mammalia: racoon, elk and weasel
- Amphibia: frog, round frog

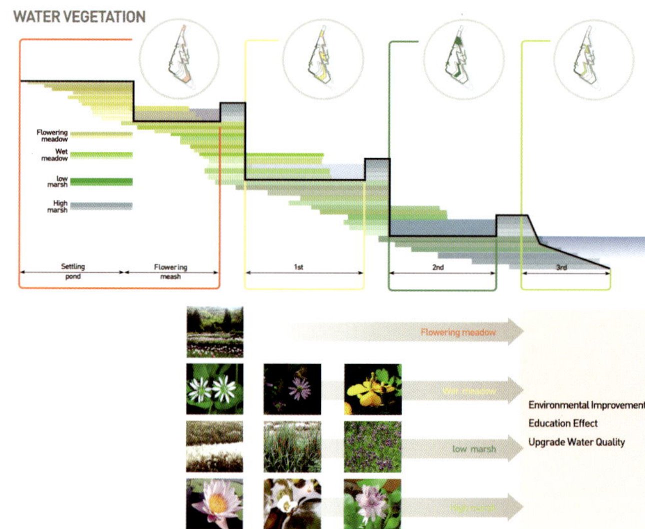

URBAN SCAPE

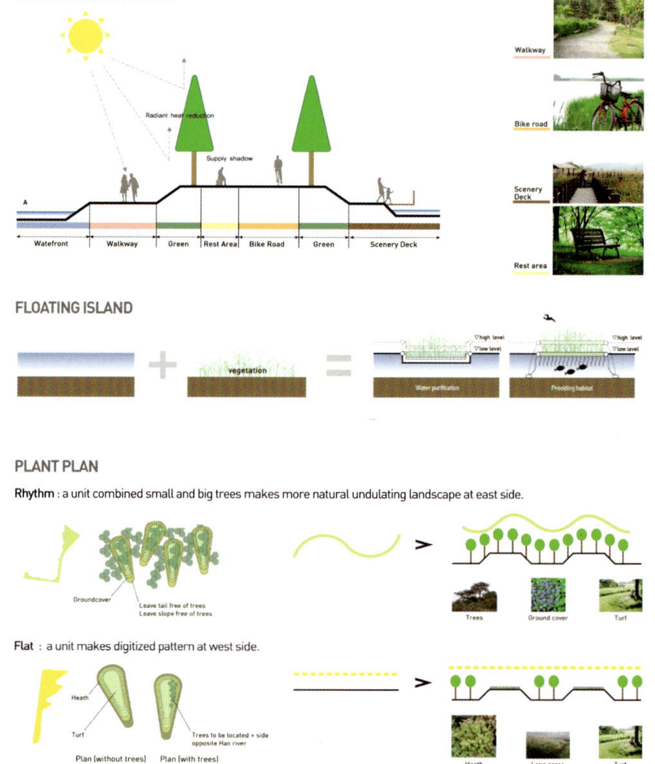

WATER CYCLE

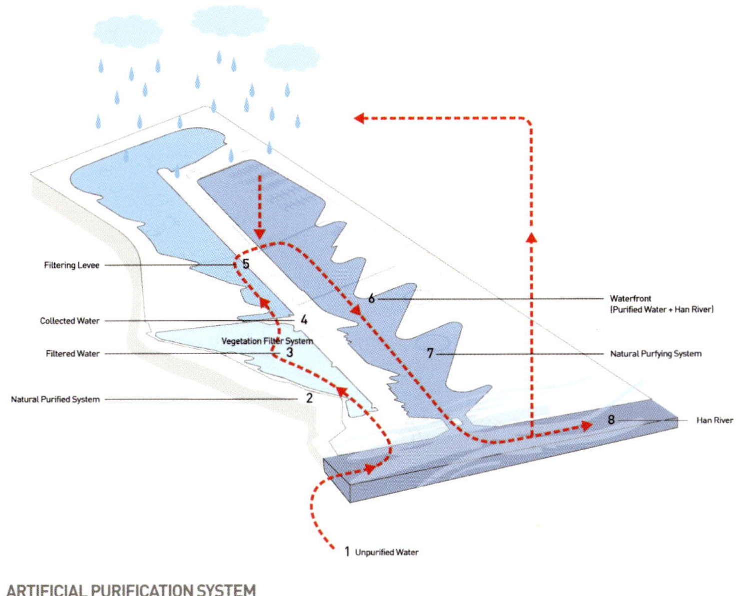

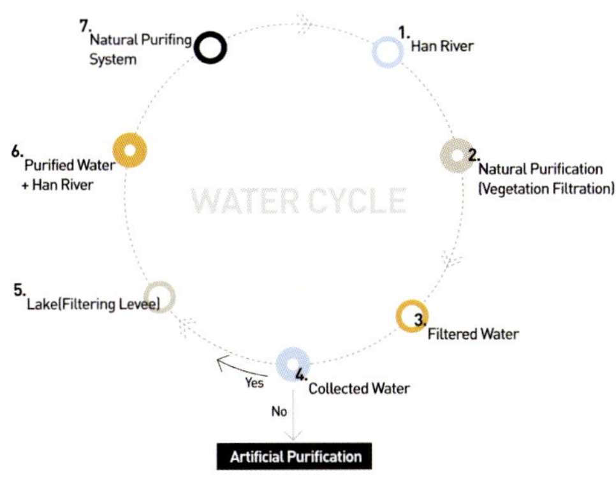

As the heart regenerates and sustains the body, the M-park regenerates the water for the Han River. The multiple bodies of water act as natural filters for the water being pumped or directed through the site. River water is cleaned as it passes from body of water to body of water and eventually back to the Han River. Rain water run-off from the surrounding area is diverted though filtering pools which will cleanse the water while it passes through to the Han.

These systems are accessible to park visitors for activites such as swimming, fishing and canoeing and so that they might understand how these systems fit into the greater natural and urban environments. In a similar manner to how these bodies of water clean the river water, these recreational activities can be as restorative or regenerative for the people of the Magok and greater Seoul.

ARTIFICIAL PURIFICATION SYSTEM

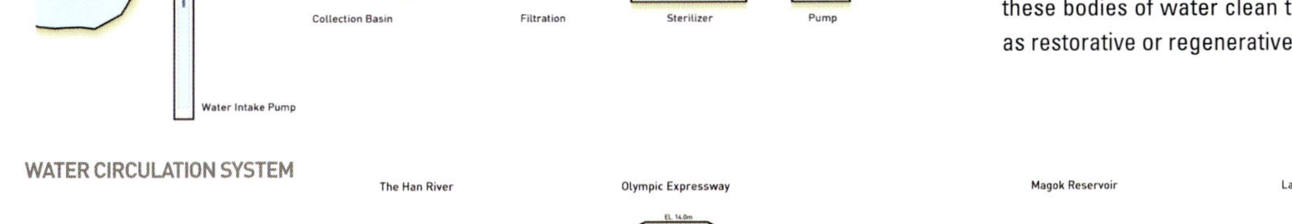

WATER CIRCULATION SYSTEM

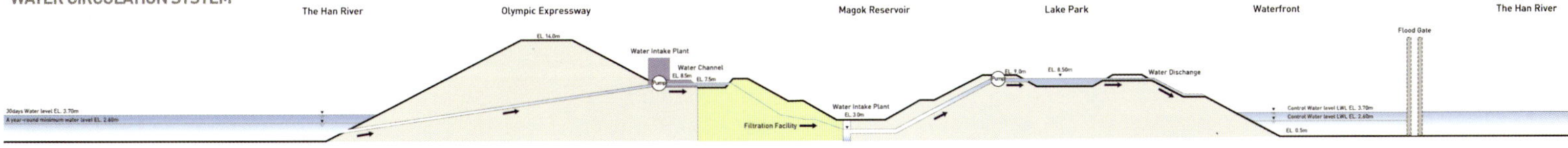

FLOOD GATE

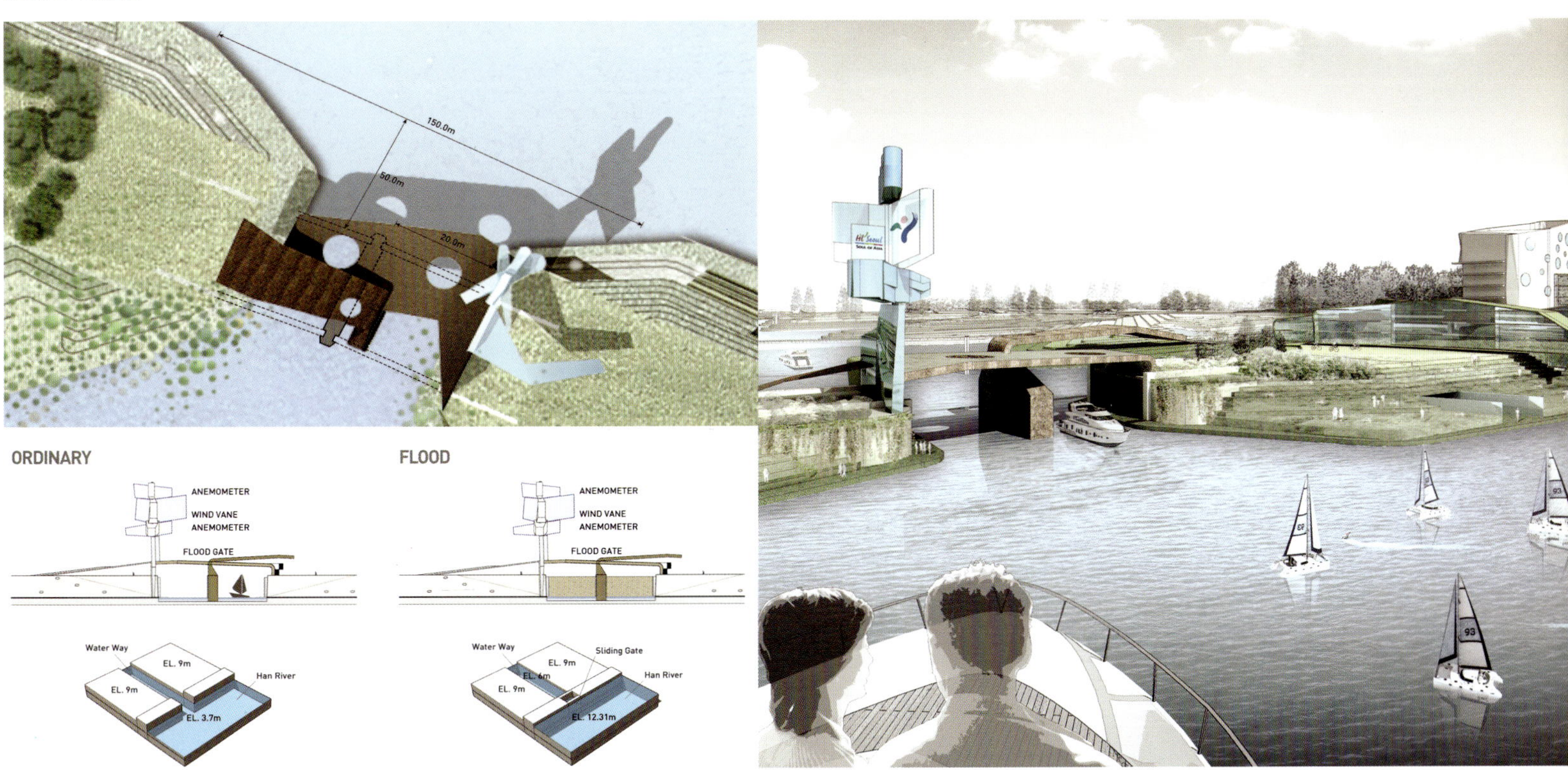

2rd Prize

Disappearing Landscapes

Eun Dong Shin 은동신 · Architect, Korea
Partners _ Park, No Wook 박노욱 + Goettsch James

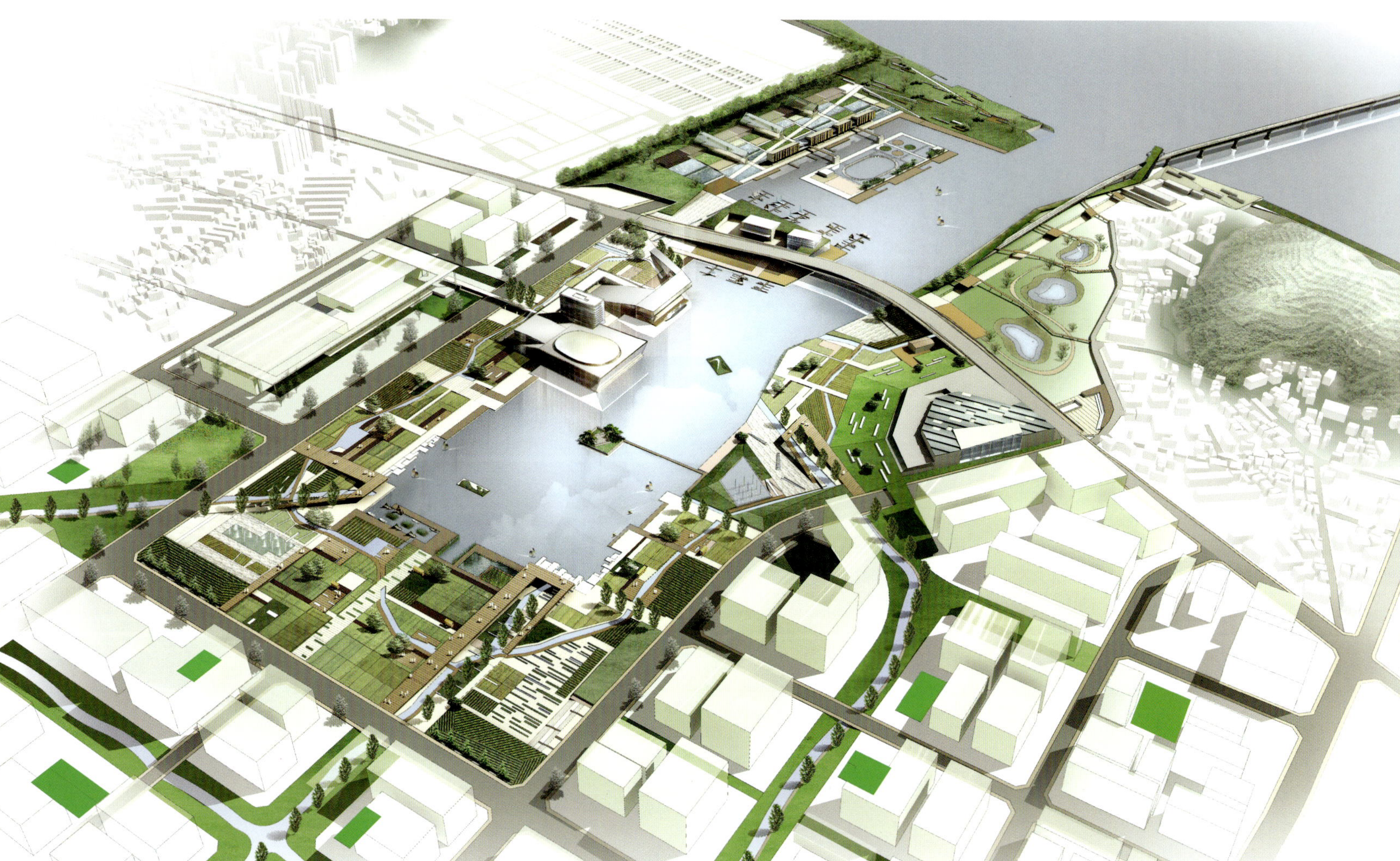

What does the Magok Waterfront connect values of the Han river to the park?
What is the meaning of the Magok Waterfront site in Seoul?
What should be preserved and sustained?
What does make the Magok Waterfront unique to other waterfronts?

We find out the clues from the vanished scenery of the Han river at the past and the sight of farmland at present.

The interactive inflow of the Han river to the urban park to rehabilitate the various activities seen around the Han river at past makes it to be the infrastructure of cultural activities. The suggested water quality management method used by a number of different water resources and those function fundamentally to preserve the critical ecology basis of the lake park.
The total farmland in Seoul is 478ha and has been decreasing caused by the expansion of the urban area. This means not just the reduction of the productive farmland in Seoul but also the reduction of the ecological infrastructure of the farmland in the city. The farmland will be rehabilitated as the complexed cultivated land of the new possibility including the ecological, environmental, social and economical merits urbane cultural activity in our design.
Seasonal change of the farmland and participatory programs will be the rejuvenator as a cultural activity space makes the new urbane scene. And the farmland functions as a sustainable ecological infrastructure, connecting with the artificial lake linked to the Han river intimately.
The Magok Waterfront will be re-birth as the new urbane park of the future, retaining the cultural and ecological possibility through the various activities around the Han river at the past and the farmland including the vitality of the land at present.

마곡 워터프론트가 한강 주변의 공원으로 연관되는 가치는 무엇인가?
서울시에서 마곡 워터프론트가 갖는 의미는 어떤 것인가?
무엇이 보호되어야 하고 지속되어야 하는가?
마곡 워터프론트가 다른 지역 및 나라의 사례와 비교했을 때, 어떤 특징을 가지고 있는가?

우리는 과거의 한강에서 사라진 주변환경의 모습과 현재의 모습을 연구하여 실마리를 찾아보았다. 한강에서 볼 수 있는 모습들은 도시 공원에 영향을 미칠 수 있는 유입요소들로서 과거 한강 주변에서 볼 수 있었던 다양한 문화적 활동을 재건하기 위한 문화활동의 기본적인 구조가 될 수 있다. 다양하고 각기 다른 수자원을 통하여 제안된 수처리방법과 이렇게 파생된 기능들은 기본적으로 호수 공원에서 생태계의 중요한 기반을 형성할 수 있다. 서울시 총 농경지의 면적은 478ha로 도시지역의 확충으로 인하여 매년 줄어들고 있는데, 이는 단지 서울주변의 비옥한 토지가 줄어들고 있는 것만을 의미하는 것이 아니라 도시속의 생태구조 또한 줄어들고 있음을 의미한다. 농경지는 우리가 설계한 디자인에서 생태학적, 환경학적, 사회학적, 경제학적인 장점이 내포되어 새로운 가능성을 지닌 다중복합적인 토지로 재건될 수 있다.

사계절의 뚜렷한 변화와 이와 관련된 기능적 프로그램들은 하나의 문화적인 장을 형성하게 되어 새로운 도시적인 장면을 연출하게 된다. 마곡 워터프론트는 미래지향적인 도시속의 공원으로 탈바꿈하여 과거 한강주변에서의 다양한 문화적, 생태학적인 가능성을 유지하면서 활기를 불어넣어줄 것이라 믿고 있다.

WATER AREA

Urban infrastructure + facility

rain-water culvert

Water transportation circulation

marina expansion if needed
marina
passenger
ferry

Pedestrian + Bike circulation

Olympic Expressway is designed below the water level to allow for the pass of water transporation
Yang-cheon road is designed in a streo graphic way
ventilating opening designed as a floating object to connect the circulation of water space

Connection to greens

Kang-seo ecological swamp park
Mt. kung

Ratio of water permeability ground

release of nonpoint pollution source
rain-water culvert
rain-water gathering facility
rain-water reservoir
connection to the rain water reservoir in the district

impermeable ground (15%)
permeable ground (85%)
rain-water reservoir
rain-water gathering facility

LAND AREA

Urban suggestion of the Magok district

Connection with the waterfront park

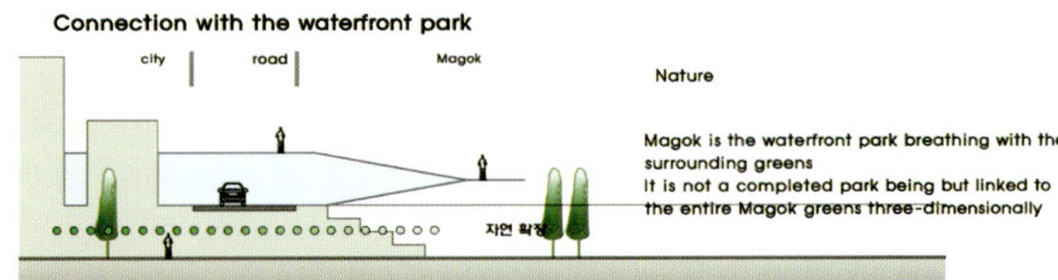

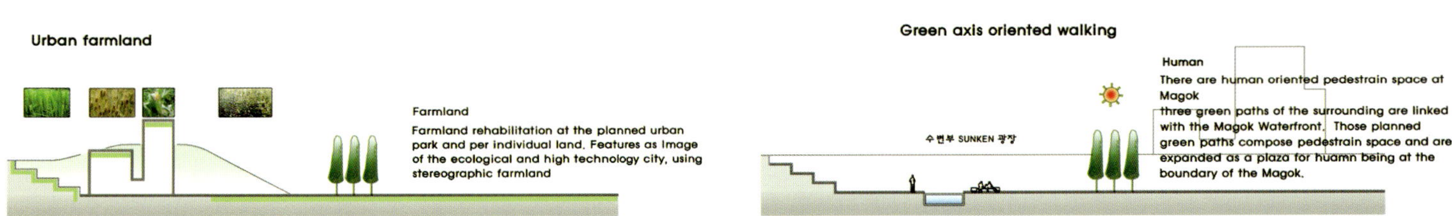

ZONE A _ Flood Gate

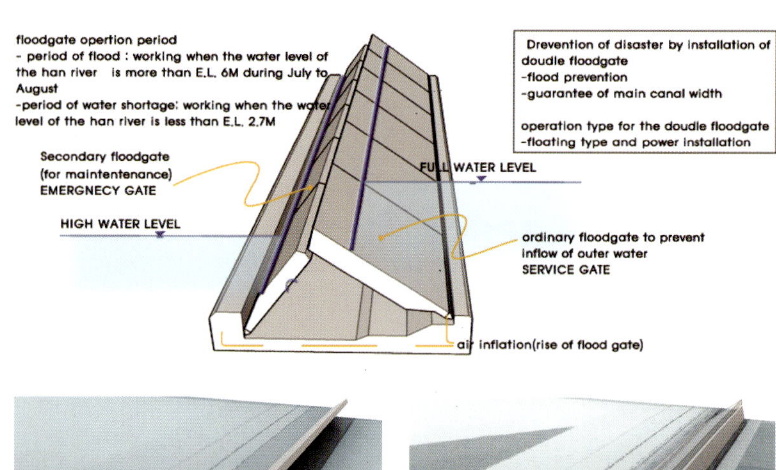

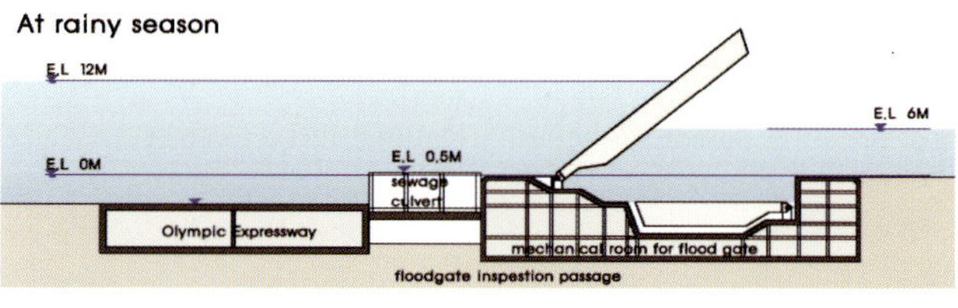

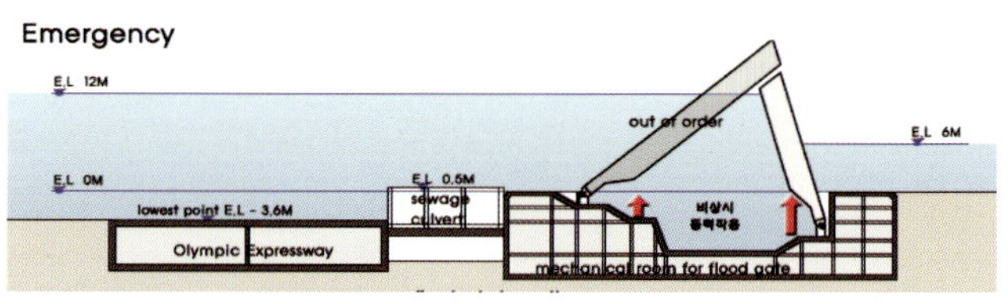

ZONE B _ Waterways / Marina

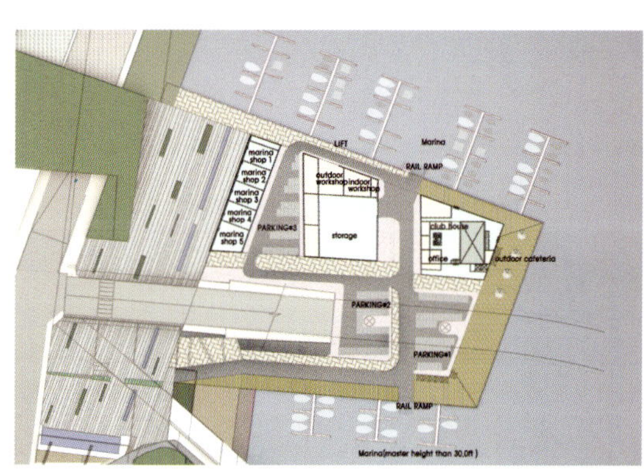

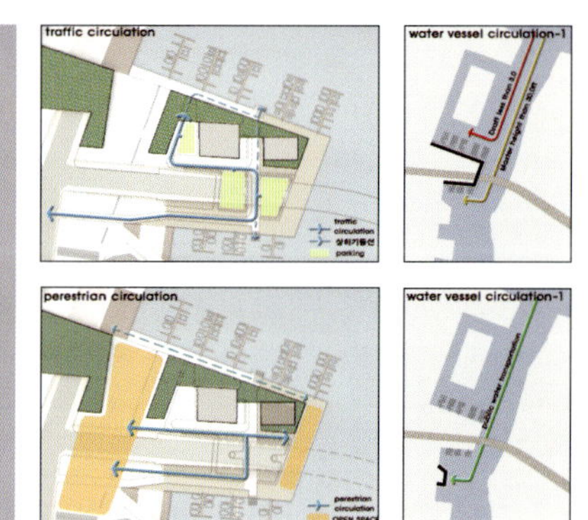

ZONE B _ Detention Reservoir

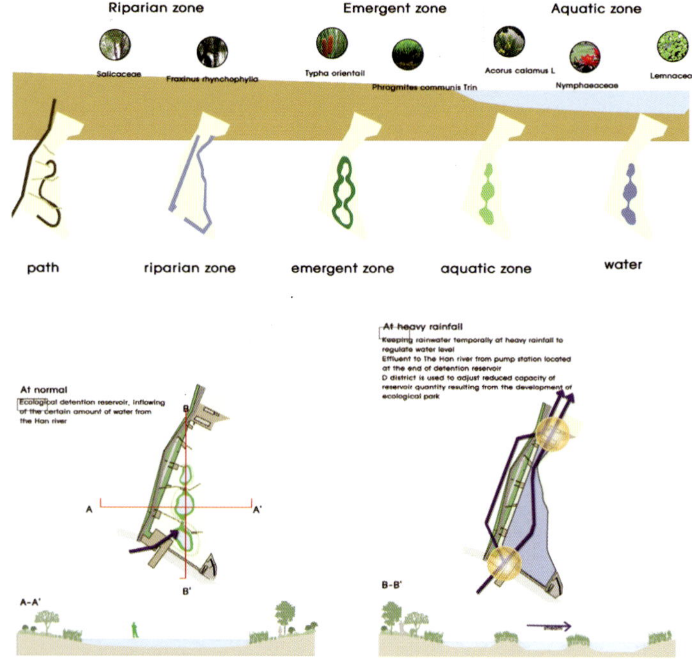

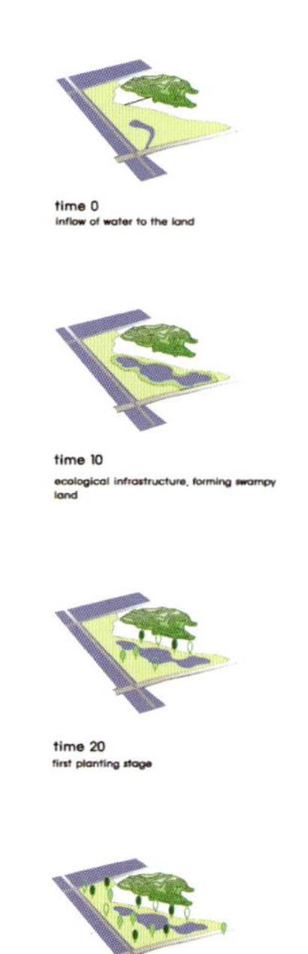

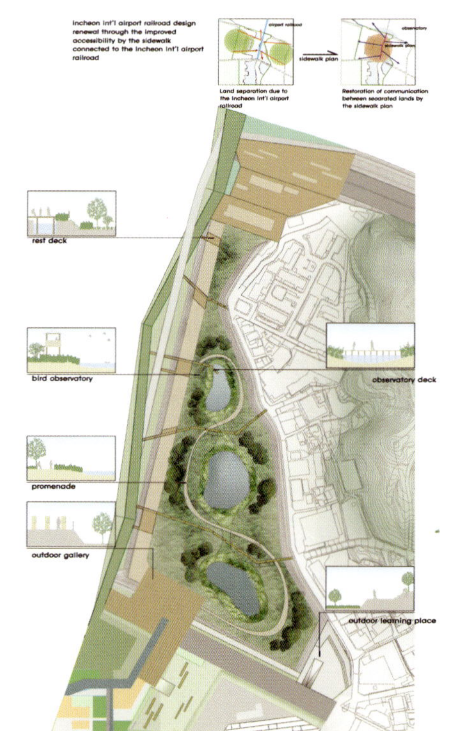

MASTER PLAN
Han River
Ferry
The Han River water front park
Observatory
Seonam Sewage
Treatment Plant
Marina
subsidiary facility
Greenhouse for
for water plant
Leports facility
Promenade on riverbank
Incheon Int' l Airport railroad
Soakru(Pavilion)
Reservoir / Ecological Park
Mooring Facility
Observatory on the deck
Yangcheon Hyanggyo
Gallery of Guem-jae jeong seon
Marina Plaza
Plaza
Yangcheon-gil Crossing
Observation Course
Gallery for Han River
(The old drain pump station)
Ferry for
a passenger ship
Cogeneration
Facility
Bridge
Covention Center
Exhibition facility Complex
Garden in water
Winter Praza
Winter Praza
Ferry
Plant in water Deck
Fountain plaza
Cultivated land Plaza

PLANTING PLAN

Spring

Summer

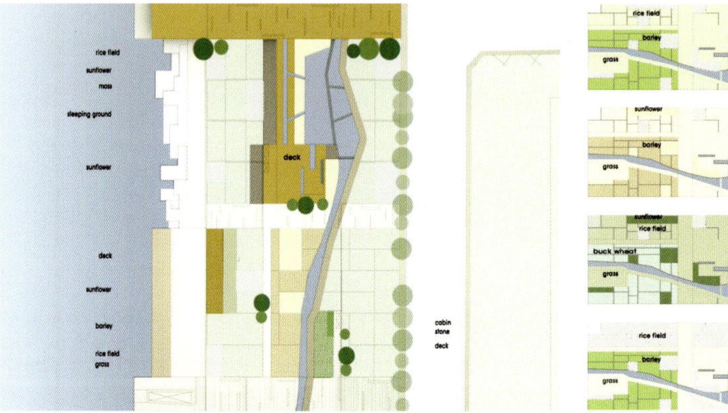

Fall

Winter

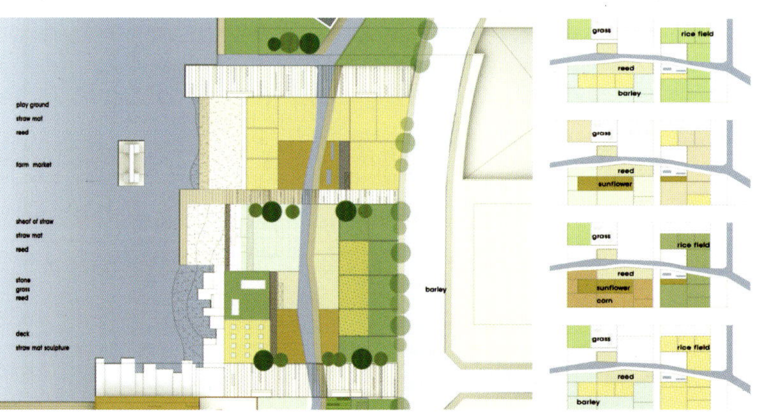

ZONE C _ Detention Reservoir

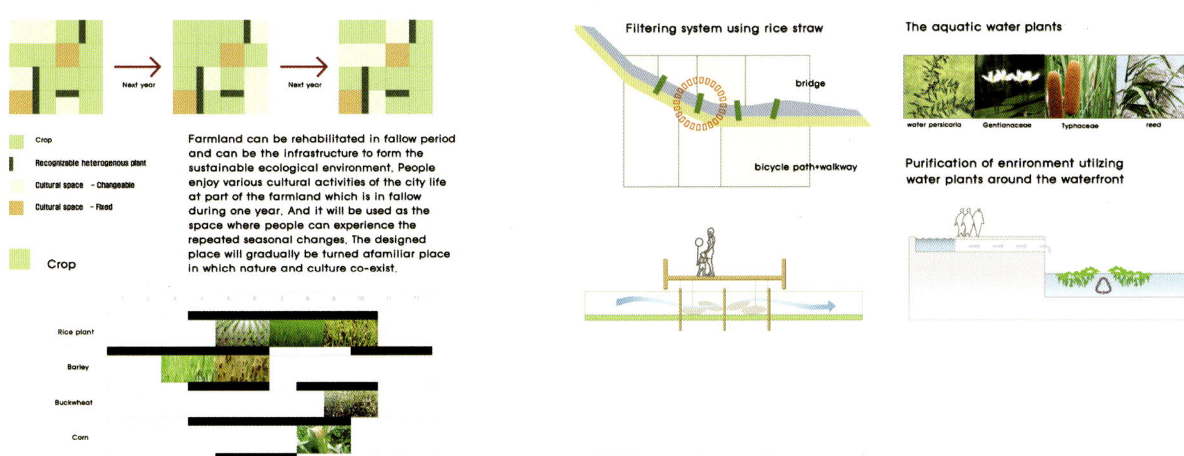

Farmland can be rehabilitated in fallow period and can be the infrastructure to form the sustainable ecological environment. People enjoy various cultural activities of the city life at part of the farmland which is in fallow during one year. And it will be used as the space where people can experience the repeated seasonal changes. The designed place will gradually be turned afamiliar place in which nature and culture co-exist.

Recognizable heterogenous plant

Cultural space - Changeable

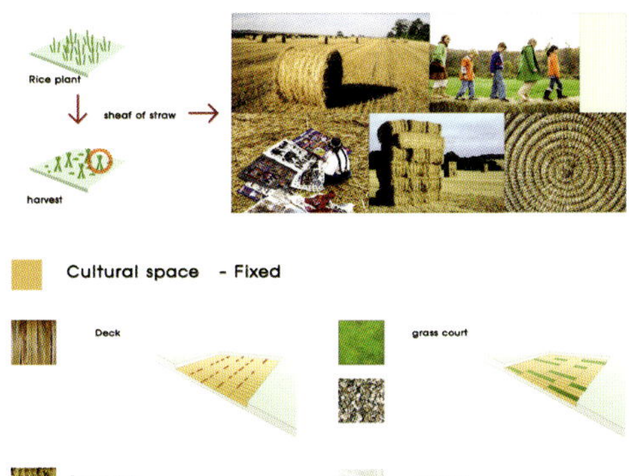

Cultural space - Fixed

ZONE C _ Ferry. Commer Store

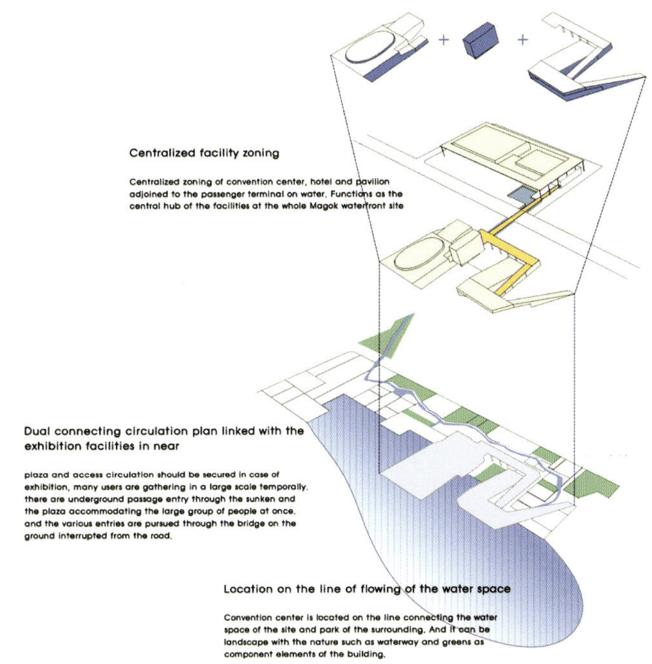

Centralized facility zoning

Centralized zoning of convention center, hotel and pavilion adjoined to the passenger terminal on water. Functions as the central hub of the facilities at the whole Magok waterfront site

Dual connecting circulation plan linked with the exhibition facilities in near

plaza and access circulation should be secured in case of exhibition, many users are gathering in a large scale temporally. there are underground passage entry through the sunken and the plaza accommodating the large group of people at once. and the various entries are pursued through the bridge on the ground interrupted from the road.

Location on the line of flowing of the water space

Convention center is located on the line connecting the water space of the site and park of the surrounding. And it can be landscape with the nature such as waterway and greens as component elements of the building.

3rd Prize

Magok's Botanical Tissue

David Leven(Leven Betts Architects) · Architect, U.S.A. | **Partners** _ Felipe Correa + Douglas Meffert+ John, An + Nat Oppen Heimer

The main objective of this project is to provide a clear framework for the Magok Waterfront that incorporates high-impact gestures with larger, strategic infrastructural and ecological actions. In doing so, the project aims to establish a broader operation that can turn this key area of the 'Han River Renaissance Project' into a significant dual location : a place that is at once a dynamic multipurpose program and an archi -tectural icon. On the one hand, the project operates as a primary intermodal connector that brings into the area a wide host of transport infrastructures, and enhances them with a varied brief that includes recreational ecologies, water-sports activities, retail space, research and development facilities, and a proposed con -vention center. On the other hand, through discrete, yet high-profile architectural pieces the project can establish significant visibility that will endow the Magok Waterfront with worldwide recognition and showcase, on an international scale, Seoul's public-rivate capacity to conceive and execute successful, comprehensive restruc -turing plans.

Because of the increasingly high-paced com -petition among urban centers worldwide to establish themselves as powerful economic and cultural hubs, high-profile design operations have become key elements by which individual cities achieve competitive advantage. Within this context, our project capitalizes on the strategic location of Magok in its proximity to downtown Seoul and major transportation infrastructures to enhance an urbanistic operation that establishes clear links to the city at large, while rendering Magok visible in relation to the other operations along the Han River.

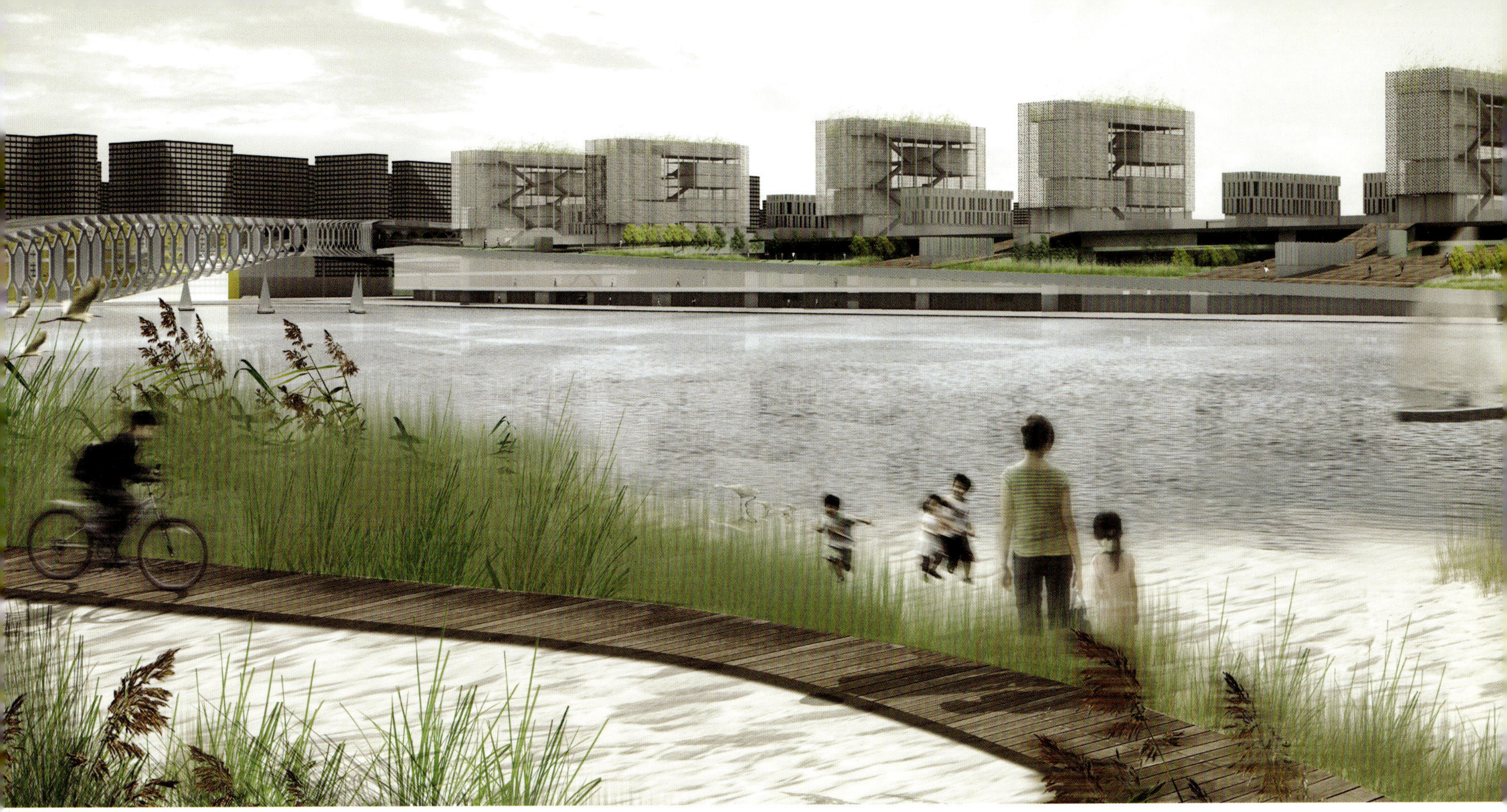

본 프로젝트의 주요 목적은 마곡 워터프론트를 체계적으로 구성하면서 생태학적으로 파장을 줄 수 있는 명확한 틀을 제시하는 것으로, '한강 르네상스 사업'의 주요 구간인 마곡지구를 다목적 기능을 갖춘 문화적 요충지 및 건축적 상징물로 구축하는 것이다. 다른 한편으로, 이 프로젝트는 대중교통시스템에 있어서 호스트 역할을 가지고 관리기능을 강화시킨다는 것에 있으며, 휴양의 성격을 띤 자연환경, 수상스포츠 활동, 상업공간, 연구 및 개발 시설, 컨벤션 시설 등을 포함한 다양한 계획으로 하부조직을 보강했다. 또한 독립적이지만 인식성이 높은 건축 요소들은 마곡 워터프론트의 인지도를 높여주어 세계적으로 건축 분야의 높은 평가를 받을 수 있는 확연한 모델로 급부상할 것이다. 보다 역동적인 경제적 · 문화적 허브를 구축하기 위한 세계 도시들간의 경쟁이 증가하고 있는 요즘, 높은 수준의 설계안과 디자인이야말로 경쟁에서 살아남을 수 있는 요소가 된다고 본다. 따라서 제시한 안은 서울시내와 근접한 주요교통의 요충지로서의 마곡의 지리적 장점을 최대한 이용함으로써 한강을 따라 진행되는 다른 사업들과 더불어 마곡지구를 돋보이게 할 것이다.

BOTANICAL TISSUE OPERATIONS

green systems

watershed

energy

shuttle bus stops

recycling

bike share

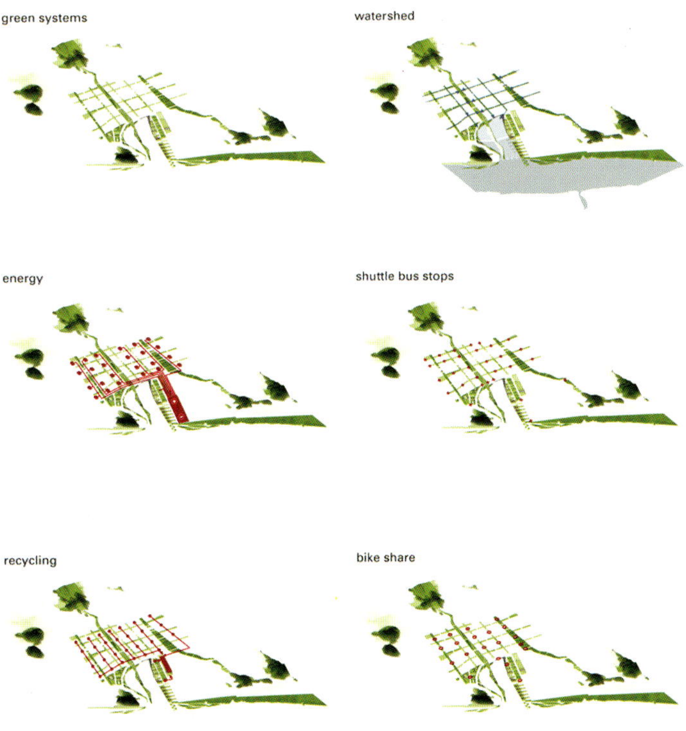

WATERSHED

Storm water run-off is channeled and filtered through green roofs, planted waterways that are located in the linear green spaces that course through the grid, and the working, recreational wetland that filters all excess storm run-off generated by the neighborhood, convention center and park.

ENERGY

Methane from the waste treatment plant will be harnessed for district heating, cooling and power production for the convention center, R & D park, recreational programs and residences of the neighbor-hood.

SHUTTLE BUS STATIONS

Connected to the local intermodal station, the shuttle bus network stations are located at the intersections of the main road and greenways. These nodes occur at approximately every 500 meters or at key road to green and green-to-green intersections.

RECYCLING

To the west of the convention center, a recycling facility is planned to process the large waste stream that a center of this size and scope will produce. Additionally, the solid waste of the neighborhood would be processed at this facility.

BIKE SHARE STATIONS

Bike share stations occur at the green intersections in the grid of the residential neighborhood. The Squid contains the main bike share station that also provides recreational transportation within the lake and wetland park.

BOTANICAL TISSUE

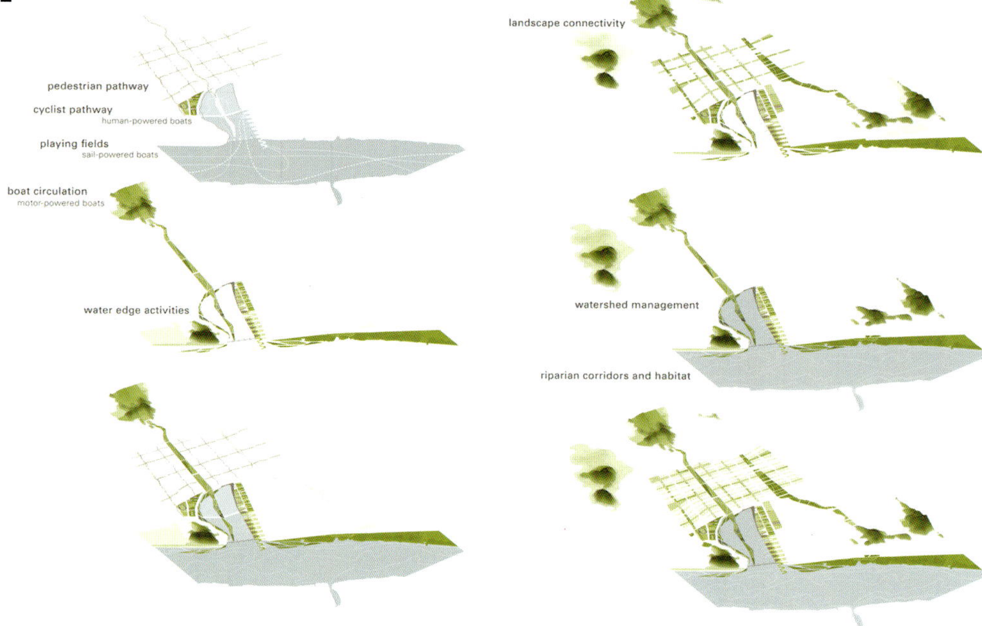

BOTANICAL TISSUE VIGNETTES

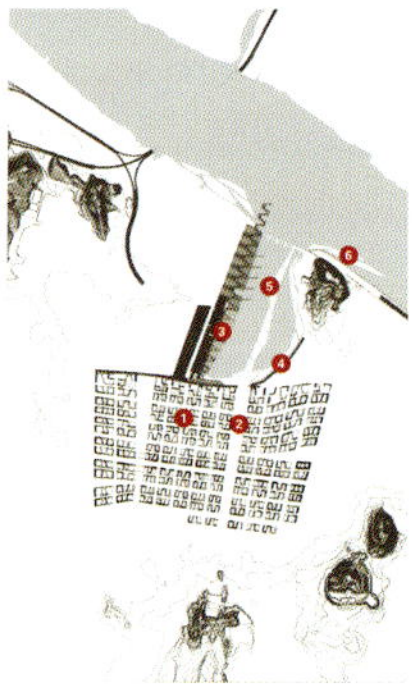

RIVER AND WETLAND ZONES

There are two broad categories of wetland that support a variety of plant and wildlife in South Korea: intertidal/tidal and freshwater. The interventions recommended in this proposal relate entirely to freshwater habitats. The zoning of these systems will include habitat for fish and migratory bird flyways, and restoration of food webs for resident species.

Freshwater wetlands are largely concentrated in the historical floodplains of South Korea's major rivers: the Han, Geum, Yeongsan, and Nakdong. For the Magok Waterfront site, we propose the expansion of an existing Han-related habitat that consists of freshwater marsh, bayous, and ponds. The expansion will enhance natural and semi-natural freshwater wetland areas in the region, many of whose viability is increasingly threatened by continued environmental degradation.

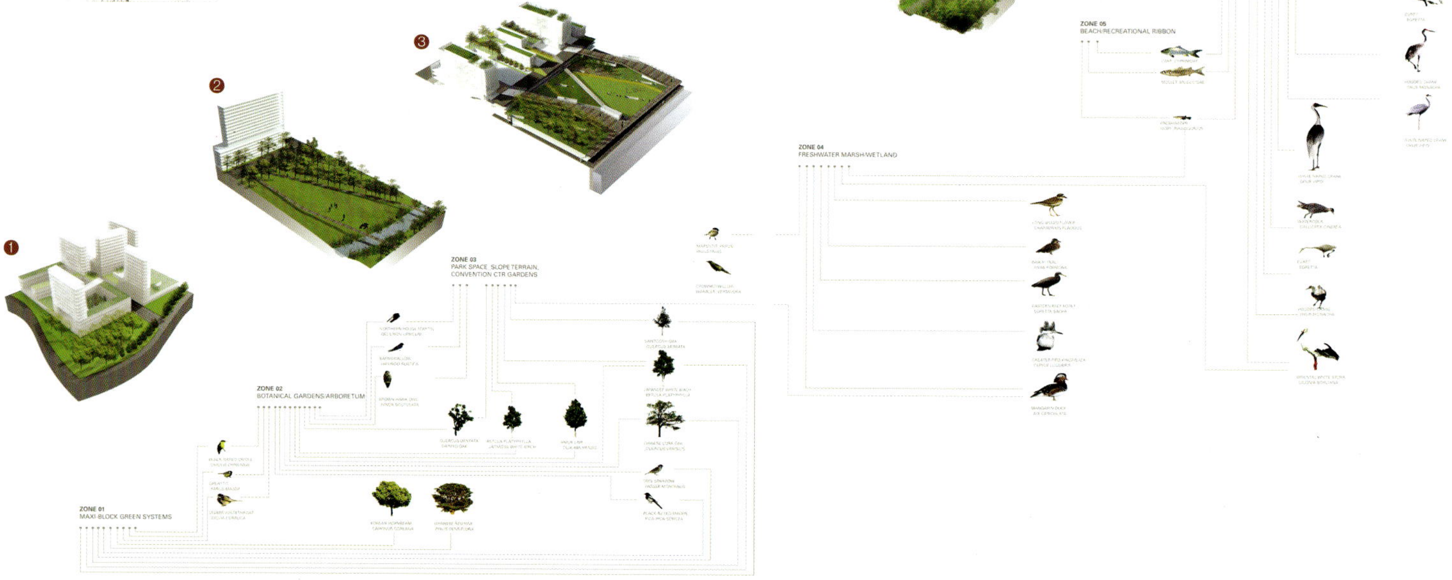

ZONE 01
MAXI-BLOCK GREEN SYSTEMS

ZONE 02
BOTANICAL GARDENS/ARBORETUM

ZONE 03
PARK SPACE, SLOPE TERRAIN, CONVENTION CTR GARDENS

ZONE 04
FRESHWATER MARSH/WETLAND

ZONE 05
BEACH/RECREATIONAL RIBBON

ZONE 06
RIVER ISLANDS

PROGRAMMATIC ANCHORS CONSTITUENT ELEMENTS

PROGRAMMATIC ANCHORS PLAN

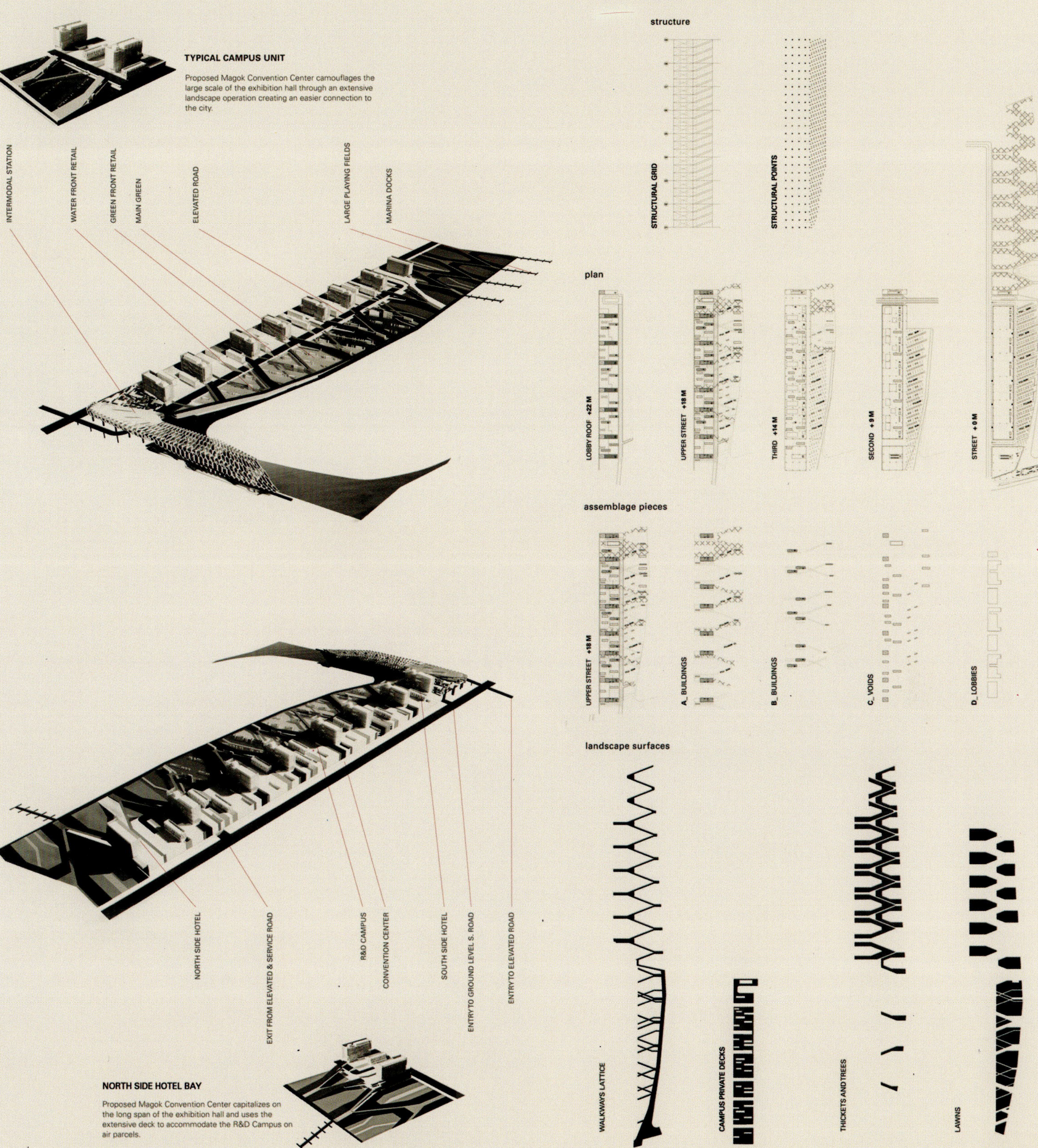

3rd Prize

Magok Harbor Park

Cho Ju Hwan 조주환 · Architect, Korea | **Partners** _ James Corner + Schwitler Craig + Si Chaisoo

A NEW HARBOR PARK FOR MAGOK :

· With the increasing homogenization of the way urban places around the world look and feel, might there be a newly designed place for Magok that is uniquely Korean, related to Seoul, the Han River and the unique topography of the region?

· What if the waters from the river—for so long barricaded from social use by levees and barriers—could be allowed to flow into the Magok site, to create a new watery interior, a new harbor for boating and water-related uses?

· And what if this new harbor could be framed by a vibrant urban edge, social, 24-hour, eventful and cosmopolitan?

· And perhaps, in the center, a new topography of hills, islands and pro-montories reminiscent of "Asian mountains in the water"- gardens of azalea, pines, flower meadows and overlooks?

· How might this new urban harbor park - with its inter-mixing of hard urban wharfside uses with soft ecological landscapes - create a unique, new form of public landscape for Magok, reconnecting the city to the river?

마곡의 새로운 항만공원

· 전 세계가 주목하고 있는 가운데, 마곡에 형성될 설계안은 과연 한국적이며 한강과 서울과도 어울리면서 지역적인 특색을 살려줄 수 있을 것인가?

· 오랜 세월 동안 쌓여있는 제방과 장애물로 인해 막혀있는 한강에서 마곡까지의 수로가 다시 흐를 수 있도록 함으로써 보트나 수상과 관련된 문화를 이용할 수 있는 수자원으로 개발하면 어떨까?

· 새로 형성될 항만공원은 과연 도시 외곽에 위치한 생동감 넘치는 곳으로 24시간 운영이 가능하며 세계적으로 알려질 수 있는 아이콘이 될 수 있을까?

· 진달래와 소나무가 어우러진 꽃동산과 정원이 형성된 '아시아의 강변언덕' 으로 지형적인 특색을 상기시켜주는 인공섬으로 조성한다면 어떨까?

· 어떻게 하면 선창 주변 건축물과 생태경관이 조합된 이 항만공원을 독특하고 참신한 마곡의 공공경관으로서 도시와 강을 연결하는 수단이 되게 할 수 있을까?

CREATE A NEW HARBOR PARK FOR MAGOK

First, bring in the water, allow it to pulse, flow and circulate with the river, while offering protection, restfulness and calm;

Then, activate the water with diverse boating activities, movement and contact—small row and paddle boats, kayaks, windsurfers, sailboats, yachts, ferry taxis, rowing vessels, even a specially designed "magok punt";

Surround the water with a vibrant urban wharfside edge, scattered with cafes, restaurants, pavilions, markets and tree-lined places to sit;

And construct new hills and islands in the water as naturalized destinations, invoking the quintessential asian image of mountains in water while establishing amazing new plantations, gardens, paths and high vista overlooks;

Finally, tunnel the main river road below grade to allow for an uninterrupted social connection to the han river, while establishing a new generous waterfront park area linked to a dramatic new water barrier feature;

And, create a rolling vehicular signature bridge across the mid-portion of the site, to lend strong identity and focal point to the center of the site.

마곡에서의 새로운 항만공원을 건립

먼저 한강에서 수로를 형성하여 물을 끌어들임과 동시에 자연을 보호하고 고요하면서도 차분한 분위기를 만드는 한편, 보트, 외륜선, 카약, 윈드서핑, 요트, 페리택시, 조정 등을 즐길 수 있는 문화적 활동을 활성화하기 위한 공원을 계획하였다. 한강 주변에는 카페, 레스토랑, 퍼골라, 시장, 벤치 등을 골고루 분포시켜 사람들이 이용할 수 있도록 하였다. 언덕과 섬을 물 속에 조성하여 물 위에 떠있는 전형적인 아시아 산지의 이미지를 연상할 수 있도록 했다.

마지막으로 공원과 한강을 연결하기 위해 비탈아래로 강변도로가 지나는 터널을 뚫어 새로운 워터프론트 공원지역으로 구축하도록 하였다. 또한 중앙부를 가로지르는 시그니처 교량을 만들어 초점부를 형성하고 강한 정체성을 부여한다.

HYDROLOGY-DRY / NORMAL

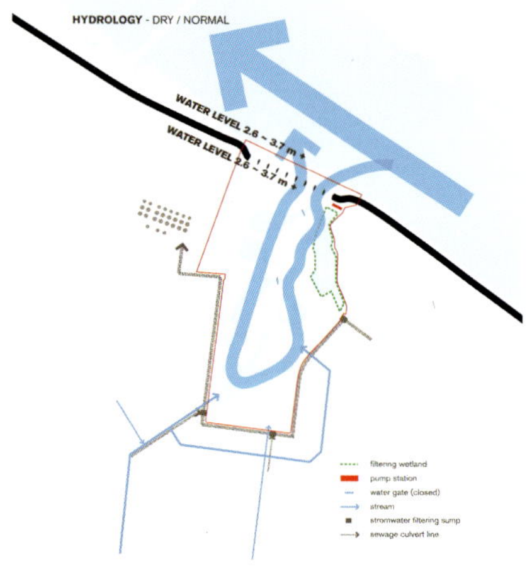

HYDROLOGY-WET

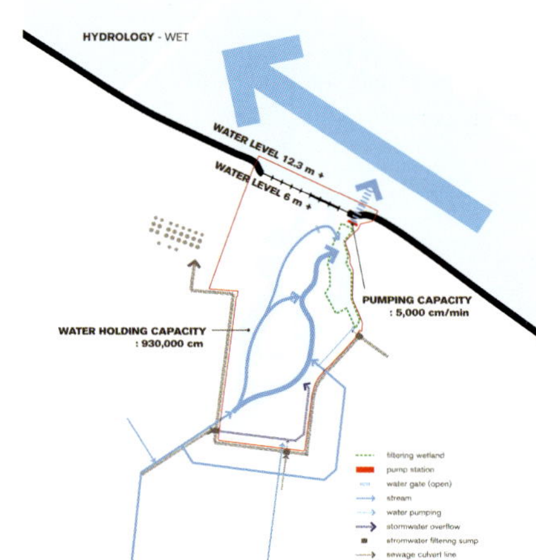

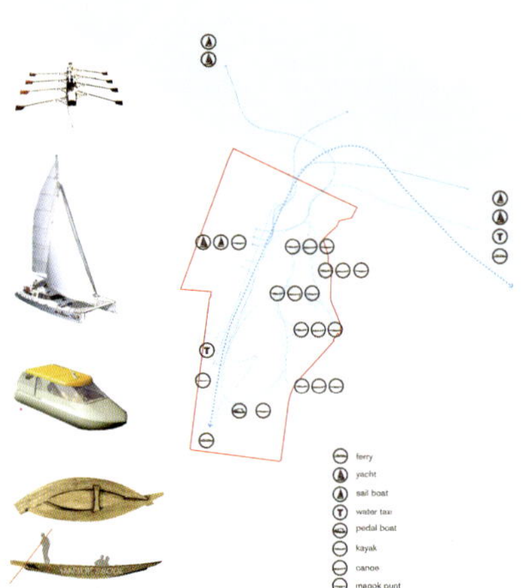

FLOOD GATE+TUNNEL

WHY TUNNEL?

The masterplan prioritizes openness into and out of the inlet. Currently, the eight lane highway bridge visually disconnects pedestrian access to the sea as well as boats to the marina. The proposed tunnel is to be built adjacent the bridge to allow traffic flow during construction.

TUNNEL + FLOOD GATE INTEGRATION

Positioning the tunnel below the flood gates offers three main advantages:

1. The foundation and structural system of the flood gate piers serves to offset the bouyancy of the tunnel at a regular interval.
2. Ventilation shafts and intermediate points of access can be accomodated within the flood gate piers.
3. Minimizes the construction footprint for flood gate and tunnel pieces.

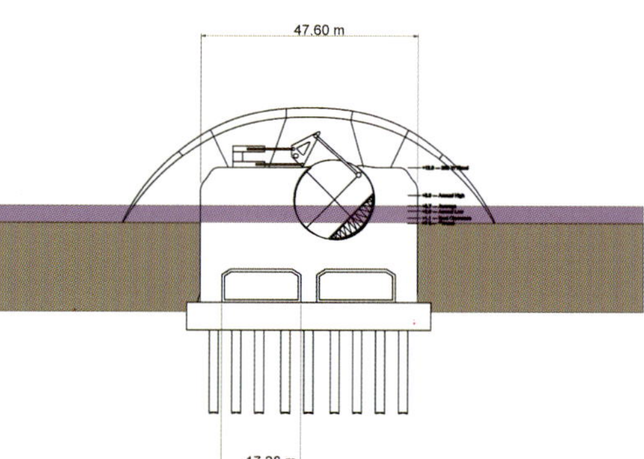

TRANSVERSE SECTION

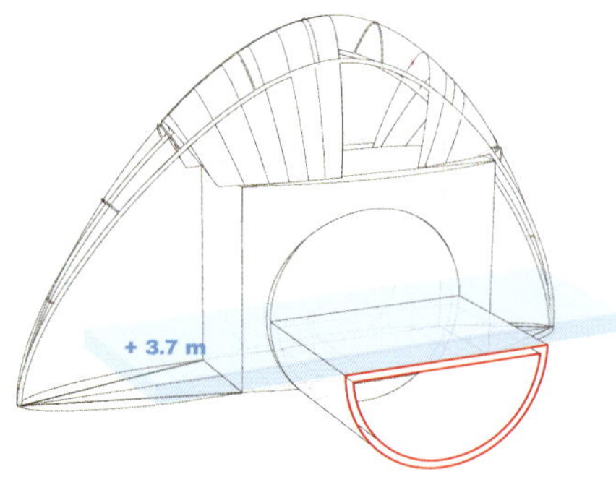

WATER LEVEL + 3.7

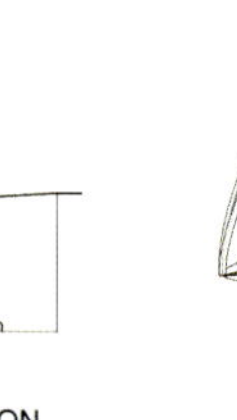

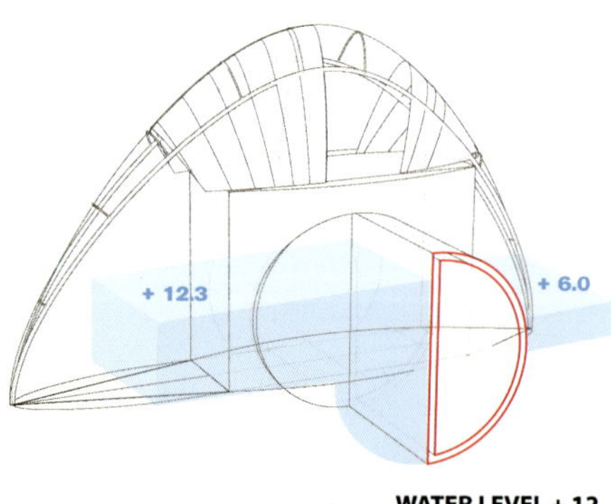

WATER LEVEL + 12

0% Grade

166.67 m | 450.00 m | 238.00 m | 450.00 m | 166.67 m

LONGITUDINAL SECTION

MASTER PLAN

WATERFRONT
01 WATERFRONT PARK
02 LINEAR PARK + LOOK OUT
03 FLOOD BARRIER
04 RENOVATED PUMP STATION

CULTURAL WHARF
05 WATERFRONT MARKETPLACE
06 ART BOSQUE
07 MIXED USE DEVELOPMENT
08 MARINA AND YACHT CLUB
09 SPORTS CENTRE
10 EVENT ROOF
11 BRIDGE
12 PUBLIC DOCK
13 BOATS DOCK
14 FERRY DOCK
15 FLOATING POOL
16 MAGOK BAY

CIVIC PLAZA
17 BAY PLAZA + FOUNTAIN
18 CONVENTION CENTRE
19 HOTEL
20 AUXILIARY BUILDING

PROMENADE
21 EVENT PLAZA
22 PROMENADE PAVILLION
23 LOOK OUT TERRACE

NATURE
24 FILTERING WETLAND
25 PINE FOREST
26 AZALEA HILL
27 ISLAND CAMPING HARBOR
28 KAYAKING HARBOR
29 TEA GARDEN
30 FISHING HARBOR
31 CANOEING HARBOR
32 WINDSURFING HARBOR
33 ROWING COURSE
34 SUN LAWN

INFRA-STRUCTURE
35 SUBSTATION
36 WASTE INCINEREATOR
HEAT FUSION GENERATOR
REUSE/RECYCLING STATION

Merit Award

Magok Harbour

COXALL PHILIP(Mcgregor partners) · Landscape Arcitects, Australia

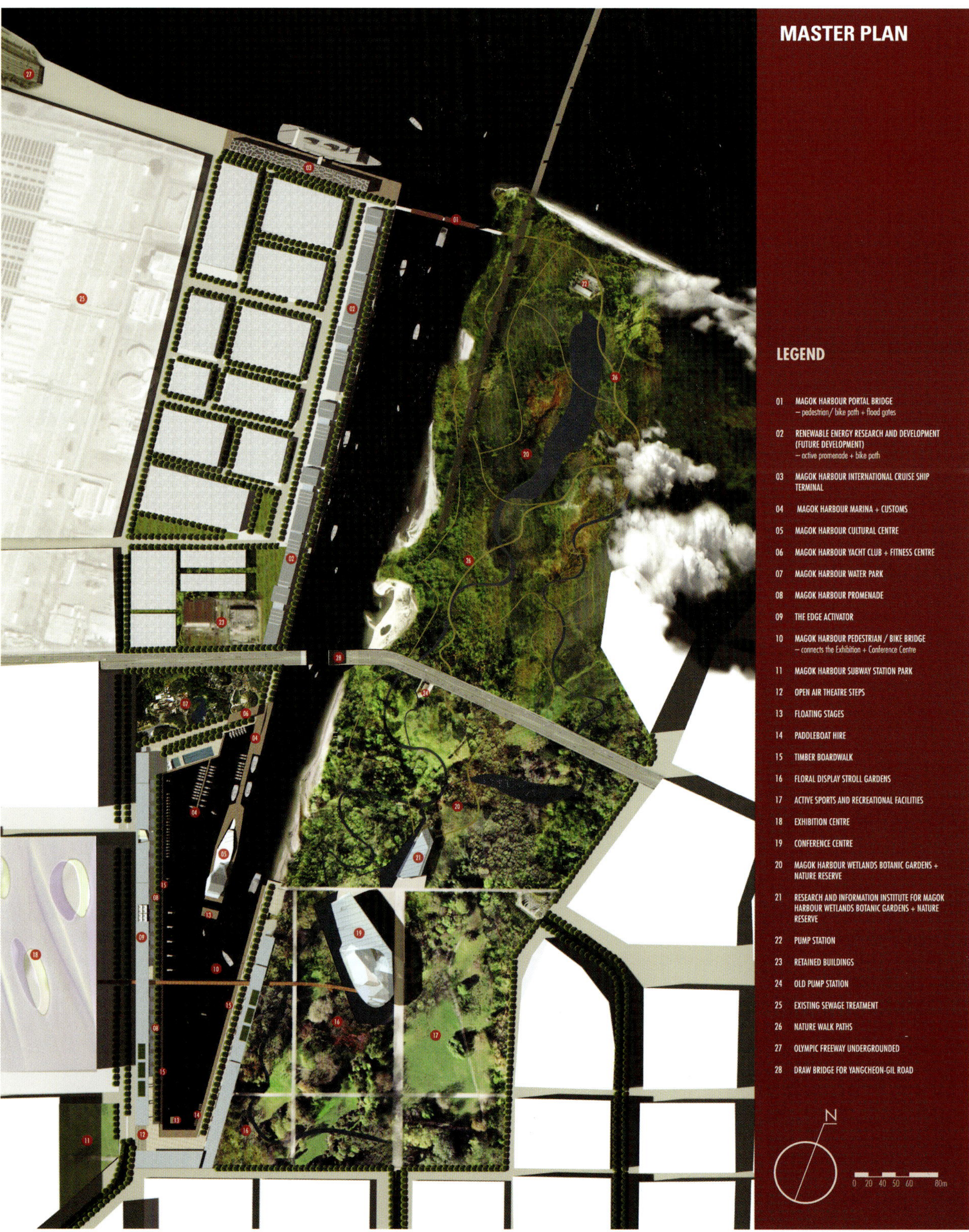

THE DESIGN STRATEGY

MAGOK HARBOUR
Bring the harbour and boats into the heart of Magok, so that the harbour and its edge become the centre of Magok. Add associated programs – Marina, Yacht Club, Customs House and Water Park. Rename the project **Magok Harbour**.

01

▲ View of the inner harbour, Activated Edge and Promenade

▲ View of the Activated Edge and Open Air Theatre Steps from the Promenade

ACTIVATED EDGE & PROMENADE
Construct a human scale commercial / residential building to the harbour edge as an activator. Ensure it becomes Magok Harbour's principal active strip by leasing the lower floors to restaurants, theatres, coffee shops and tourist facilities. Sell the top floors as high end residential apartments.

Construct a generous promenade between the Edge Activator and the harbour. Ensure it becomes the principal urban space for Magok Harbour both night and day by promoting outdoor eating, street theatre and festivals. Plant flowering trees along its length to provide seasonal colour in spring and autumn and shade in summer.

Construct a floating timber boardwalk on the harbour edge. Connect it to the Promenade with ramps, in order to provide intimate access to the water's edge. Locate a number of water facilities including paddle boat hire an harbour cruises along its length. Provide floating stages on the harbour for use during the festivals and as a base for fireworks and concerts.

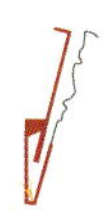

02

▲ View of the inner harbour and Floating Boardwalk abutting the Promenade Edge

CULTURAL CENTRE
Build a cultural centre in the harbour that becomes an internationally recognized icon for Magok Harbour. Make it a focus for both Korean and international cultural events. Ensure it becomes a centre for the community by providing a library and other community related functions.

03

▲ View looking back at the Marina and Cultural Centre from the harbour

PARKLANDS
Develop the eastern side of the site as multi use parklands with an ecological and flood control focus. Use the parklands as a biotope to clean and polish all stormwater from the new town prior to its discharge into the harbour. Use this polished water release to help flush the harbour.

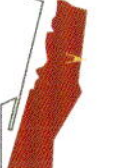

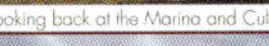

THE LEVEE
Build a levee between the harbour and parklands to manage and control water movement between the two.

04

▲ View from the Levee across the harbour to the Renewable Energy Promenade with its vertical axis wind turbines

WETLANDS BOTANIC GARDENS & NATURE RESERVE
Establish a Wetlands Botanic Gardens and Nature Reserve that reinstates and protects this key tract of land. Initiate a benchmark conservation project for the revitalisation of the wetlands. Including the establishment of a sanctuary for plants, animals and birds. Provide controlled access via raised boardwalks, for bird watchers, nature walks and educational tours. Establish breeding grounds for the migratory Oriental White Stork and make it a symbol for Magok Harbour.

THE WETLANDS RESEARCH & INFORMATION INSTITUTE
Develop a Wetlands Research and Information Centre at the edge of the wetlands, that researches, protects and manages this sensitive eco system.

06

▲ View of the Wetland, Research and Information Institute at the edge of the Wetlands Botanic Gardens

PARKLANDS & CONVENTION CENTRE
Develop the southeastern corner of the parklands into an area for active sporting pursuits including jogging and bicycle riding. Provide Flower Gardens for strolling and picnicking in the western corner. Locate the Convention Centre and hotels within these grounds with a direct Pedestrian Bridge link to the Exhibition Centre.

08

▲ View of the Convention Centre within the Active Parklands

PORTAL BRIDGE & FLOOD GATES
Underground the Olympic Expressway. Build a new pedestrian / bike bridge as an entry portal to the harbour. Integrate the flood gates into the bridge structure as a below water system similar to that used in Venice.

INTERNATIONAL CRUISE SHIP TERMINAL & RENEWABLE ENERGY RESEARCH AND DEVELOPMENT CENTRE
Develop the portion of sewage treatment works abutting the river as an International Cruise Ship Terminal and provide solar powered minibus connections to Magok Harbour Town Centre. Promote development of a Renewable Energy Research and Development Centre abutting the harbour edge. Install vertical axis wind turbines as signature structures along the building / harbour promenade.

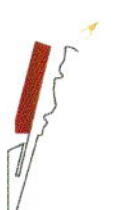

09

▲ View looking at the harbour entrance, Portal Bridge and International Cruise Ship Terminal

The importance of harmonising the apparently irreconcilable contradictions between urban growth and the needs of the environment is now universally recognised. Korea in parti-cular, after half a century of spectacular economic growth accompanied by massive urbanisation and environmental cost, is now focused on developing more appropriate models. Magok Harbour must be a beacon for this new direction.

The philosophical paradigm known in Korea as Eum and Yang expresses this contradiction and its resolution. Eum and Yang are oppo-sites. Part of Eum is in Yang, part of Yang is in Eum. Eum and Yang are interdependent. Eum and Yang consume and support each other. The design locates Eum and Yang on opposite sides of the harbour, yet each exists to support the other.

This proposal recognises the interdependency and interconnectedness of these systems and puts forward a design that shows how these objectives are compatible in ways that are sustainable and sustaining.

WATER

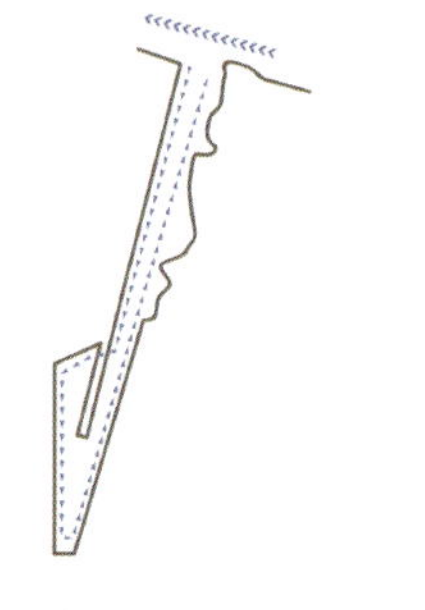
HAN RIVER TIDAL FLUSHING

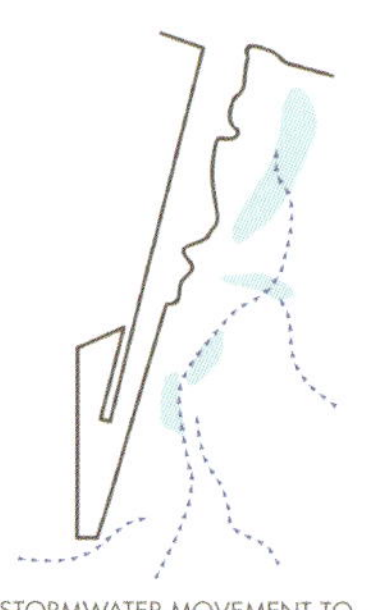
STORMWATER MOVEMENT TO FILTRATION PONDS

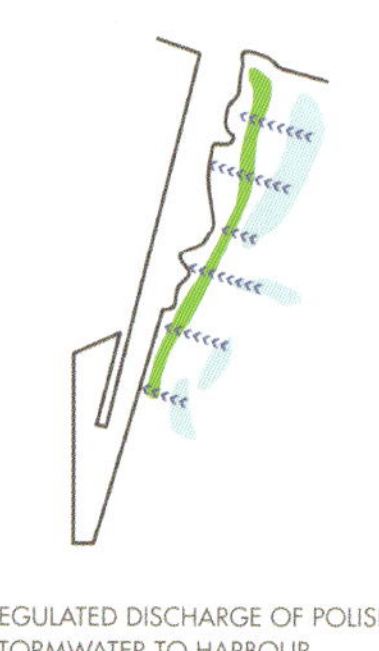
REGULATED DISCHARGE OF POLISHED STORMWATER TO HARBOUR

THE PORTAL BRIDGE

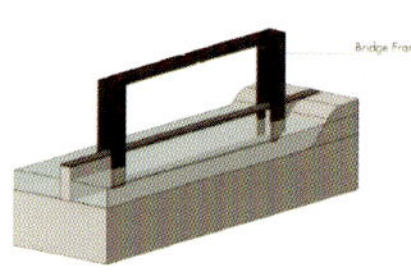
PORTAL BRIDGE IN NORMAL MODE

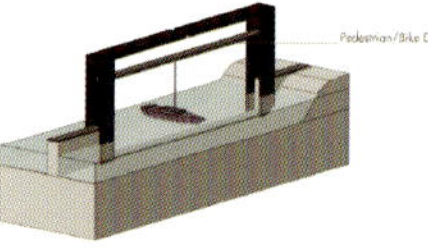
PORTAL BRIDGE WITH PEDESTRIAN/ BIKE DECK RAISED

PORTAL BRIDGE WITH FLOOD-GATE RAISED

MAGOK BOTANIC GARDENS & NATURE RESERVE

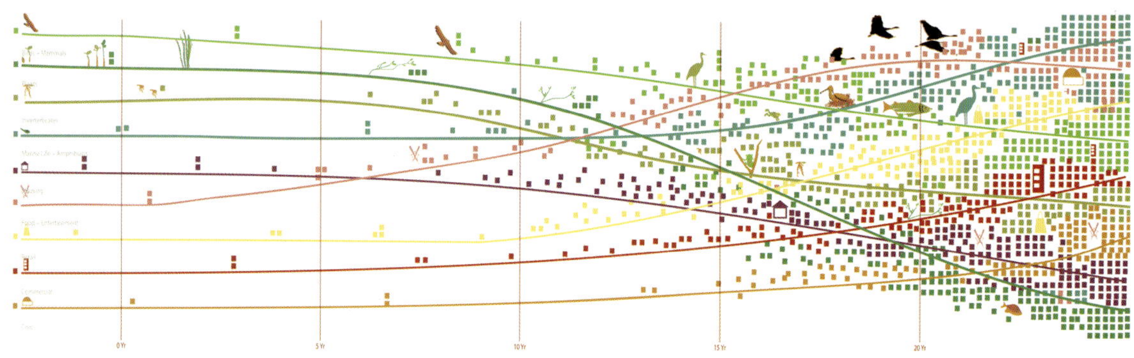
NATURAL & SOCIAL BIODIVERSITY INCREASES OVER TIME

THE LEVEE

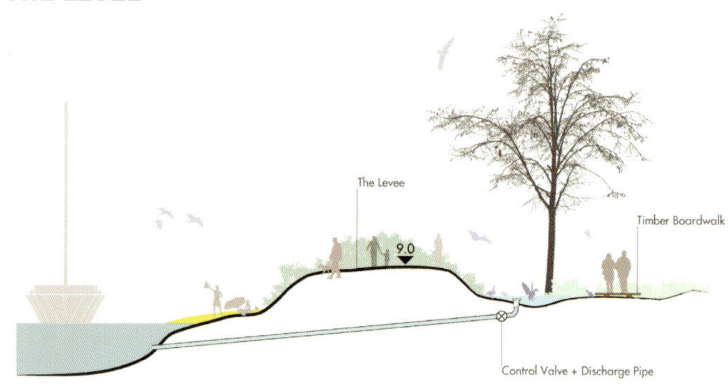

WETLAND LEVEE

ACTIVED EDGE, PROMENADE & HARBOUR INTERFACE

ACTIVATED EDGE & PROMENADE

Merit Award

New Yacht Marina Create Integration Function for All Surrounding Area

Valentin Oleynik · Architect, Russia

Proposed scheme of land use around yacht marina is not corresponding with new, and its been updated corresponding with new rectangular concept.

Main Concept
New yacht marina create integration function for all surrounding areas.

Around water zone of yacht marina grouping parks, cultural object and multi-functional complexes. All functional zones combine together in a compact rectangular scheme.

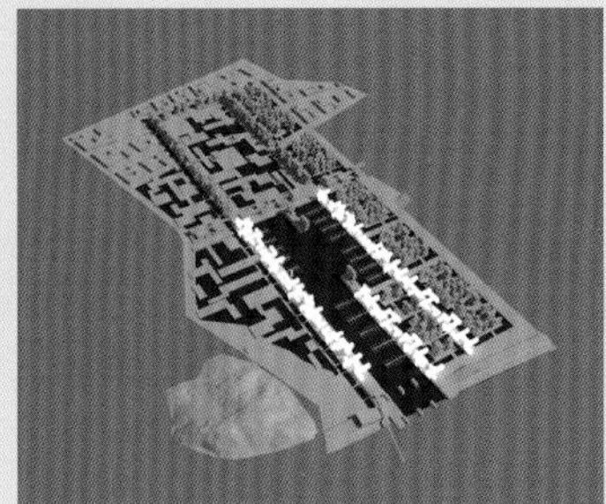

Mixed use for yachtsmen apartments, hotel, entertainment, shops in street level storages for yachts, workshops.

On the banks of yacht marina situated walking zone connected with zone of multi-functional buildings including apartments for the yachtsmens, hotel and on the first floor cafes, shops, galleries and zones with storages for yachts and workshops.

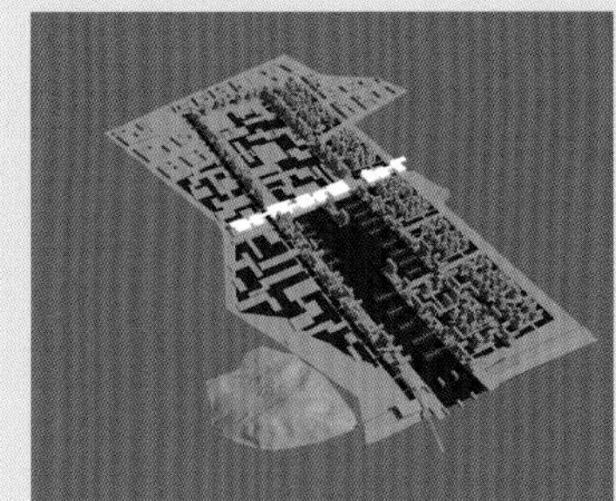

Congress and Exhibition Facilities Complex

On the border of yacht marina and cultural water park situated connected zone with complexes of congress and convention centers and wide walking zone opened to the water surface.

Also in this zone situated yacht club with hotel, entertainment facilities and observation restaurant at the top of the building.

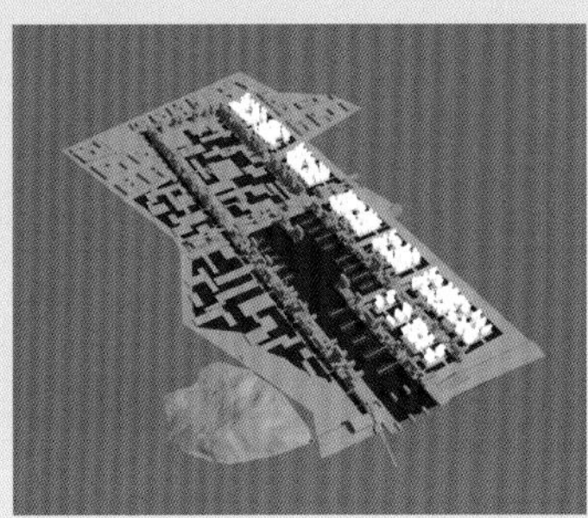

Mixed use apartments closed connected with overall rectangular scheme all houses rised from ground on pilons, giving the freedom of moving.

In the idea zone of seonam sewage treatment plant create a living complex situated on the artificial island surrounding by the channels and close connected with overall rectangular scheme.

All houses standing on the pilons that creates free space on the ground level for public and park zones.

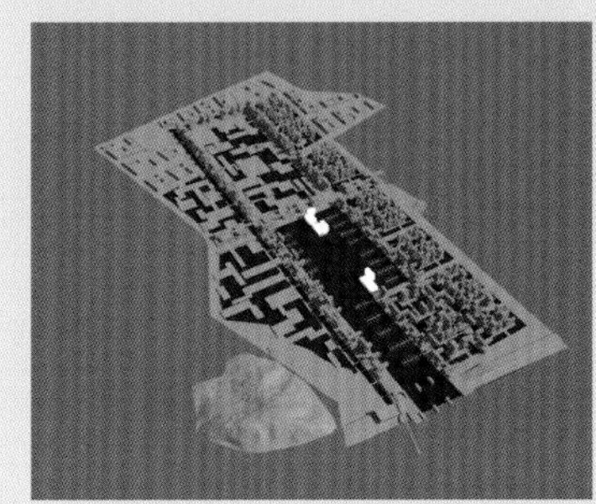

Yacht clubs

Yacht clubs, surrounded by water contain hotel, entertainment facilities and observation restaurant at the top of the building.

It must be one of the site landmarks.

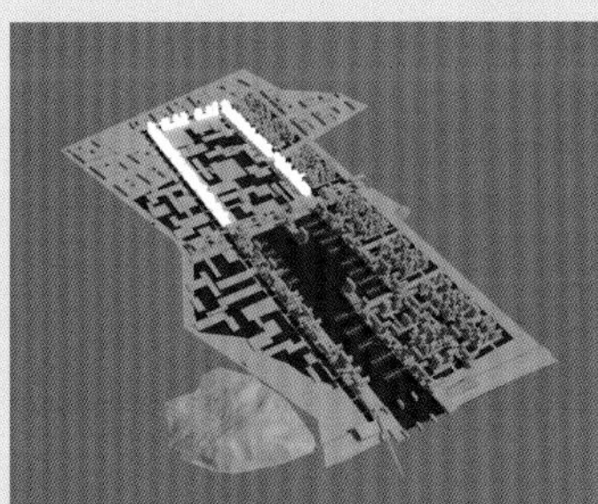

Mixed Use

Entertainment, shops in platform and office, apartments in towers.

Strip of tall mixed use towers(50~60m high) standing on the multi-functional platform.

In the border of park and towers created a wide walking area.

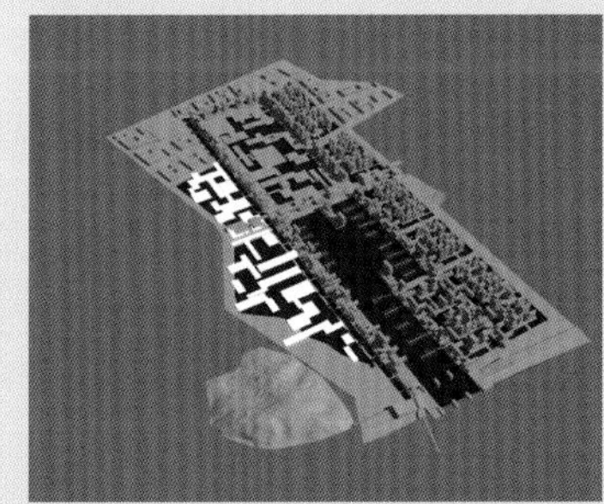

River Park

Situated on the low, flooding territiry and used this for water entertainment park with artificial rectangular island and stations for little boats.

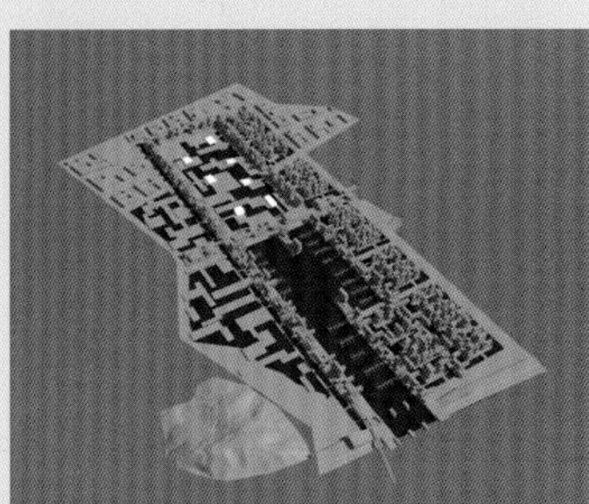

Cultural and entertainment free standing building in the park-theatres, galleries, media libraries.

Landmark buildings integrated in park environment.

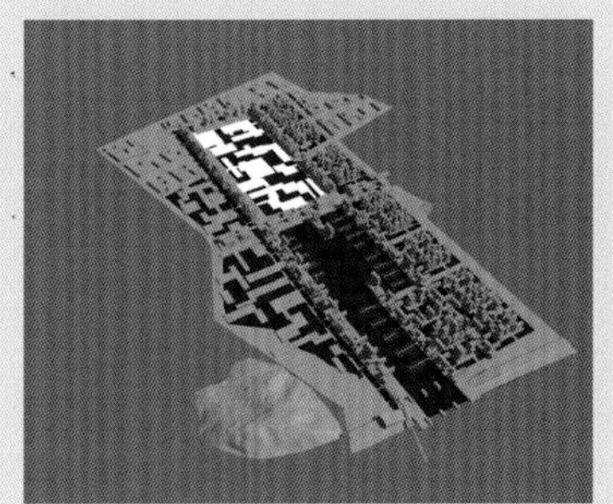

Cultural Water Park

Second park integrated in overall rectangular scheme and become an attraction place like central park in New york. This park surrounded on three sides by a strip of tall mixed use towers(50~60m high) standing on the multi-functional platform.

In the border of park and towers created a wide walking area.
In the park situated important cultural public buildings for all Magok area.

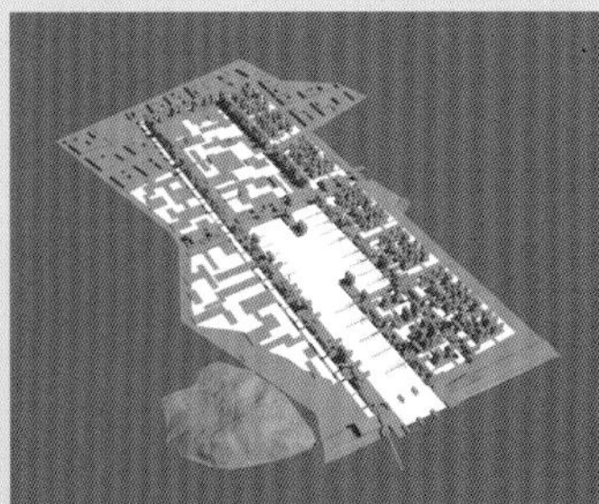

Main Concept

New yacht marina create integration function for all surrounding areas.

Around water zone of yacht marina grouping parks, cultural objects and multi-functional complexs all functional zones combine together in a compact rectangular scheme.

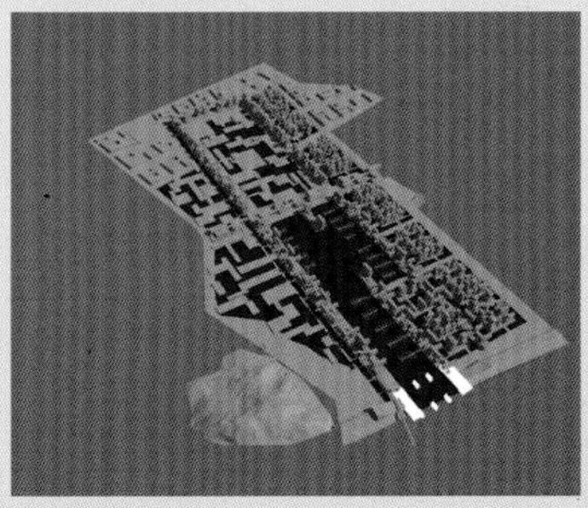

Floodgate Facilities

Olympic Expressway getting to the underground tunnel when crossing the area of floodgate facilities to avoid enter toll ships to the marina area.

The scheme of the floodgate facilities is similar to the floodgate facilities on river thames near london, it's a few rotating segments, closed when water level is high and become a two layered dock.

OVERALL CONCEPT SCHEME

OVERALL CONCEPT SCHEME

MIXED USE
ENTERTAINMENT,SHOPS IN PLATFORM
OFFICE, APARTMENTS IN TOWERS

MIXED USE
ENTERTAINMENT,SHOPS IN PLATFORM/ OFFICE,
APARTMENTS IN TOWERS

MIXED USE FOR YACHTSMEN
APARTMENTS, HOTEL, ENTERTAINMENT, SHOPS IN STREET LEVEL
STORAGES FOR YACHTS, WORKSHOPS

MIXED USE FOR YACHTSMEN
APARTMENTS, HOTEL, ENTERTAINMENT,
SHOPS IN STREET LEVEL
STORAGES FOR YACHTS, WORKSHOPS

MIXED USE
APARTMENTS

CULTURAL AND ENTERTAINMENT FREE STANDING BUILDINGS IN THE
PARK - THEATRES, GALLERIES, MEDIA LIBRARIES

MIXED USE
APARTMENTS

CONGRESS AND EXIBITIONS FACILITIES COMPLEX

CONGRESS AND EXIBITIONS
FACILITIES COMPLEX

CULTURAL WATER PARK

YACHT CLUB/ HOTEL

YACHT CLUB/ HOTEL

LAKE PARK 1

RESEARCH AND DEVELOPMENT FACILITIES

RESEARH AND DEVELOPMENT FACILITIES

CONNECTED WATERWAYS

CONNECTED WATERWAYS

YACHT MARINA

FLOODGATE FACILITIES

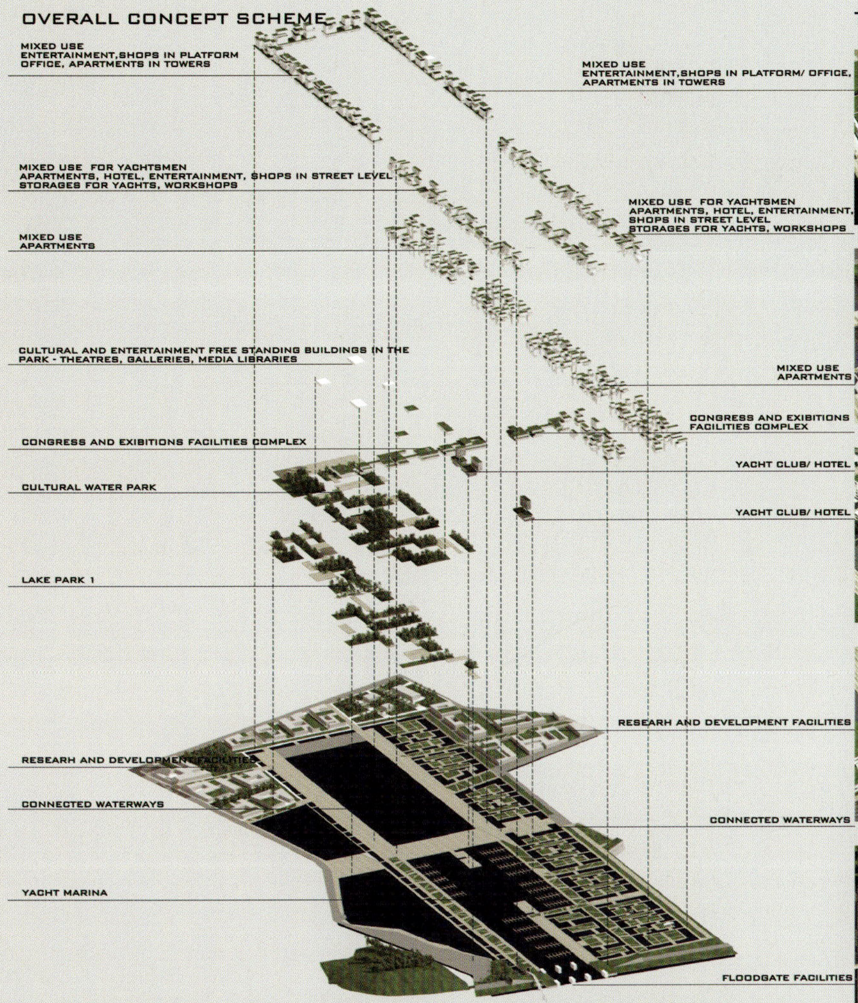

TYPOLOGY OF BUILDINGS

MIXED USE FOR YACHTSMEN
APARTMENTS, HOTEL, ENTERTAINMENT,
SHOPS IN STREET LEVEL
STORAGES FOR YACHTS, WORKSHOPS
MAXIMUM HIGH 15 - 30 m

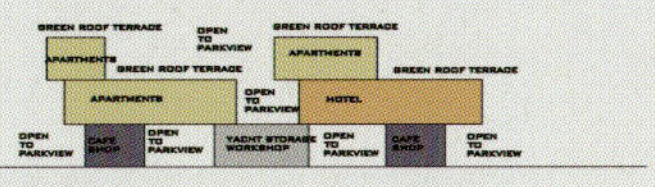

MIXED USE
ENTERTAINMENT,SHOPS IN PLATFORM AND
OFFICE, APARTMENTS IN TOWERS
MAXIMUM HIGH 50 - 60 m

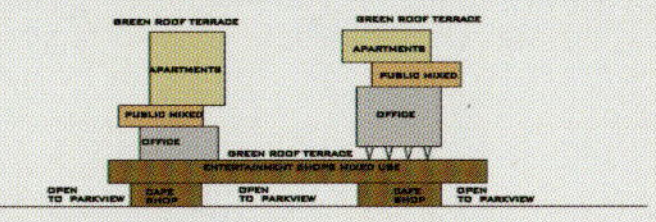

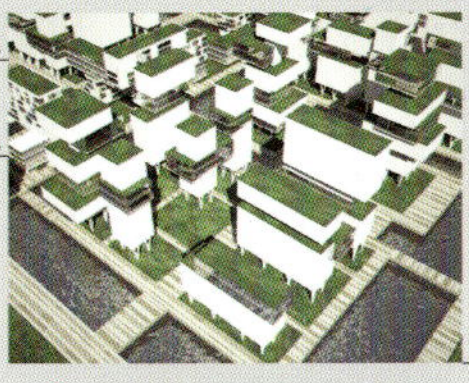

MIXED USE APARTMENTS
ALL HOUSES RISED FROM GROUND
ON PILONS, GIVING THE FREEDOM
OF MOVING
MAXIMUM HIGH 40 - 50 m

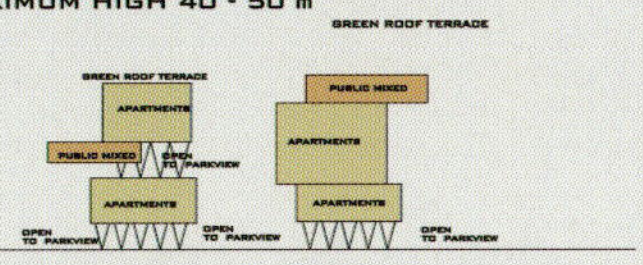

CONGRESS AND EXIBITIONS
FACILITIES COMPLEX
MAXIMUM HIGH 15 - 25 m

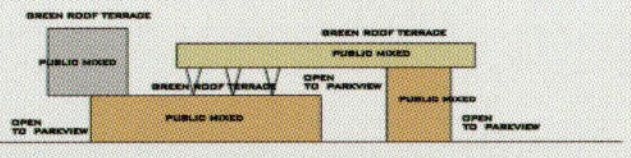

CULTURAL AND ENTERTAINMENT
FREE STANDING BUILDINGS IN THE
PARK - THEATRES, GALLERIES, MEDIA
LIBRARIES

LANDMARK BUILDINGS INTEGRATED IN
PARK IVIRONMENT
MAXIMUM HIGH 15m

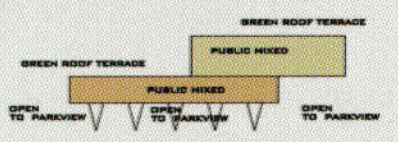

Merit Award

Cultivating the Edge

Lee Jeff(Lee+Papa and Associates) · Landscape Architects, U.S.A.

The design of the Magok waterfront will surface the significance of the site's existing groundplane and strengthen the form of its horizontality.

Lying in the floodplain of the Han river, this 117ha site is thin in profile but possesses a richly layered groundplane. The design of a shallow harbor will fold and unfold this thin profile, pulling elevations up and down to maximaize the potential of existing hydrological and soil substrate conditions. Measures are taken to preserve soil fertility and to emphasize the strong visual relationships bettween land, water, and sky.

A lock and floodgate system where the inlet meets the river functions as a mechanism to reveal water horizons. This horizon vacillates based on times of natural flooding and drought, but it is also controlled by the floodgate. The locks displace boats to meet the water level of the site when river level is low. The water level is preserved at a minimun of 4m during drought season and 6.5m during flood season, averaging a level of 4m the majority of the year. Grading of the harbor creates sufficient depth for Magok Canal, a navigable waterway that circuits through the outlying development. Ramps and terraces help bring the grade up to 9m to meet the outlying Magok New Town development.

EMERGING HORIZONS

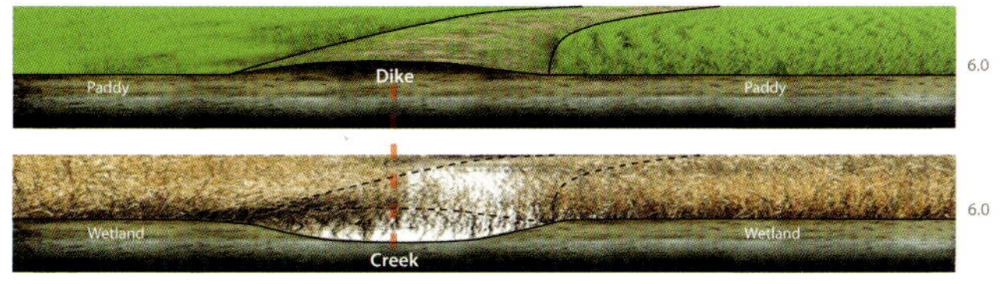

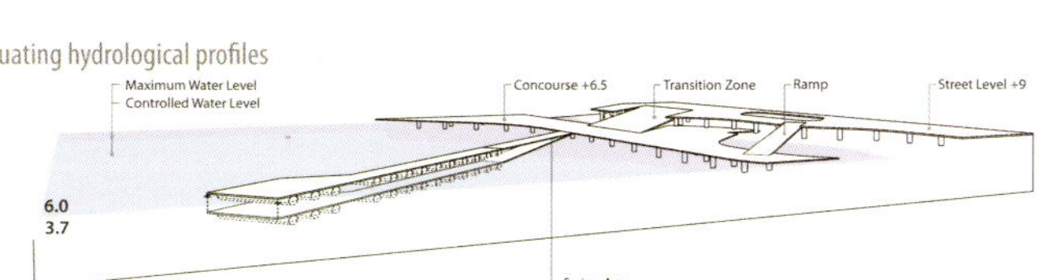

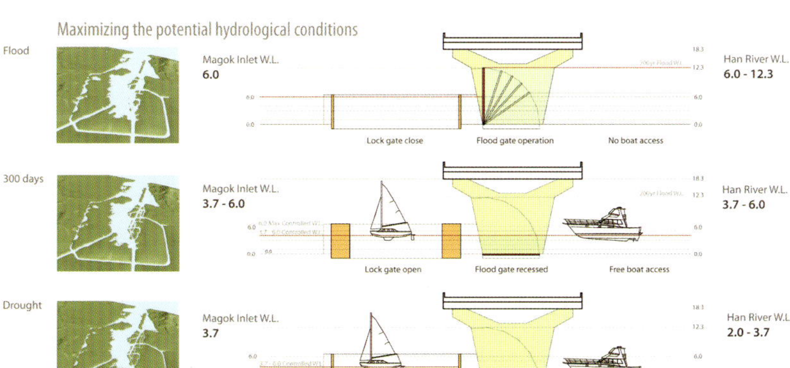

BRIDGING EXPERIENCES

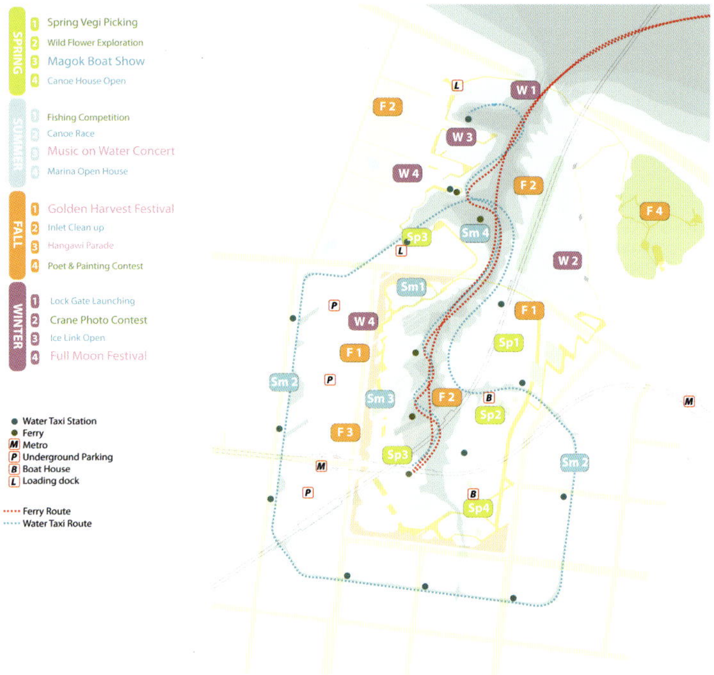

SPRING
1. Spring Vegi Picking
2. Wild Flower Exploration
3. Magok Boat Show
4. Canoe House Open

SUMMER
1. Fishing Competition
2. Canoe Race
3. Music on Water Concert
4. Marina Open House

FALL
1. Golden Harvest Festival
2. Inlet Clean up
3. Hangawi Parade
4. Poet & Painting Contest

WINTER
1. Lock Gate Launching
2. Crane Photo Contest
3. Ice Link Open
4. Full Moon Festival

- Water Taxi Station
- Ferry
- M Metro
- P Underground Parking
- B Boat House
- L Loading dock

- Ferry Route
- Water Taxi Route

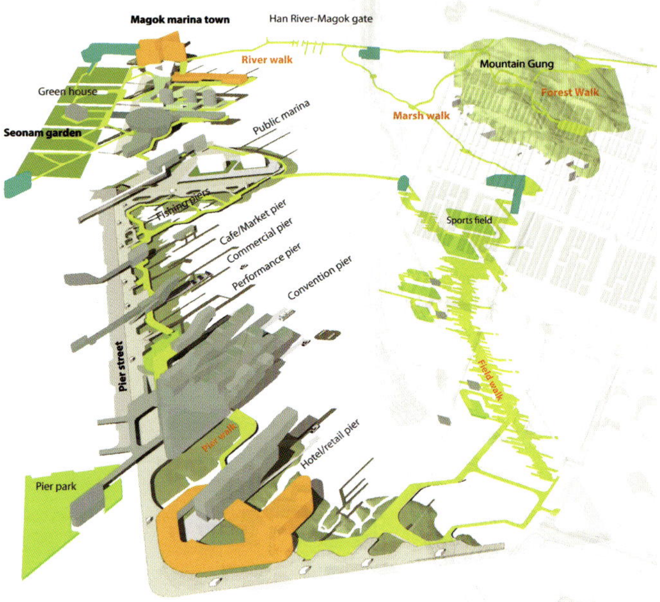

Magok marina town
Han River-Magok gate
River walk
Mountain Gung
Forest Walk
Marsh walk
Green house
Seonam garden
Public marina
Sports field
Fishing piers
Cafe/Market pier
Commercial pier
Performance pier
Convention pier
Pier street
Hotel/retail pier
Pier park

Jury Comments

서울과 같은 국제도시에서 국제설계경기가 열릴 때마다 전 세계의 계획가들이 관심을 갖고 참가하듯이 이번 설계경기에도 전 세계적으로 상당히 높은 수준의 아이디어를 보여주는 많은 작품들이 출품되었다. 특히 이번 마곡 현상공모는 입지와 프로그램이 복잡하고, 대규모인 만큼 많은 분야의 전문가들의 공동작업이 요구되었다. 다행히 참여한 대부분의 팀들은 기술적 측면과 리더십에서 균형잡힌 모습을 보여주었다.

신도시를 계획하는데 있어서 아무것도 없는 상태에서 시작되는 법은 없다. 설계자는 주변의 풍부한 도시조직과 복잡한 교통시스템, 그리고 무엇보다도 거대한 자연을 바탕에 깔고 작업을 시작해야 한다. 마곡은 한강이라는 주요 하천과 공항, 고속도로변에 근접하여 위치함으로써 국제적 관문으로서 손색없는 입지특성을 갖고 있다.

보통의 현상공모에서처럼 한 전문분야가 주도적으로 팀을 이끌어 나갈 수 없는 것도 이번 현상공모의 특징이다. 이런 경우에는 '랜드스케이프 어바니즘Landscape Urbanism' 이라고 하는 조경과 도시설계가 혼합된 다소 생소한 새로운 분야의 전문가가 계획의 방향을 제시해야만 한다. 마곡은 이러한 새로운 전문분야와 전문가가 가로계획, 토목구조물, 생태기반구조 등을 결정함에 있어 디자인 리더십을 발휘할 수 있는 완벽한 대상지이다.
세계적으로 볼 때, 이렇듯 복잡한 속성을 가진 프로젝트들은 많다. 마곡 르네상스 프로젝트는 주변 지역의 구조를 보완하고, 도로, 간선도로, 고속도로, 공익시설, 교량, 홍수조절용 수문 등을 잘 결합할 수 있는 매우 독특한 기회를 제공하고 있다. 이러한 프로젝트를 잘 할 수 있는 유일한 방법은 주변지역까지 확대하여 계획을 수립하는 것이다. 다시 말해서 도시의 조직이 프로그램을 이끌어가도록 하는 것이다.

그러므로 이 현상공모에서 참여자들로 하여금 많은 이슈들을 다루도록 요구하는 것은 어쩌면 당연한 일이기도 하다. 마곡 프로젝트는 한강 르네상스 프로젝트 전체에서 매우 강한 상징성을 갖는다. 동시에 오픈스페이스가 부족한 이 지역에 충분한 오픈스페이스를 제공할 뿐만 아니라 마곡 신도시의 심장으로서의 기능을 가져야만 한다. 따라서 이 프로젝트는 인근 업무지구와 R&D지구가 원활하게 작동할 수 있도록 충분한 어메니티를 제공해야만 하며 더 나아가, 미래 한국의 공원 설계의 모델이 될 수 있는 질적 수준을 확보해야 한다.

이번 공모에서 참여자들은 하수처리, 교통, 수질관리, 홍수조절 등 모든 기술적 문제들을 해결해야만 했다. 이 모든 문제들을 해결하기 위해서 당선자는 건축가, 생태전문가, 조경전문가를 비롯하여 많은 엔지니어들을 동원했음이 분명하다.

이상의 것들을 종합해 볼 때, 우리 심사위원들은 위에 기술한 복잡하고 다양한 현상을 이해하고 반영한 안을 당선작으로 선정하였다. 이 과정에서 설계지침을 준수함에 있어서 다소 미흡한 점이 있더라도, 제안한 내용이 창의적이며, 미래의 조경가, 계획가, 도시설계가, 엔지니어와 건축가들을 고취시킬 수 있는 리더십 원칙이 발견된 작품의 경우 당선안에 포함시키는데에 주저하지 않았다.

As might be expected of an International Competition that attracted designers from all over the world to address the needs of a global city like Seoul, a great deal of work was produced and a huge amount of quality ideascontributed. What may be unique however is that the site and the program complexities required an integrated, fully multi-disciplinary team. Most of the teams possessed this balanced composition of skills and leadership.

It is no longer possible to create new towns from scratch. Designers must work with a rich urban fabric, complex transportation systems and critically important chunks of nature. Magok is an exceptional location on a Han River and airport, highway, uniquely qualifying it as an international gateway. For this reason the submittals have to be complex. The usual professional disciplines cannot have the leading voice.

A new professional leader in landscape urbanism must provide the direction. The Magok is a perfect district to demonstrate this new synthesis and essential form of integrative design which must include conceptual design leadership on the alignment of roads, civil engineering structures, as well as ecological infrastructure. Around the world there are many opportunities of this complex nature. The Magok Renaissance Project is unique an offering this opportunity to complement the fabric of the surrounding city and to integrate the roads, arterials and expressways, the utilites, bridges and floodgates, etc. The only way to do these projects well, will be to extend the project boundaries to include the adjustment surrounding areas. inother words the typology of the urban fabric needs to lead the program. Therefore it was quite appropriate in this competition that the participants were asked to deal with many different issues. Magok park has to be ideologically a symbol of the whole Han River Renaissance Project. At the same time, it should provide plenty of open space to the much needed region as well as serve as the heart of Magok New town, so that it has to supply lots of amenities to the adjacent R&D activities. Furthermore, the park must show the quality that it can be the model design for future park design in Korea.

The participants were also requested to solve all the technical problems involving sewage system, storm water drainage, transportation, water quality control as well as flood gate system. To solve all the issues raised, the winner must be grouped with architects, ecologists, landscape urbanists, as well as many engineers.

Therefore, considering all the above, and when deemed appropriate, we, the jury members, have decided to giverecognition to a few schemes in spite of their short comings in meeting the design guidelines, due to the creative quality of the proposed concepts and the leadership principles they contributed to future landscape architects, planners, urban designers, engineers and architects.

Gangbuk Large Park

강북대형공원 마스터플랜 국제현상설계 공모전

서울시는 지난 4월 24일 국제현상공모를 통해 강북에 조성키로 한 대형공원의 기본계획이 될 최우수 당선작으로 (주)씨토포스와 미국 IMA의 작품 '개방Open Field'을 최종 선정했다고 밝혔다. 이번 국제현상설계 공모전은 서울시가 지난 2007년 10월 발표했던 강북구 드림랜드를 포함한 90만㎡의 부지를 대형 녹지공원으로 조성하겠다는 계획에 따라 세계적인 수준의 공원 조성을 위해 국내 · 외 전문가의 창의적인 아이디어와 폭넓은 디자인계획을 얻고자 시도한 것이다. 공모의 대상 범위는 사업구역도 1 · 2단계 구간으로 90만㎡에 이르며, 1단계의 경우 드림랜드 부지와 주변 32만㎡ 등 66만 2천627㎡를, 2단계로는 2013년까지 나머지 24만2천1㎡를 공원화할 계획이다.

Location Beon-dong, Gangbuk-gu, Seoul, Korea	위치 서울특별시 강북구 번동 산 28-6
900,000㎡ **Area**	면적 약 900,000㎡
Promoters Governments of Seoul Metrocity	공모주최 서울특별시
Schedule Announcement of Park Plan 2007. 10. 16 Announcement of Design Guideline 2008. 1. 15 Final Submission 2008. 4. 7 Announcement of Winner 2008. 4. 23	공모일정 강북대형공원 조성발표 2007. 10. 16 공모 및 설계지침 공고 2008. 1. 15 작품제출 2008. 4. 7 당선작발표 2008. 4. 23

심사위원

조경_ Tashiro Yoritaka · 위원장, 일본 지바대학교 교수, 일본 조원학회 회장
Nick Smith · 영국 Aldar Laing O'Rourke Design & Technical Manager
Kim, Young Dae 김영대 · 영남대학교 교수
건축_ Zheng Shiling · 중국 동제대학교 교수 | Too, Kerl 유걸 · IARC 대표
문화_ Yun, Ho Jin 윤호진 · 단국대학교 교수
생태_ George D. Gann · 미국 지역보전협회 관리자

Gangbuk Large Park is...

Seoul city hosts 'International Master Plan Contest for Gangbuk Park' for Gangbuk area which has insufficient green zones. This site has caused the place to become a slum, losing the role of a Park. In this competition, the 1st prize winning 'Open Field', The proposal was so nice to show the Korean cultural sense of rural landscape which was quite compatible with face water system of ponds and water ways with soft landscape of existing forests and grasses. The proposal of cultural facility represented by the art center, because of its urban sense of attractiveness, was evaluated as of the symbolic existence of this park in the future which attracts many people to visit here.

Jury Comments

I think that parks and green spaces in large cities today are thought as one of most valuable environmental settings because they take roles of contributing toward global warming and creating healthy lifestyles because of their green and cultural potentials. A city who has many large and appropriate urban parks within it shall be thought as most refined city. In this context, to have many parks which all of them are located in a well balanced manner and accessed by all equally. To live with parks is an expression of highly elegant lifestyles. Every large city, historically speaking, have tried to do so under different progress of development. After heated and serious discussion we came to a happy conclusion to choose one work as the best from this board. The proposal was very much superior in the following contexts: We have recognized that this proposal was so nice to show the Korean cultural sense of rural landscape which was quite compatible with the proposed new urban landscape of the central facility of °∞urban and cultural center°±. Rural landscape was proposed with successful surface water system of ponds and water ways with soft landscapes of existing forests and grasses. The proposal of Cultural facility represented by the arts center, because of its urban sense of attractiveness, was evaluated as of the symbolic existence of this park in the future which attracts many people to visit here. This well organized plan of urban and rural landscapes with enough cultural and recreational environment by existing natural basis may be the best solution of the presenter for the culture park. In any way, I hope this proposal, when arranged appropriately by the city government of Seoul under distinguished professional team, will surely be the unique model of new large urban park in Seoul. Well balance of nature and urban context is a question of a park to be considered, and 'void and build' balance of this park will be the best solution to the question of what is a new park in this region of the city of Seoul. I believe that this park will take an important role to realize the good lifestyles of all who live there. _ Tashiro Yoritaka | Chiba Univ, Japan

강북대형공원 국제현상공모의 심사위원장이라는 역할은 내게 흥분되고 어려운 일이었다. 국제적으로 조경, 생태, 건축, 문화의 전문분야에서 초청된 심사위원들과 함께 주의 깊고 신중한 토의를 통해 공원에 대한 우수한 계획을 제안하고 있는 작품을 찾아내야 했기 때문이다. 대도시의 공원과 녹지는 지구온난화와 건강한 생활방식에 직접적인 영향을 미치는 가장 가치 있는 환경중의 하나라고 생각한다. 다수의 크고 적절한 도시공원을 갖고 있는 도시는 가장 세련된 도시로 간주될 수 있을 것이며, 다수의 공원들은 균형되게 위치하고 동시에 쉽게 접근될 수 있어야 한다. 공원과 함께 생활한다는 것은 상당히 고상한 생활방식의 표현이다. 역사적으로 보아도 모든 대도시는 개발 과정에서 공원조성을 시도해 왔음을 알 수 있다. 우수한 작품들에 대한 열띤 토론을 거친 후 심사위원들은 한 작품을 선정하는 결론에 이르렀다. 심사위원들은 이 작품에 대해 '도시와 문화적 중심' 으로 계획된 새로운 도시 조경의 중심시설물과 함께 상당히 조화로운 전원풍 조경으로 한국의 문화적 감각을 보여주는 것이 좋았다고 평가했다. 존재하고 있는 숲과 잔디의 유연한 조경과 함께 연못들과 수로의 지표수 체계를 계획한 점이나 아트센터로 대표된 문화시설물의 계획 등은 방문하는 많은 사람들에게 매력을 줄 수 있으며 미래에 이 공원의 상징적인 존재가 될 수 있으리라 생각되었다. 또한 부지 내에 충분한 문화와 함께 도시와 전원풍 조경의 계획을 체계화했으며, 기존의 자연환경 기반에 의한 레크리에이션 공간이 문화공원을 위한 제안자의 최선의 해답으로 보인다. 서울시에서 능력있는 전문가팀을 구성해 시행해 나간다면 이 작품은 틀림없이 서울의 새로운 대규모 도시공원의 독특한 모델이 되리라 생각한다. 자연과 도시환경의 만족스러운 균형은 고려돼야 할 공원의 쟁점이며, 이 공원의 '비움과 채움' 의 균형은 이 지역의 새로운 공원이 무엇인가라는 쟁점에 최선의 해답이 될 것이다. 이 공원은 주변에 거주하는 모든 사람들에게 풍족한 생활방식을 체득시키는 중요한 역할을 할 수 있을 것이다. _ 타시로 요리타카

1st Prize

Open Field

CTOPOS (주)씨토포스
Partners _ Sigan Architecture (주)시간건축 + David, Choi 최신현

History

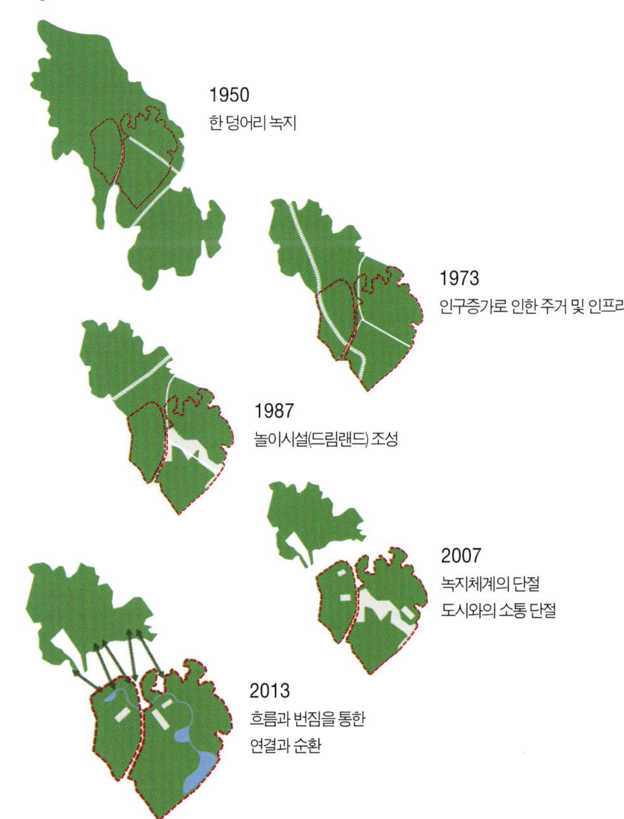

What is Open Field?

Open Field is useful emptiness

The Open field is a much needed open space in the crowded city. It will provide a breathing room and opportunity for diverse recreation, cultural events and various outdoor venues for the people of the city. The park is planned as the city's 'Madang', a traditional courtyard in Korean architecture, with flexible and multiple uses. The Open field's design maximizes the connections to the surrounding communities and the adjacent redevelopment areas. The park design focuses on different opportunities to connect to the city.

· Ecological - The Open field maintains a vital link to existing open space preserves that stretch from the valley to the mountains. It connects the critical ecological systems and waterways in the part of the city.
· Social - The park will physically connect and open up the previously closed circulation flow to the city. It will expand the existing network of walking trails, bike paths and roads to link the adjacent neighborhoods and the city beyond.
· Cultural - The Open field celebrates the past history of the site and to bring new cultural venues for the city within the park.

채움보다는 비움을 통해 공간의 잠재성을 확보함으로써 불확실한 도시의 변화에 대응하는 '도시의 여백' 을 제공한다. 또한 땅의 의미를 드러내기 위해서 장소의 기억과 재생을 위한 기존의 인공지형을 활용한 시설배치 및 비워진 땅의 의미를 확장하여 공간 및 시설배치에 중점을 두었다. 이런 시설배치기법에 있어서 고유의 자연적인 요소를 도입하며 절제된 형태와 방식으로 조성하고자 한다.
전통적인 한국적 공간의 유형에서 추출한 요소들 즉, 기존 도시의 '가로' 와 '마당' 의 개념을 현대적으로 새로이 해석하여 공원의 조직을 구성하였으며, 가로의 활용으로는 한국적 공간의 위계를 통한 공간경험의 다양화를 추구하였다.
우리의 마당은 일본의 석정, 서양의 가든 및 코트와는 달리 공간의 비움을 통해 작업의 공간, 행사의 공간 등으로 그 기능적 성격을 가진다 할 수 있다. 이것이 그들과의 가장 큰 차이점이라 할 수 있으며 그 중심에는 비어있으되 쓰임이 있는 공간의 특성이라 할 수 있다.
강북공원의 Open Field 또한 마당의 확장된 공간영역을 의미하며 비워진 중심공간의 기능 역시 변화하는 도시구조의 다양한 요구와 변화에 능동적으로 변화 혹은 진화해 갈 것이다. 단순한 비움이 아닌 쓰임을 위한, 자생적 생성의 장을 위한 비움인 것이다.

The development of the cities in Korea differs from the western countries. The speed of chaotic growth due to burgeoning population in the city occurred much more rapidly during the last few decades. The cities in Korea developed haphazardly without careful planning process and grew based on rapid changes in economic, social and political changes. The long term planning was not the priority, the immediate current stimulus drove the growth. Therefore, lot of the urban ills were created, including lack of open spaces for the growing population in the city. The Open field and similar parks are needed to satisfy the needs of the people of the growing city in the years to come.

강북공원은 도시흐름의 변화에 어떠한 자세를 취할 것인지의 기로에 서있다 할 수 있다. 이러한 도시 문제의 해결책으로 '흐름을 통한 순환과 소통' 의 역사를 제안하고자 한다.

Disconnection with urban structure(Park as island)

The current lack of easy access into the park area almost isolates it like a green island. The neighborhood surrounding the park does not have many physical connections to the park to be able to use it.

주변 도시구조와의 단절로 인한 밀집된 주거지의 다양한 요구를 수용할 수 없는 공간구조를 가지고 있으며 도시 속의 섬처럼 떠다니고 있다.

By connecting the surrounding communities with numerous access points into the park and linking the hiking trails and bike paths, the city's needs for the open space and recreation is far better satisfied than before. This allows the use of the valuable assets that is the existing forest and open space.

중요한 자원요소인 숲을 공원의 한 요소로 생각하고 적극적으로 주변 도시구조와 연결하였으며 도시와 공원의 경계를 열고 확장하여 도시민들의 다양한 요구를 수용하고 변화에 유연하게 대처할 수 있는 공원구조를 제안하고자 한다.

드림랜드의, 시설들이 사라지고 남은 땅의 모습은 평평하게 정지된 인공지형의 모습으로 나타나고 있다. 그래서 남겨진 공간의 빈터를 개발하기보다는 끊어진 자연의 흐름을 연결하는 복원의 공간으로 계획하여 과거 시설중심의 계획이 가져온 참혹한 실패를 반복하지 않도록 하고, 비우고 또 비워서 미래에 유연하게 대처할 수 있는 쓰임이 있는 여백의 공간으로 계획하고자 한다. 기존의 눈썰매장으로 쓰여진 경사지에 위치한 테라스 형식의 인공지형은 이미 만들어진 땅의 형태를 살려 전망이 가능한 땅의 기억을 살린 문화재생의 공간으로 계획한다. 개발이 용이한 가용지 중심공간이 아닌 버려지고 소외된 계곡부 혹은 경사지에 관심을 가짐으로써 다양한 공간의 활용을 가능하게 하는 새로운 공원의 패러다임을 추구하고자 한다.

Macro condition & Visual structure

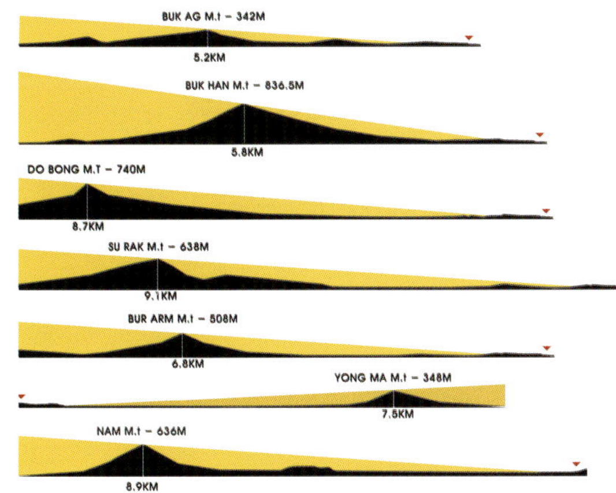

전통적인 서울의 산악경관요소의 도입과 차경 그리고 조망축의 형성

산업사회의 계획 조건에서는 배제되어 오던 자연을 중요한 고려의 대상으로 삼고자 한다. 우리나라의 산이나 구릉지는 도시경관의 지배적 요소로서 랜드마크의 역할을 담당했다. 전통의 경관개념은 주로 풍수지리설에 입각하여 산을 숭배하는 사회문화적 분위기에 의해 형성되었다. 예를 들어 서울 남산의 경우, 서울 시가가 가장 잘 보이는 자리로서 서울의 전통적인 랜드마크 역할을 담당해 왔다. 이러한 대표적인 대상지 주변의 자연적인 경관요소들을 내부경관에 도입하기 위해 차경의 기법을 도입한다.

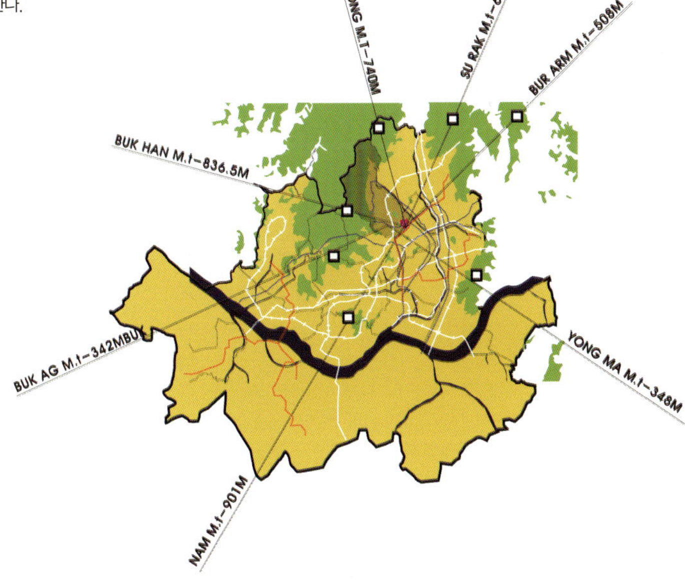

2개의 지하철, 경전철, 버스 등 다양한 대중교통 중심의 공원 접근체계를 구축하였으며 유동적인 공원 이용인구가 아닌 주변의 재개발구역들 장위, 길음, 미아 등의 정주인구를 고려한 공원 프로그램을 구성하였다. 광역녹지체계에 있어서는 생태적 징검다리 역할을 수행하는 패치의 기능을 가지며 인근의 소규모 녹지 패치들과의 연계는 에코터널 등과 같이 직접적인 연결을 제안하였으며, 우이천-중랑천-한강으로 이어지는 광역수계상의 상류부 수계의 성격을 지니고 있다. 공원의 성격 또한 시설중심이 아닌 경관중심으로 계획하여 채움보다는 비움을 통해 사방으로 둘러싸인 서울의 대표적인 산악경관을 차경하여 공원 경관의 핵심으로 계획하였다.

Conceptual Structure(Blue-Blur-Blossom)

Blue ; The water flows and connects the nature and the city.
물, 그 푸른빛으로 흐름과 연결이 되다.
푸른빛은 이제 강북으로 흘러 연결의 매개체가 된다.
물이 흐르는 골짜기를 따라 옛 지형이 되살아나고 그 곳에 스며드는 푸른 빛은
풍경을 채우는 Blue, 생태를 살리는 Blue, 문화를 만드는 Blue로 재탄생하게 된다.
Eco Blue _ 생태와 만나는 물, Urban Culture Blue _ 도시의 문화적 감성
물은 상징적 연결을 넘어 푸른 자연과 더불어 살아가는 도시의 문화를 만들어낸다.
Landscape Blue _ 경관을 그리는 물은 풍요로운 시각적 경험으로 작용하고
도시, 숲, 하늘로 열린 경관과 어우러진다.

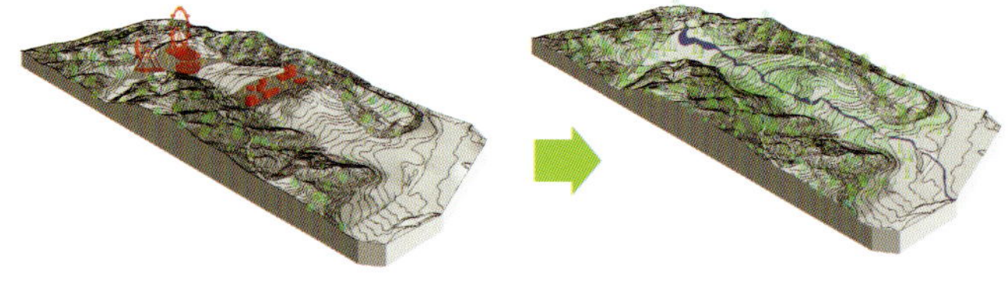

Blur ; Link a ink blot that spreads on a tissue, the park spreads and seeps into the city to create a new environment.
화선지에 떨어진 먹이 번지듯 우리의 삶도, 자연의 여유도, 도시의 기억도 공원에 번져 스며든다.
경계가 없어진 공원은 강북이라는 도시 깊숙이 스며들어 새로운 시스템으로 거듭난다.
땅의 비움과 채움은 움직임을 담음으로 자연스럽게 만들어진다.
정해지지 않은 활동은 영역의 제한 없이 번져 열린 공간으로 도시문화에 스며들고자 한다.
Edge blur _ 녹색의 섬이었던 공원은 경계를 허물고 도시로 번져나간다.
Activity blur _ 활동의 영역은 경계의 틀을 무너뜨리고 전체가 열린 공간이 된다.
Memory blur _ 옛 기억을 되살리고 오늘을 기억하는 장소가 된다.

Blossom ; Creation and progress of the park that blooms in the open field.
만개, 전성기, 번영, 발전, 생산
강북의 open field, 만개하다.
complex, diversity, bleak ···
도시의 삶을 꽃을 피우고 문화라는 결실의 열매를 맺는다.
살아 숨쉬는 도시의 새로운 아이콘 "강북 Open Field"를 만든다.
이제 도시가 그리고 공원이 변하고자 한다.

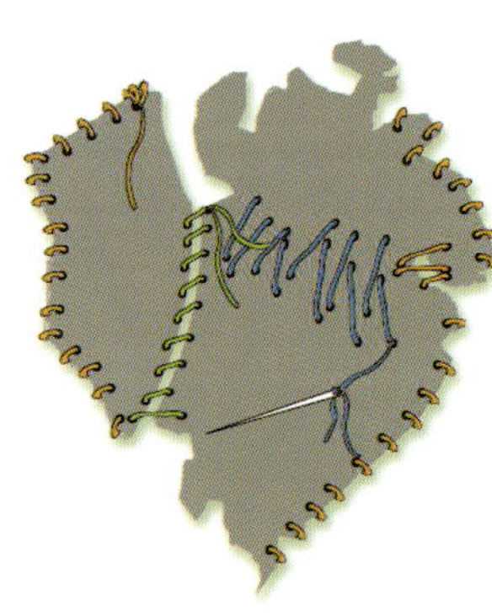

Connectivity 1(Niddle & Thread)

The two open spaces that were severed by the road is sewn back with the ecological needle and thread. The eco-tunnel is the starting point for the reconnection that will revitalize and restructure this city space. Like the stitching process, it is not a fast machine like process. It will take time to reconnect and establish the ecological and city fabric as desired.

도로로 인해 단절되었던 두 개의 녹지공간을 바늘과 실로 기우듯이 생태적으로 연결한다. 또한 공원의 경계부와 도시구조를 한땀한땀 꿰어 새로운 도시재생의 공간을 만들고 동서로 나누어졌던 주민들의 도시 커뮤니티 구조를 새롭게 창조하고자 한다. 바느질의 함축적인 의미는 빠른 시간에 끝내버리는 기계적인 작업이 아닌 속도는 느리지만 시간에 따라 천천히 변화해야 하고 다양한 요구에 유연하게 대처해야 하는 공원의 미래상을 인간의 수작업에 비유하고자 함이다.

Connectivity 2(Traditional Village Approach)

Much like the traditional Korean village layout, the common public spaces are located along the pedestrian circulation and the water. These common areas provide venues for various functions and activities. The design follows the logic of our fore fathers' wisdom and methods and applied it to our modern day design. This is an expression of Korean tradition and methodology and a symbolic point for the park.

산수를 고려한 마을공간의 구성으로 주거지 내부로 유입된 물길은 상징성과 실용성을 갖춘 수체계를 이루고 있으며, 길을 통한 공간구성형식으로 안길을 따라 마을의 공동시설을 배치하고 이 공동시설은 마을의 진입시 장소성을 나타내는 상징공간으로 조성되었다. 강북공원의 open field 또한 자연스럽게 흐르는 물길을 통해 공간의 프로그램들이 연계되어 있으며 생태적인 수로의 기능도 수행한다. 또한 자연의 흐름에 맞추어 내부가로에 배치된 시설들은 원의 중추적인 프로그램을 담고 있다. 이와 같이 옛 선인들의 공간구축방식을 현대에 재해석함으로써 한국적인 또 다른 표현이자 공원을 대표하는 상징성을 찾아가고자 한다.

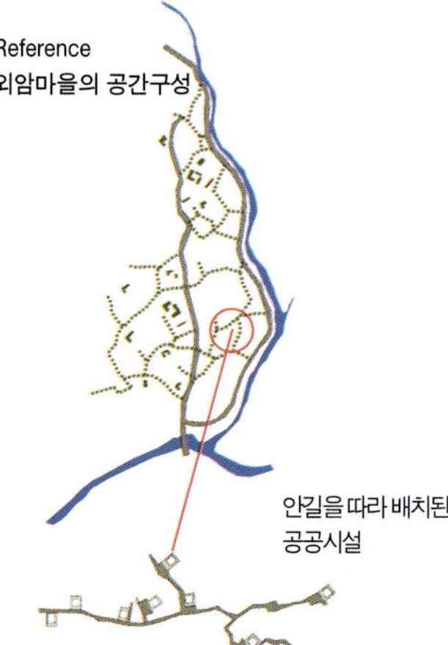

Process

주변상황과 단절된 섬과 같은 공원의 구조 _ 우이천, 중랑천과의 광역적 블루네트워크가 요구됨	도시구조의 다양한 관계들 _ 다양한 도시의 힘들을 공원구조에 연결하기 위한 도시구조의 탐색	경계 흐리기를 통한 도시구조와 공원의 결합 _ 도시의 Edge와 공원의 영역들이 중첩되어 공원의 일부를 형성	중심공간의 여백 _ 채움보다는 비움을 통한 다양성과 불확정적인 도시구조에 유연하게 대처	중앙부의 여백공간을 순환하거나 혹은 관통하며 공원을 체험할 수 있는 동선체계의 수립	네트워크공간 _ 서로 관계 맺으며 성장 혹은 자생적으로 생성되는 프로그램공간의 형성

단절된 도시구조와 공원의 영역을 흐리게 하여(blur) 서로 관계 맺게 하고 대상지 전체를 관통하는 물(blue)이라는 흐르는 요소를 통해 프로그램의 확산 혹은 생성을 유발하게 되고 공원문화를 꽃 피우게 되는 촉매 기능을 수행한다. 또한 이 물들은 불확정적 프로그램의 수용을 위해 크게 비워진 대상지의 중앙부를 관통하며 이 공원의 주요한 경관요소로 활용된다. 계획된 각각의 시설공간들은 단기간에 모두 형성되는 것이 아니라 자생적 혹은 주변상황의 요구에 의해서 점진적으로 개발될 것을 고려하여 전체적인 공원조성의 틀을 수립하고 그 틀이 지속적으로 성장, 진화하는 전략을 수립한다.

Spatial Structure(S.M.L.XL & Thematic walk)

The basic spatial organization of the park is series of rings that focus on the center that radiate outward towards the edge of the park. The existing topography is basically intact and the traditional Korean spatial? organization was used to layout the park. The interconnected network of trails and paths link major and minor spaces and their various functions and programs. The inner most and lowest elevation ring of paths connects primary program venues for the park. The middle level ring connects all the minor open spaces like a spider web and functions as jogging/hiking trail and exercise areas. The highest elevated outer ring trails provide views of the park and a venue to enjoy a relaxed stroll through the forest.

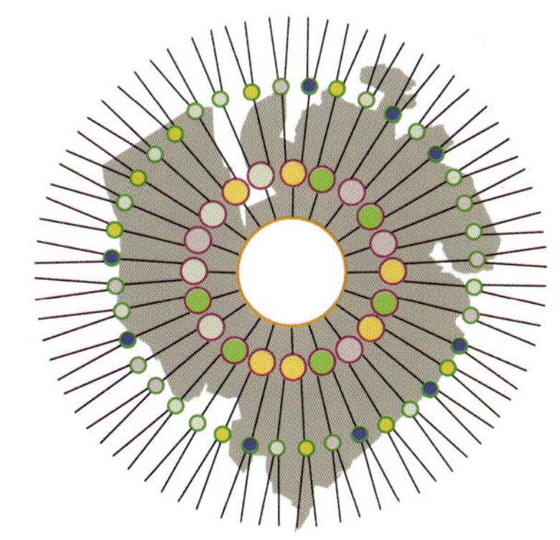

Spatial structure 공간구조 다이아그램

공간의 기본적인 구성체계는 외연에서내부로 자연스럽게 연결되는 동심원 형식의 체계로 장소중심의 사고를 통해 나타나는 대지의 조직방식을 통하여 도시의 다양성을 수용하는 공간, 지형에 순응하는 동선의 흐름, 반위계적이며 분산화 된 공간체계를 반영한 한국적 공간구조체계를 수립하였다. 또한 땅의 의미를 재해석하고 구성하고자 하였으며 자연요소들을 내부공간에 도입하여 본질적인 구조를 바탕으로 하는 공간체계를 수립하였다.

Structure by scale & location 규모와 위치에 따른 공간구조

공간의 크기와 위치에 따라 4종류의 유형으로 구분하였으며, 공원 프로그램들을 중심공간이 아닌 주변의 공간들로 분산함으로써 공원의 중심공간을 시설이 아닌 경관 중심의 탄력적이고 유연한 공간 배치를 가능하게 한다.

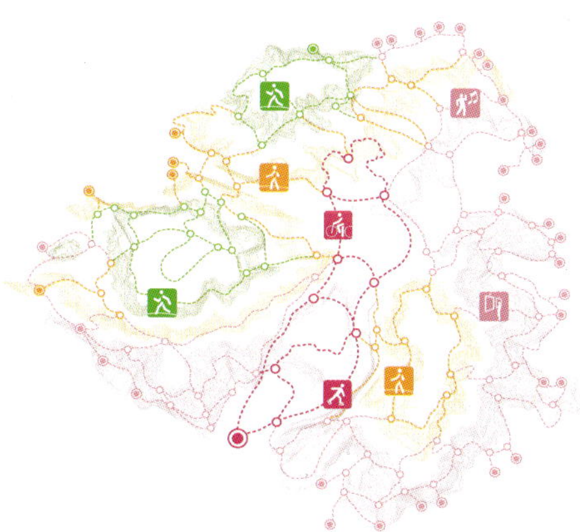

Structure by thematic walk 동선의 고저차에 의한 공간구조

동선의 고저차에 따른 주제가 있는 보행의 계획 및 공간의 연결로, 평탄지에 위치한 하단부의 중심동선은 공원의 주프로그램들을 연결하는 주동선으로 설정하고 중간레벨인 언덕에는 거미줄처럼 얽혀있는 소규모 시설 프로그램들을 연결하는 동선과 동시에 운동공간으로 활용한다. 상부의 산책로는 전망과 함께하는 산림욕을 즐길 수 있는 산책로로 계획되어진다.

Master plan

Small - Edge
1 Identity plaza
2 포켓공원
3 쌈지공원
4 Entrance Garden
5 Children Play Ground
6 Street park

Medium- Patch
1 수림산책로
2 시인의 길
3 인라인 스케이트장
4 한국야생화 정원
5 Sports park

Large - Node
1 습지원
2 보태닉 가든
3 생태체험관
4 음악정원
5 이벤트의 길
6 한국정원
7 단풍대

XLarge - Field
1 문화센터
2 전망타워
3 미러가든
4 지하주차장 지역
5 노을언덕
6 멀티극장
7 레스토랑
8 대형잔디
9 월광폭포
10 월영지
11 아이콘갤러리
12 미술관

1. Small - Edge

- 공간체계 중 가장 작은 단위규모로서 공원의 경계부에 위치
- 전이공간 성격과 지역주민의 가장 사랑받는 시공간적 친밀도가 높은 공원을 의미
- 공간의 위치에 따라 포켓파크, Identity플라자, 쌈지마당, 가로공원들의 다양한 유형의 공간으로 진화해 간다.

2. Medium - Patch

- 공원이 지녀야 할 근린적 프로그램을 분산 수용
- 중앙의 Great-Lawn의 시설 프로그램 설치부담을 줄이는 역할
- 자연에 순응하는 길 중심의 공간 배치를 통해 Wellbeing, Sports 등의 활동형 공간 프로그램을 수행한다.

3. Large - Node

- 시설지들이 공원의 주동선과 접하여 배치
- 2km 길이의 순차적 이동경로를 통해 경관 체험과 다양한 문화 활동의 참여
- 자연 공원 속에 살아있는 문화 / 역사 이벤트 참여공간

X Large(Field)

X Large - Performing Art & Exhibition Center

지형의 축적

기존의 눈썰매장은 '지형의 복원'을 위한 하나의 대상지이며, 또한 새로운 건축물 배치를 위한 별도의 지형훼손이 필요없는 최적지이다. 대지 조건에 맞는 형태와 프로그램을 적용하여 대지의 재활용도를 높이고 방문객이 자연스럽게 누적된 또 하나의 지형을 인식할 수 있도록 하는 것이다. 주변 지형 높이와 차이가 거의 없는 지붕선의 불규칙적인 흐름은 건축물과 지형이 접하는 면과 선을 최대한 확보하여 방문객이 다양한 경험을 할 수 있는 더 많은 기회를 제공한다.

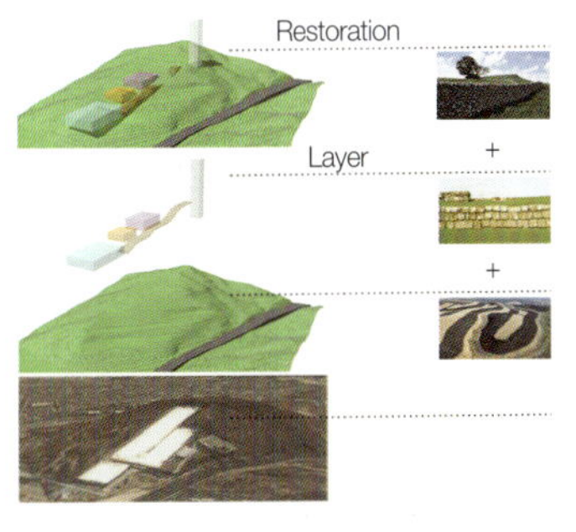

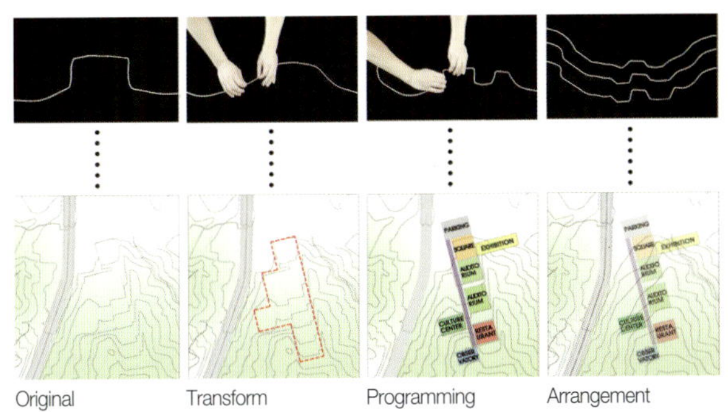

층별에 따라 각각의 주제로 구성

지형의 높이 변화에 따라 다양한 옥외 조경공간이 제공됨

노을지는 하늘과 풍요로운 숲을 편안하게 전망할 수 있는 퍼포밍 아트센터

Poetic Landscape(詩景)
One day in Gangbuk's Great Open Field

As I gaze beyond toward's In Wang Mountain, I entered the park greeted by array of fountains dancing in the water. It is very pleasant as I stroll along the stream towards the lake that looks so refreshing. Eating picnic lunch on the great lawn and enjoying the warm sun is a simple pleasure that is rare these days. I hurry towards the gallery to avoid the spring rain. There is a Monet exhibition today. His paintings are very much like this park. I head towards the plaza after a cup of tea as the dusk settles in. As I head towards the parking area, theme lighting from the floor entertains me. I could hear the joyful screams from the children ahead of me.

멀리 인수봉을 바라보며 공원으로 들어가는 길은 정겹기만 하다. 입구의 오밀조밀하게 배치되어 있는 소담한 분수들이 손님맞이를 한다.
구불구불 시원스레 흐르는 계류를 따라 복원된 옛 선인들의 쉼터가 또 다른 감흥으로 다가온다. 모두들 사진 찍기에 분주하다.
옛 정취에 빠져 잠시 걸었다 싶더니 어느새 커다란 호숫가에 도착한다.
넓은 잔디밭에서 따사로운 볕 아래 도시락 먹는 즐거움이야말로 진정한 都市樂이 아닌가…. 잠시 호숫가를 산책하며 옛 고향의 정취에 빠져든다.
아, 옛 사람들은 이런 곳에서 꼭 시로서 아름다운 경관을 노래 했다 했던가. 이런 저런 상념 또한 내겐 오랜만의 즐거움이다.
갑자기 떨어지는 봄비를 피해 미술관으로 몸을 옮긴다. 아 오늘은 모네의 그림전시가 있는 날이다. 정원그림들이 이 공원과 무척 닮아 있음을 가슴으로 느끼게 된다.
차 한잔을 마시고 어둑어둑해진 공원을 나서려 Performing Art Center가 있는 광장으로 향한다.
광장에는 벌써 어둠이 짙다. 바닥에 설치된 형형색색 변하는 인공조명이 아름답다. 밟을 때마다 무게에 감응해 그 색이 변하나 보다.
아이들의 즐거운 함성이 이곳 저곳에 메아리 친다.

2nd Prize

THE OLD GATE An Invitation to the Lost Poetical Lyricism

Synwha Consulting (주)신화컨설팅 | **Partners** _ Won Yang 원양건축 + HASSEL

To Koreans, a mountain is a spiritual space that realizes the ultimate world sought for by a city and the essential value of life 'Man dwells upon this earth poetically.' An urban park is a space of alchemy that transformsa prose into a poem. We propose a space inspired by Bachelard, a park as the matrix of the earth, which is full of the poetic sentiment, imagination and the dreams of childhood days.

Park with Stories that Communicate with Scenery

산 + 하늘바라기 + 이야기

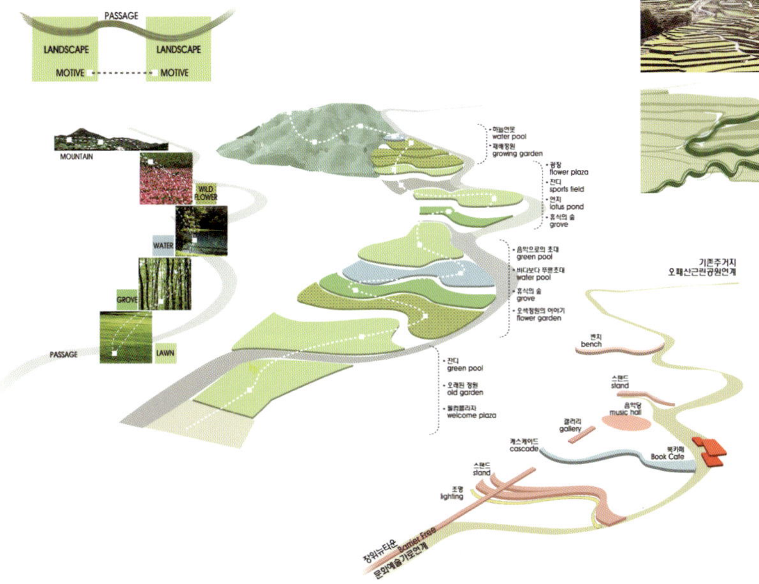

Design motive

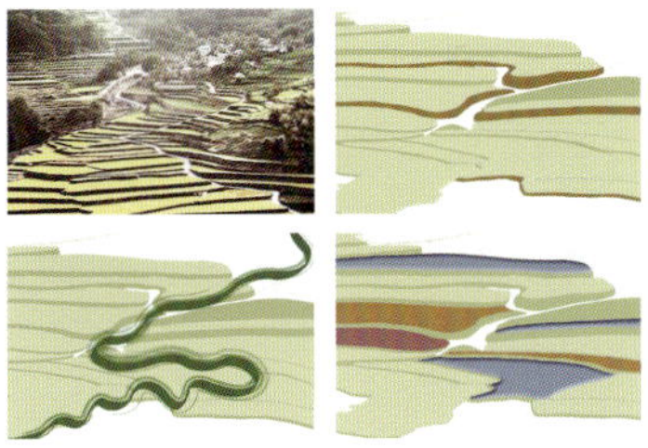

한국의 풍경을 디자인하다 _ 산 + 하늘바라기

세 개의 산이 자리 잡고 있는 대상지의 경관 특성을 고려하여, 비산비야의 우리네 풍경을 디자인한다. 산에서부터 자연스럽게 흘러내리는 (농촌의) 정겨운 풍경은 도시 속의 명소로 거듭나, 수많은 사람들의 이야기를 담아내는 넉넉한 장소가 되기에 충분하다. 산과 어우러지는 풍경은 계곡 자락을 흘러내리는 천수답(하늘바라기)을 모티브로 디자인한다. 천수답은 자연과 더불어 사는 삶의 방식과 이념을 대변한다. 속도 지향의 사회 대신 환경과 조화롭게 사는 방식을 기반으로, 도시의 속도와 방향에 반하는 느림의 미학을 추구한다. 도심의 바쁜 일상 속에서 고향 같은 낙원으로서의 강북 대형공원이 되길 희망한다.

풍경과 교감하는 이야기가 있는 공원 _ 산 + 하늘바라기 + 이야기

도시 속의 낙원을 만들기 위해 네 가지 키워드로 접근한다. 산과 이어지는 천수답의 흐름Landscape과 이를 연계하는 구불거리는 길Passage을 만든다. 여기에 이야기의 화두가 되는 요소motive들을 공원의 곳곳에 배치하여, 이용자들로 하여금 자신만의 이야기story telling를 지어가면서 새로운 감흥을 느낄 수 있는 장소를 계획한다.

잃어버린 기억의 매체로서 공원이 일상 가까이에 놓이며 문학과 예술, 한국인의 신명나는 풍류가 분출할 수 있기를 바란다. 그것을 일깨우는 우리 주변 160㎢의 조각난 산들을 소중한 시선으로 다시 볼 수 있기를 희망한다.

2nd Prize

Gangbuk Cultural Eco Park

GROUP HAN Associates | **Partners** _ HanSeol Green (주)한설그린 + Dawool 다울건축 + Harvard Univ.(Niall G. Kirkwood)

The present appearance of the site for Gangbuk Park was an example of a short sighted vision that failed to foresee the future and was not sustainable in the aspect of environment. Thus, we placed the focus of the newly built Gangbuk Park on transforming the image of present Gangbuk area and creating a environmentally, culturally and socially healthy and active space to set up a model for 21st century Seoul urban park. Gangbuk Park will be perceived as a symbol of restoration by the residents with its elevated quality of the green spaces, relaxing areas and the surrounding communities.

지속가능성이라는 개념은 이미 대부분의 한국 도심공원에서 광범위하게 채용되고 있다. 그러나 비단 좁은 의미의 지속가능성에 대한 측면 뿐만 아니라 시민사회적, 환경적, 문화적, 사회적, 경제적, 그리고 정치적 견지의 총체적 지속가능성에 대한 혁신적 디자인을 실현하는 데 진정한 난관이 있다. 우리는 다음 네 부분의 지속가능성에 대한 명확한 판단과 방향을 제시한다.

시민사회적 지속가능성 _ 새로운 공원에 의해 창출되는 자원소비와 도시 어메니티 간의 연계에 주목할 것이다. 우리는 서울이 녹지공간을 보다 생산적이고 지속가능하게 이용함으로써 얻을 수 있는 기회를 제시할 것이다.
환경적 지속가능성 _ 대지의 건강성과 주변 커뮤니티와의 관계에 대한 급진적인 변혁안을 제시함으로써, 자연자원의 소비와 국지적/지역적 규모의 녹지 공간의 혜택에 대한 연관성을 밝힐 것이다.
문화적 · 사회적 지속가능성 _ 새로운 공원은 미학적 틀로서의 경관에 대한 새로운 시각을 제시할 것이다. 모든 연령대의 방문객이 향유할 수 있는 자연의 아름다움과 살아있는 문화요소들을 운용하는 방법을 제안한다.
경제적 · 정치적 지속가능성 _ 개방적이고 투명하며 민주적인 녹지공간을 실현하기 위해 우리는 기존 경관의 재활용에 주목한다. 국지적이며 도시적인 수준에 부합하기 위해 부지가 재정비 되고 나면 주변의 커뮤니티에 정치경제적 활력을 재충전하는 기회가 될 것이다.

건강한 미래공원을 위한 본 제안은 기존의 공학, 원예, 재프로그래밍 등에 한정된 제안들에 의존하지 않으며, 환경적인 투자의 측면에서 접근한다. 시민들과 전문가들에게 강북공원은 현재의 사회, 환경, 커뮤니티의 이슈의 해결이 가져오는 혜택을 보여줄 것이다. 버려진 땅에서 실현가능한 명확한 아이디어를 제안함으로써 우리는 강북공원이 시민사회의 많은 거주자와 방문자들에게 사랑받을 수 있는 곳으로 거듭나길 기대한다.

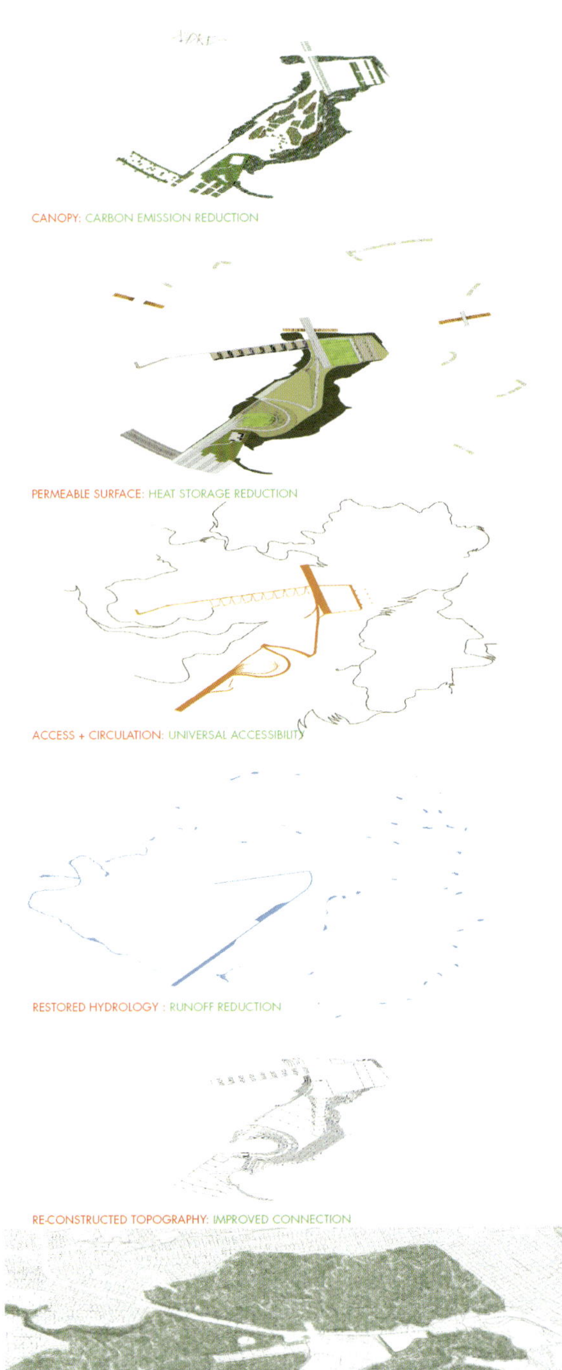

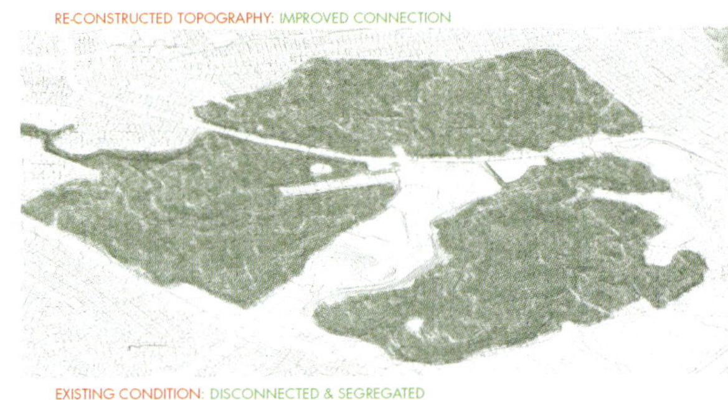

2nd Prize

Gangbuk Park the Responsive Land

L.E.T Landscape Architects Associate | **Partners** _ Dong Woo 동우건축 + Martha Schwartz + Southern Green

Dreamland was left as an isolated island due to the cultural centralization into urban areas. In Dreamland, which was confined to a space by a physical boundary of Euclidian geometry, the time flow seemed to bear no relation to that of the outside environment. And it was the very factor tha weakened the organic combination between the inside and outside of Dreamland. The closed ness of a dream bounded by Euclidian geometry is now being brought down to be exposed to positive and continuous stimuli. Here, we make researches into the responding land that can be organically combined with the fierce stimuli of time.

1. 드림광장 (Dream plaza)
2. 잔디주차장 (Green parking)
3. 장위폭포 (Jangwi waterfall)
4. 안개분수 (Mist fountain)
5. 창포원 (Iris garden)
6. 상징조형물 (Symbol sculpture)
7. 비지터센터 (Visitor center)
8. 오픈갤러리 (Open galley)
9. 꽃향기언덕 (Flower hill)
10. 보존림 (Existing Forest)
11. 어번비치 (Urban beach)
12. 해담못 (Sunny pond)
13. 카페테리아 (Cafeteria)
14. 드림폭포 (Dream waterfall)
15. 태양열집열판 (Solar panel)
16. 억새언덕 (Pampas hill)
17. 물억새습지 (Pampas grass swampy)
18. 재료테마원 (Material garden)
19. 다이나믹물놀장 (Water playground)
20. 축제의너른뜰 (Festival yard)
21. 농구장 (Basketball court)
22. 들꽃언덕 (Wild flower hill)
23. 꽃무리비탈 (Flower hill)
24. 기억의정원 (Memorial garden)
25. 강북문화테마원(Gangbuk cultural garden)
26. 거울연못 (Reflecting pond)
27. 야외스탠드 (Outdoor stand)
28. 솔향기언덕 (Pine Hill)
29. 숲속프롬나드 (Promenade in forest)
30. 스카이워커 (Sky walker)
31. 산마루터 (Top of mountain place)
32. 릴렉스 가든 (Relax garden)
33. 숲속꼬마놀이터 (Forest of children)
34. 환상슬라이드 (Exciting slide)
35. 재활용놀이동산 (Recycling playground)
36. 숲속빈터 (Clear space in forest)
37. 동구마당 (Village Plaza)
38. 실버정원 (Silver garden)
39. 웰빙정원(Wellbeing garden)
40. 미꾸리연못 (Loach pond)
41. 소리의 정원 (Sound garden)
42. 오픈스쿨 (Open school)
43. 물방개연못 (Diving beetle pond)
44. 곤충관찰원 (Insect observation garden)
45. 나비원 (Butterfly garden)
46. 산림비오톱관찰원 (Biotop observation garden)
47. 가로소공원 (Porket Garden)
48. 대숲정원 (Bamboo Garden)
49. 프로그램주차장 (Parking)
50. 직업테마관 (Job experience center)
51. 전망타워 (Gangbuk Tower)
52. 숲문화체험놀이터(Children playground)
53. 배드민턴장 (Badminton court)
54. 생태통로 (Eco bridge) – 제안
55. 숯문화체험원(Charcoal experience garden)
56. 도토리관찰원 (Archon observation garden)
57. 숲속공연장 (Open stage)
58. 구릉지식생천이관찰원 (Succession Forest on Hill)
59. 잔디언덕 (Grass Hill)
60. 기존전망대 (Existing observatory)
61. 풍력발전기 (Wind generator)
62. 벽천 (Water fall)
63. 장미원 (Rose garden)
64. 화목원 (Flower plants garden)
65. 철쭉원 (Royal azalea garden)
66. 연못 (pond)
67. 숲속옹달샘 (Small pond in forest)
68. 문화재보존지 (Cultural assets)
69. 종합예술센터 (Arts hall) – 제안

감응하는 대지의 발견

우리가 대상지를 바라봄에 있어 주목한 것은 주변 환경의 끊임없는 변화와 그에 대한 대상지의 반응형상이다. 공원은 조성 후 완결이 아닌 외부와의 관계에 의해 생성, 진화되는 과정 자체이며, 시간의 흐름에 반응하는 초유기체로 인식된다. 유기체는 시간에 대한 유연한 반응을 통해 영속성을 지닌다. 즉, 대상지가 영속성을 유지하기 위해서는 변화하는 시간에 유연하게 대처할 수 있는 유기체적 감응 시스템이 필요하다. 감응 시스템을 통해 공원은 스스로 사람들이 더 좋아하는 것을 알기 시작하고, 공원과 사람들의 관계는 일방이 아닌 상호적인 것이 될 것이다. 그리고 우리는, 새로이 창조될 강북공원을 'RESPONSIVE LAND' 라 정의한다.

탈유클리드 기하학

공원은 시간의 흐름 속에서 변화를 일으키는 동적인 몸체를 지니고 행동양식에 따라 유동적인 형태로 표출되어야 한다. 탈유클리드 기하학적 형태는 기존의 무응답의 관습적 구조와는 달리, 이용과 에너지의 흐름을 발생시키며 즉각 반응과 변화가 가능하다. 탈유클리드 디자인에 있어 주목한 것은 대상지의 유동요소 중 하나인 기류의 형상이다. 숲에서 생성된 찬공기를 데우지 않고 도시로 공급하고자 하는데서 대지의 형상이 도출된다. 공원의 경계는 모호하여 외피와 내피의 구분이 어려운 구조로, 외부 에너지를 내부 에너지로 변형하는 능동적 힘을 지니게 될 것이다.

감응하는 대지

우리가 주의를 기울인 것은 외력에 무감각하며 경계의 힘으로 형상을 이루던 드림랜드에 감응의 기작을 투입하여 다양한 외부환경에 반응하는 골격을 만듦에 있다. 공원의 형태는 문화적 행동양식에 따라 유동성을 보이며, 변화와의 감응을 통해 도시와 보다 유기적인 관계를 형성해 갈 것이다. 이 창조적인 대지 시스템은, METABLEND / METAMULTIPLICATION / METASTABILITY라는 3가지 메커니즘에 의하여 도시변화에 지속적으로 감응하는 대지를 실현할 것이다.

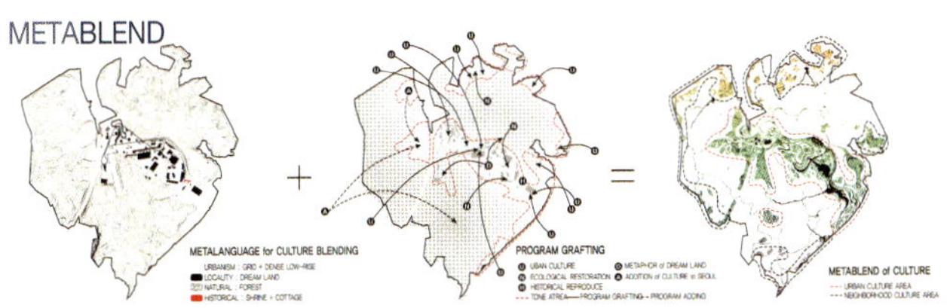

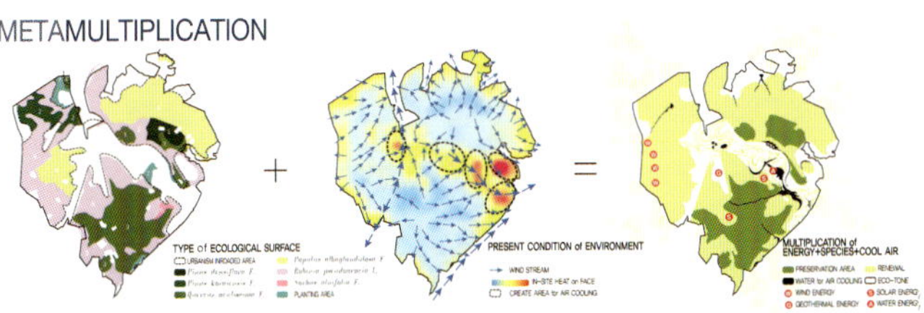

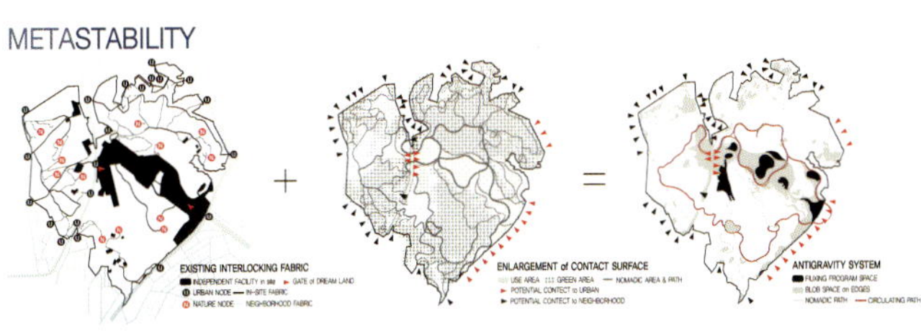

2nd Prize

FoRestinus

HaeAhn Architecture, INC. 해안종합건축사사무소 | **Partners** _ H Associates + EWoo Environmental Design 이우환경디자인

One of the district features of the Gangbuk Park proposal is that diverse programs of the park are connected as a network through the exciting trails and newly added paths to allow the visitors to go and find the programs to their taste, thus to crate the experience in this park of their own and enjoy their liberal journey. Along with the four big cultural platform installed on the former Dreamland site, the small activity pads runnning along with the trails in the mountain are largely divided into three : learning, relaxing and playing. The layered paths of these programs are linked by nodes and are maximized in three theme buildings. For example, if you went to the park to use the mountain bike path for leisure and exercise yesterday, you might want to seek for therapy of meditation or diversion today. And you may be deciding to experience Korean culture by drinking Korean tea in a Korean traditional garden. 'Korean Field and Forest' of Gangbuk Park is a positive open park that is endowed with hundreds of characters and thousands of experience to fulfill all these demands.

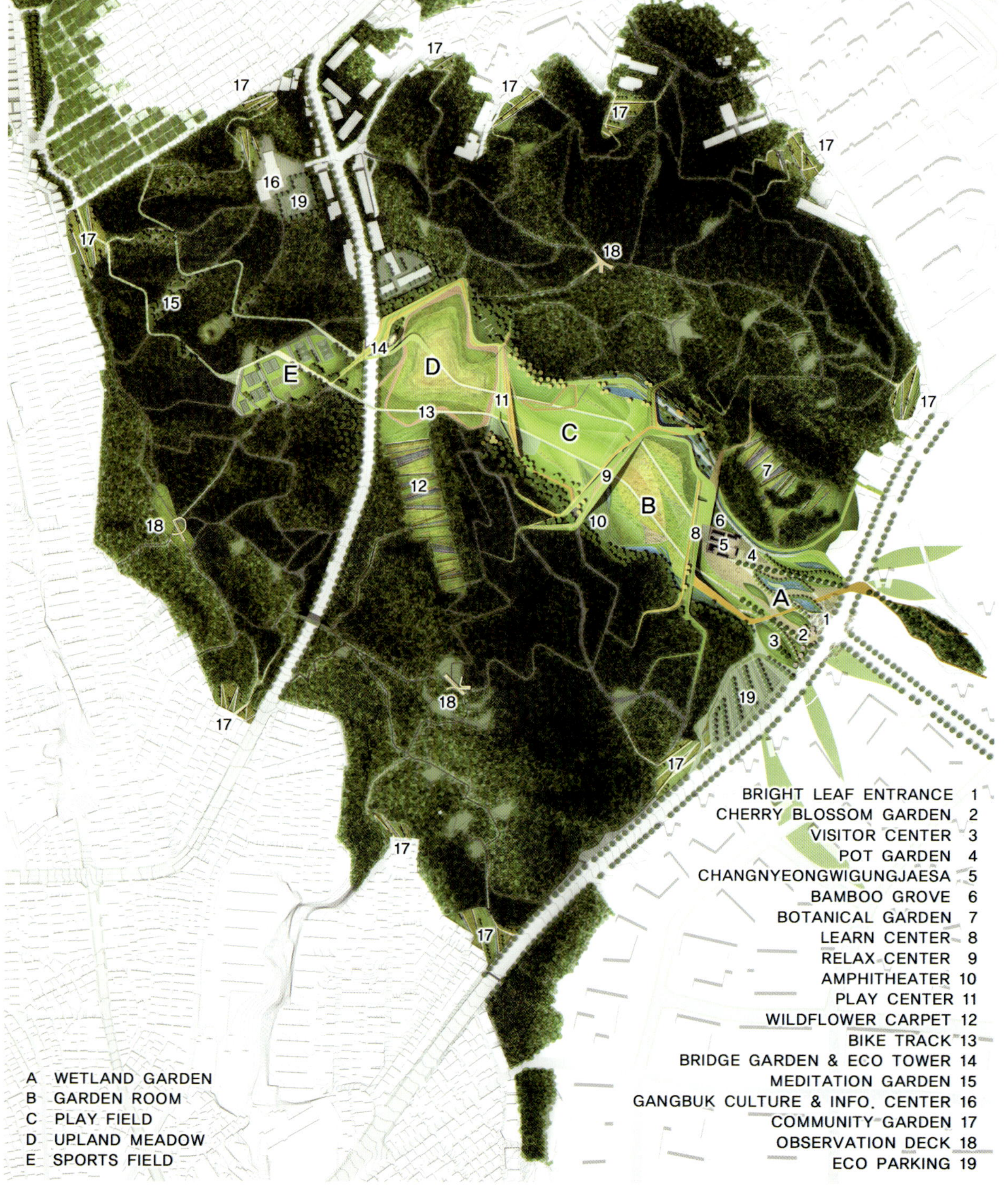

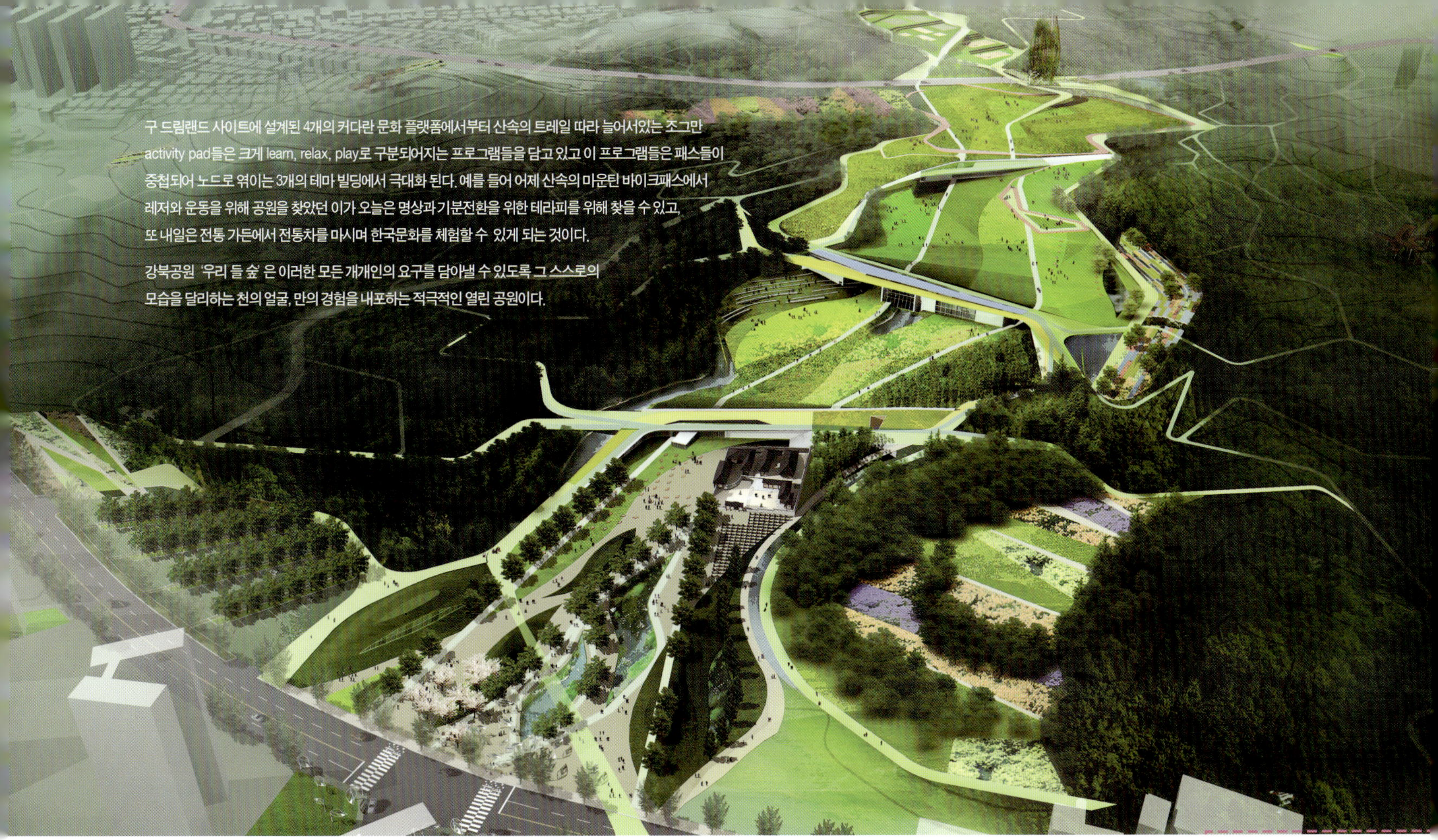

구 드림랜드 사이트에 설계된 4개의 커다란 문화 플랫폼에서부터 산속의 트레일 따라 늘어서있는 조그만 activity pad들은 크게 learn, relax, play로 구분되어지는 프로그램들을 담고 있고 이 프로그램들은 패스들이 중첩되어 노드로 엮이는 3개의 테마 빌딩에서 극대화 된다. 예를 들어 어제 산속의 마운틴 바이크패스에서 레저와 운동을 위해 공원을 찾았던 이가 오늘은 명상과 기분전환을 위한 테라피를 위해 찾을 수 있고, 또 내일은 전통 가든에서 전통차를 마시며 한국문화를 체험할 수 있게 되는 것이다.

강북공원 '우리 들 숲' 은 이러한 모든 개개인의 요구를 담아낼 수 있도록 그 스스로의 모습을 달리하는 천의 얼굴, 만의 경험을 내포하는 적극적인 열린 공원이다.

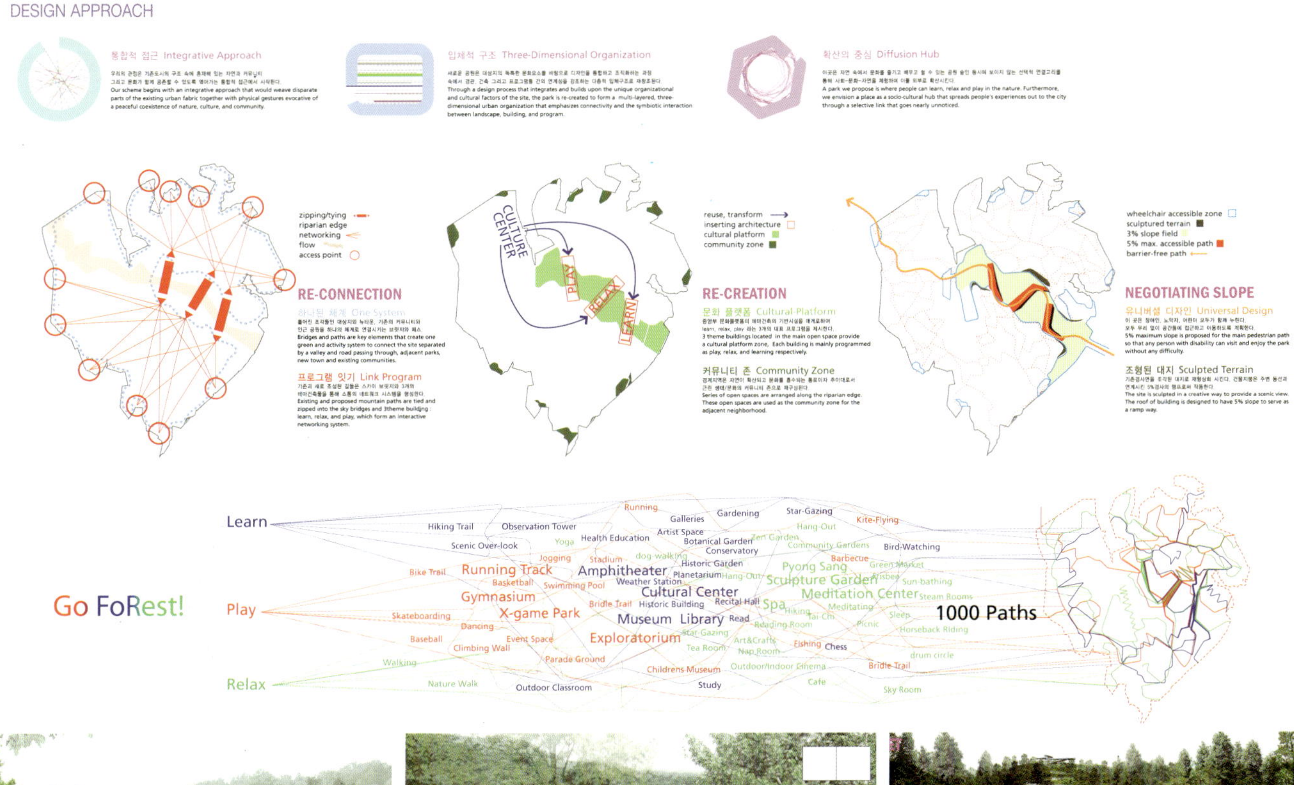

Amphitheater

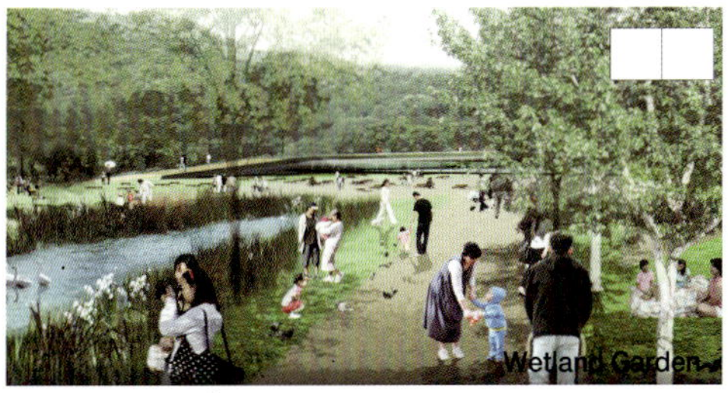

Wetland Garden

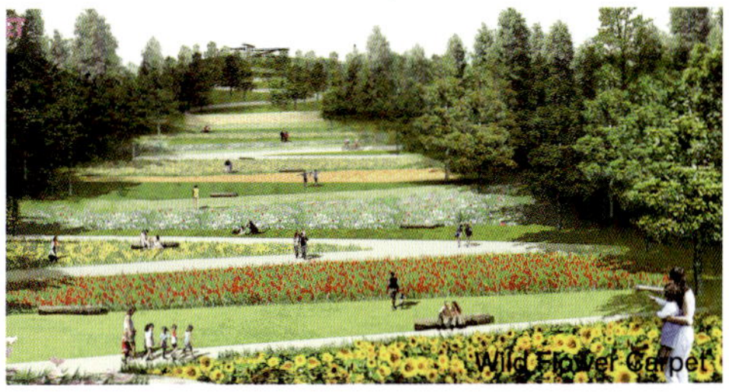

Wild Flower Carpet

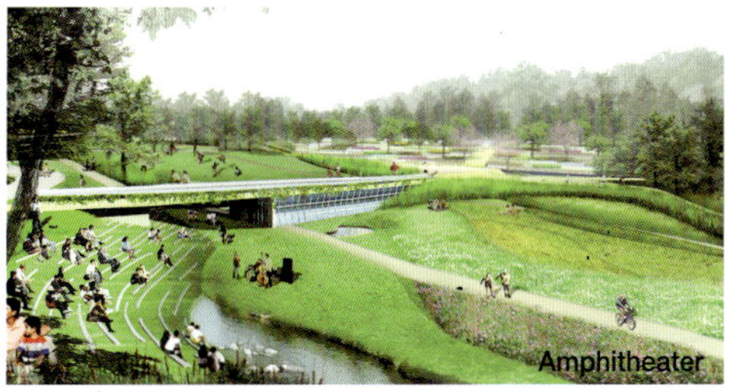

Growing Gangbuk Large Park

GAONE Landscape Design Group (주)가원조경기술사사무소 | **Partners** _ 도시건축연구소가원 + Jung, Wook Joo 정욱주 + SWA GROUP

우리는 이 공원의 이름에서도 드러나는 '대형'의 개념을 정립하는 목표와 전략을 제시하고자 한다. 그것은 단순히 단일공원의 크기가 문제가 아니라 공원의 다양성과 확장성의 문제이다. 먼저 공원 대상지 내부에서 인근 생활권부터 광역생활권까지 분포되는 다양한 스펙트럼의 프로그램을 보유함으로서 대지의 역량을 극대화 하였다. 이는 단일 사이트로서의 강북대형공원의 위상을 정립한다. 뿐만아니라 물리적으로 공원의 증식, 확장을 통해 우이천, 중랑천, 서울숲, 한강으로 네트워크화되는 대형공원권을 제안하였다. 이는 단일한 공원이 아닌 도시의 기반시설로서 실제로 작동하고 확장가능한 강북대형공원권을 구축한다. 다양성, 확장성은 강북대형공원의 잠재력이며, 이 두 가지의 구현을 통한 공원 활성화는 결과적으로 지역의 활성화를 유도할 것이다.

PARK ORGANIZATION

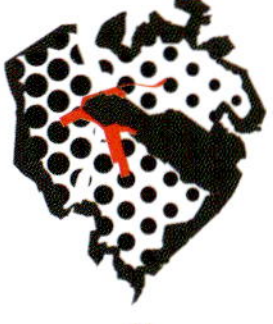

강북대형공원 GANGBUK LARGE PARK

우리가 제안하는 강북대형공원은 개념적으로 정원, 광장, 공원, 숲의 합으로 구성된다. 공원의 경계부는 정원의 성격을 부여하고 있으며, 인접생활권을 위한 근린체육, 휴게시설, 커뮤니티 가든 및 공원의 진입부가 형성되는 공간이다.

=

근린생활권 공원만들기 EDGE

공원의 경계부는 정원의 성격이 부여된 곳으로, 인접생활권을 위한 근린체육, 휴게시설, 커뮤니티 가든 및 공원의 진입부가 형성되는 공간이다. 대상지 경계부의 대부분은 숲의 가장자리로 과거 무분별한 이용으로 생태적 기능이 많이 저하된 상태에 있으므로 이 공간들은 훼손으로부터 숲을 보호하는 역할을 담당하기도 한다.

+

지역문화 HUB만들기 LATCH

과거 드림랜드의 후문 부지는 지역문화교육센터, 다목적홀, 수목문화센터를 조성, 적극적 문화프로그램의 집결지로 작용한다. 세 개의 건물 중심에 위치한 광장은 오현로와 입체교차하여 1단계 대상지와 2단계 대상지를 물리적으로 연결하고 야간에도 활발한 이용이 가능한 프로그램과 조명계획을 수립한다.

+

열린 공원 만들기 WEDGE

구 드림랜드 부지는 일반적인 공원의 역할을 수행한다. 넓은 잔디밭과 아름드리 나무들, 경사를 이용한 갤러리 건축물, 과거 수영장 시설을 재활용한 연못과 바닥분수 등이 과거의 드림랜드를 덮어쓰기 하였다.

+

도시 규모의 유인시설 만들기 FOREST

3개의 숲에는 각각 수목원, 산지형 스포츠프로그램, 생태라는 키워드를 부여하여 획일적이지 않도록 조성방향을 제시한다. 특히 대상지내 최정상부가 있는 생태의 숲에는 강북권의 파노라믹 뷰를 조망할 수 있는 전망시설을 설치한다.

GREEN DROPS

Dongsimwon Inc. ㈜동심원조경 기술사사무소 | **Partners** _ OL Consulting 오엘컨설팅 + YeChang 예창건축 + Yoshiki Toda Landscape & Architect Corporation

공원은 도시의 새로운 힘을 불어넣는 시작점이자 촉매이어야 한다.
주변 도시의 쇠락과 함께 점점 우리의 기억에서 잊혀지고 있던 드림랜드를 시간이 흘러도 시민들과 함께 변화할 수 있는 자생력을 가지는 공원으로 다시 만들고자 한다.
물방울이 떨어져 대지를 따라 흐르며 스며들 듯 공원을 조성하면서 발생하는 흐름이 확장되어 도시 전체에 활력을 불어넣는다.

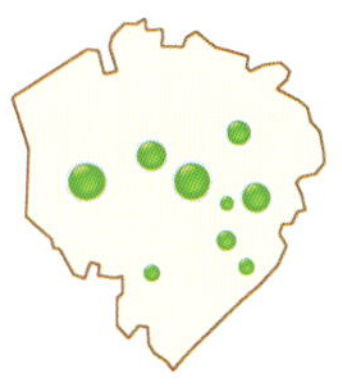

Drop of Green and Spreading _ 공원은 단순히 자연을 접하는 기회를 제공하는 것이 아니라, 지역주민의 삶과 밀접해야 하며 환경에 대한 학습 및 예술적 체험이 가능하여 문화를 창출할 수 있어야 한다. 이를 통해 도시의 공간구조 속에서 강북지역과 함께 성장할 수 있는 상징적 공원으로 거듭난다.

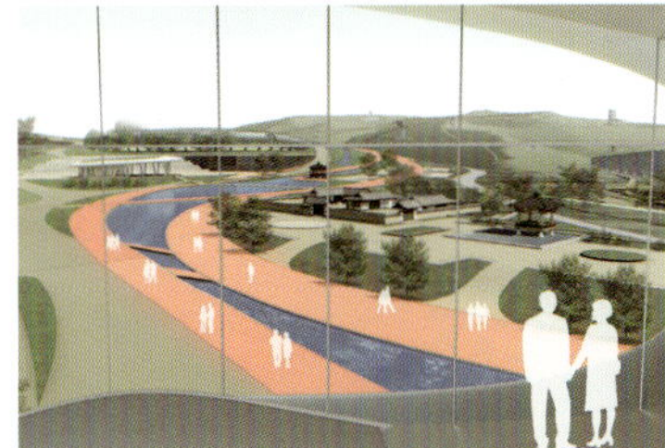

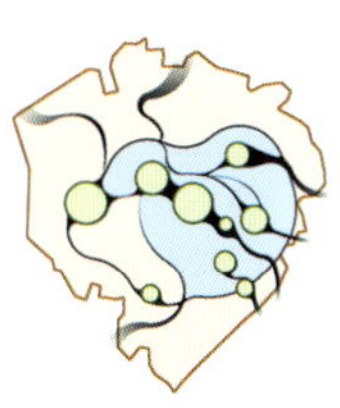

Flowing & Growing _ 공원의 모든 문화공간들은 물이 흐르듯 리듬감 있는 4가지 길을 통해 예술, 자연, 학습, 놀이체험의 기회를 제공한다. 이런 길들은 문화를 발생하는 길이며, 길들의 결절부는 실질적인 문화공간들이 배치되어 다양한 활동들을 증폭시킨다.

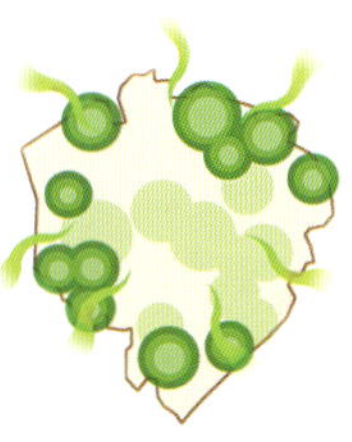

Expansion and Participation _ 이용자의 활동포인트에 배치된 30개의 Green Cell은 고정된 시설이 아니라 아트갤러리, 숲속이벤트 등 자생적 활동을 유발하는 가변시설이다. 이는 부지 경계부의 휴식, 놀이, 재배 등 시민들의 일상을 위한 패키지화된 정원과 함께 주민들이 공원을 만들어나가는 참여의 기회를 제공하고, 도시와 긴밀하게 네트워킹하게 한다.

Seoul Collage Scape

SeoAhn Total Landscape (주)조경설계 서안 | **Partners** _ Leega Plan Architecture 이가플랜건축 + IBI Group

공원에서 예술과 문화가 자연과 만나는 장소를 상상해 보라. 세계에서 가장 복잡하고 활기 넘치는 도시에 있는 숲 속에 자리잡게 되는 강북공원에서는 자연과 예술과 문화에 대한 새로운 경험을 하게 된다. 강북공원은 공간과 이벤트와 감정의 콜라주이며 과거와 현재와 미래의 콜라주이다. 강북공원은 우리들에게 숲에서의 수렵 활동에서 시작해서 인간의 본질을 예술과 음악, 춤과 같은 문화로 표현하는 것을 배워갔던 인류의 뿌리에 대해서 생각하게 해준다.

강북공원은 새로운 유형의 공원이다. 이 곳은 도시가 컬처노믹스를 인식하게 되는 문화적 실험의 장이 될 것이다.
강북공원은 장소이고 공간이고 이벤트이고 쇼 그 자체이다. 강북공원은 가족, 친구들이 모여서 놀고 젊은 연인들이 데이트를 즐기고 예술가들이 영감을 얻고, 음악가들이 곡을 만드는 곳이다. 나무 사이에서 춤을 추고, 어린이들은 뛰어 놀고, 자신 만의 장소에서 명상을 하고, 예술작품을 설치하고, 패션쇼를 하고, 숲 속에서 보물찾기를 한다.
강북공원은 모두에게 영감을 주는 생태적이고 친환경적인 장소가 될 것이다. 그리고 미래의 문화 발전을 위한 캔버스와 같은 역할을 담당할 것이다.

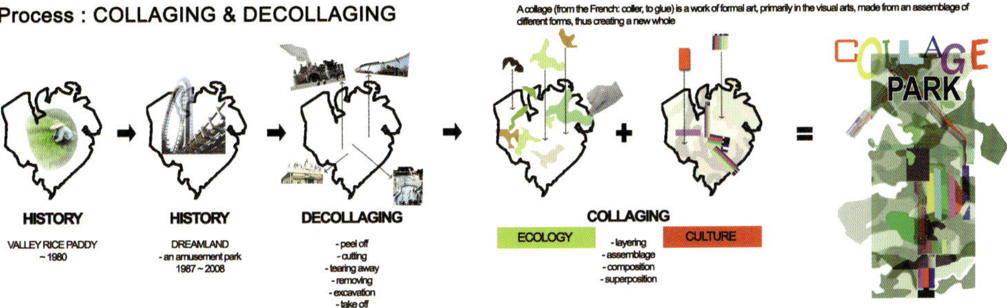

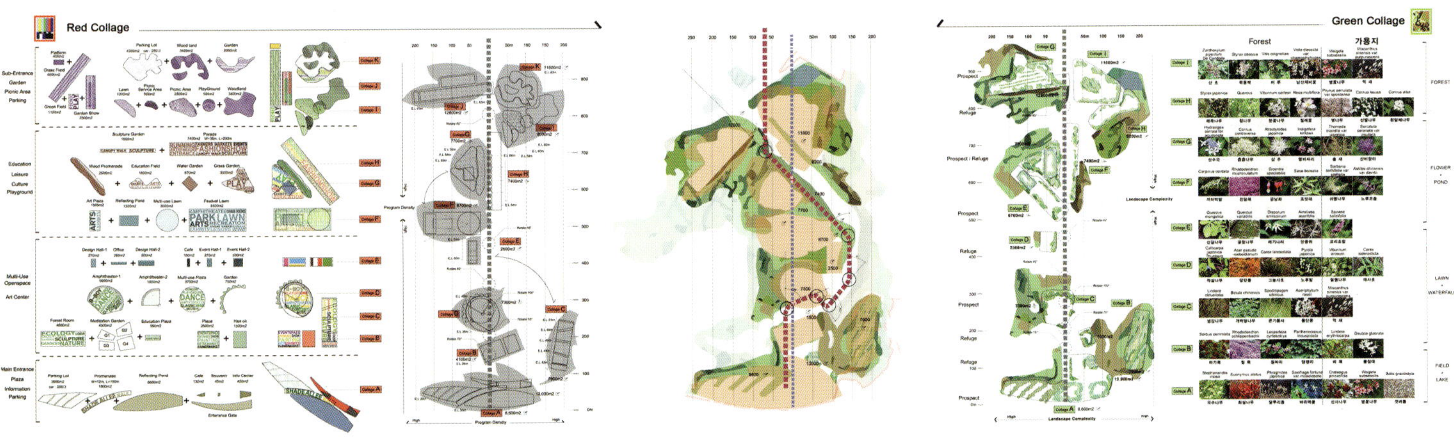

N-heart

YURIM LANDSCAPE CONSULTANTS 유림조경기술사사무소 | **Partners** _ P.I.A 피아이에이건축 + Candy Lau

'N'은 정해지지 않은 임의의 수이다. 심장은 활동에 맞는 심박수를 가진다.
'N'의 값에 따라 결과값이 달라지듯 심장은 'N' 에 의해 운동성을 달리하고 그에 맞추어 뛰고 활력소를 공급한다.
잠자고 있는 심장에 임의의 'N' 을 대입함으로서 강북의 심장은 조용히 움직임을 가지기 시작한다. 미지의 'N' 은 무한한 가능성을 기대하며, 공원은 모두의 기대치를 만족시키는 N-heart를 꿈꾼다.

계획개념 _ 각기 독립된 공간을 드림랜드의 부지와 함께 유기적인 관계를 맺도록하여 4가지 부분으로 나뉘어져있던 커다란 대상지를 하나의 공간으로서 소통시킨다. 이 공간안에 다양한 이용자의 요구를 충족시키는 소테마들을 배치시켜 이용자가 직접 주제를 선택함으로써 공간의 활용도를 높이도록 조성한다.

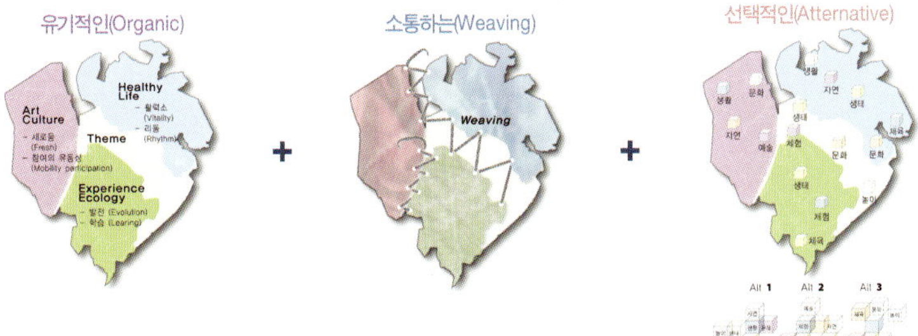

기본계획 _ 주변 환경을 고려하여 3개의 각기 다른 특성을 지니고 있는 구역을 각각의 독립된 주제를 부여, 연계되도록 계획한다.

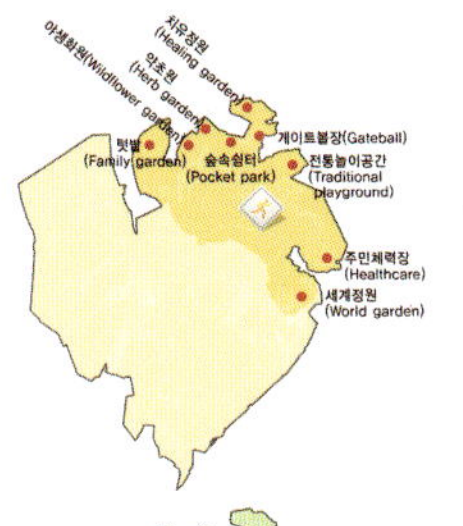

· 강북생활공원
주민들을 위한 생활공원 조성
모든 연령층을 고려한 테마공간
사회교육, 학생예능실습장 등 다양한 활동공간 제공

· 강북생태체험공원
숲을 활용한 교육이 있는 체험의 장
문화 예술 공간과 연계한 친환경적인 공간조성
전망대, 숲 탐방로, 테마체험길 조성

· 강북예술문화공원
문화와 예술을 체험하고 느낄 수 있는 예술공원 조성
순수예술과 디지털 아트 등 테마가 있는 예술품 전시
숲과 조각품이 어우러진 일상속 예술 · 문화공간 조성

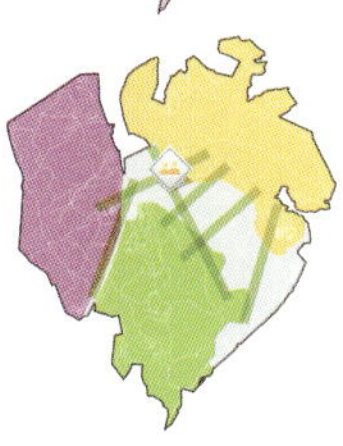

· 강북테마공원
드림랜드 부지를 이용하여 테마공원으로 계획
문화, 예술, 생태, 학습, 전통 등의 다양한 공간 조성
각 공원을 연계하는 중심적인 공간 조성

E-RUBAN PARK

TOWMOON Engineering & Architects Co.,LTD 토문엔지니어링 종합건축사사무소 | **Partners _** (株)昭和設計(沼田亘)

전체 대상지 중 개발이 가능한 지역으로 북서쪽이 높고 남동쪽이 낮은 계곡형 완경사지를 설정하여 북서와 남동을 연결하는 축을 시간 경과의 흐름, 지형이 연결되는 흐름, 주변 공원들과의 생태환경적 흐름의 축으로 규정한다. 이 흐름의 축을 중심으로 도시의 자연환경 요소와 인문사회수요가 서로 영향을 주고 받으며 변화하는 현상을 나선의 형상으로 표현하여 시간과 공간의 변화를 체험하는 도시Urban 속의 전원Rural 공간을 창출한다.

지형 나선의 흐름 _ 대상지의 지맥(땅의 역사)을 해석하고, 반복되는 역사의 흐름, 진보하는 도시문화의 흐름 등을 '나선'의 형태로 형상화하여 디자인 모티브로 사용한다. 공간의 형태 및 건축물, 시설물, CI 등에 '나선' 형태의 디자인 모티브를 적용하여 통일감 있는 경관을 계획하여 강북공원에 대형공원으로서의 아이덴티티를 부여한다.

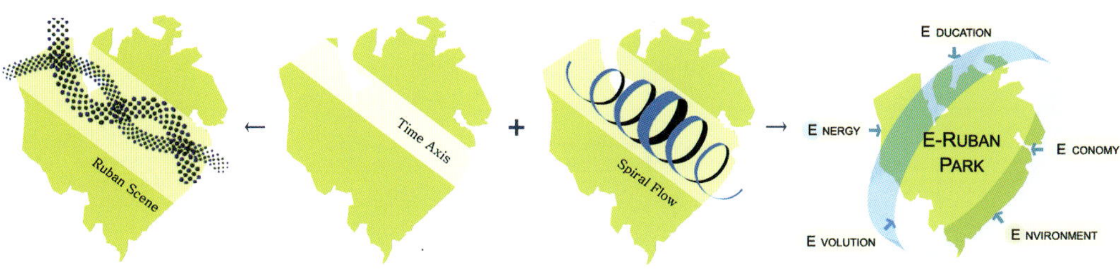

생태환경 보전, 재생 우선의 법칙 적용 _ '생태환경'을 주테마로, 생태계의 시간적 개념(가변성 천이)을 도입하여 안정된 생태환경을 구현하며, 체험형 박물관, 그린어드벤처, 농업체험 등의 운영프로그램을 도입하고 기능융합형 체험공원을 계획한다.

공간틀 디자인방침

- 숲, 마을, 거리로 구분되는 각각의 영역은 이용자들의 이용행태 및 연령대를 고려하여, 공간을 세분화하여 처리한다.
- 이용자들의 필요와 요구에 따라 디자인, 소재 등의 변화를 꾀한다.
- 자연소재에서 인공 소재에 이르는 다양한 소재를 활용하여 다채로운 경관을 창출한다.
- 색채와 사운드 스케이프를 고려하여 오감이 만족하는 공간을 제시한다.

조닝계획 _ 전체공간을 이용목적 및 환경영역에 따라 '숲의 영역', '마을의 영역', '거리의 영역', '문화서비스영역'으로 구분 숲, 마을, 거리영역의 구획은 시간의 축과 지형나선의 흐름을 매개로 하여 문화창출, 환경친화, 계층간 화합이 이루어지도록 계획

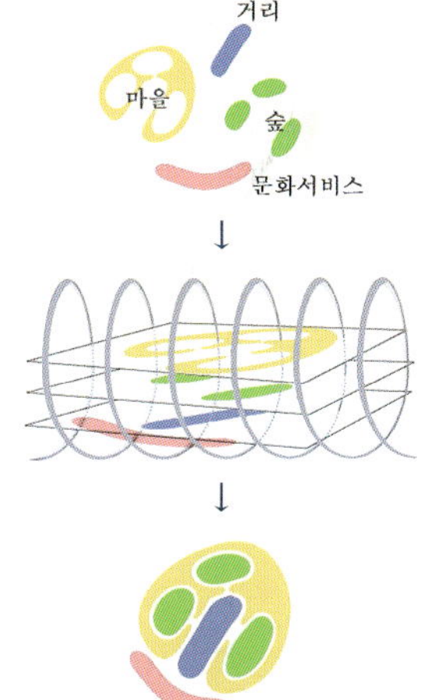

Central Open Space in Multi-functional administrative City

Yeongi-gun, Chungcheongnam-do, Korea **Location**	**위치** _ 충청남도 연기군 장남평야 일대
6.982㎢ **Area**	**면적** _ 6.982㎢ (약 200만평)
Promoters	**공모주최**
Multi-functional Administrative City Construction Agency	행정중심복합도시건설청
Korea Land Corporation	한국토지공사
Schedule	**공모일정**
Registration 2007. 3. 9~3. 24	참가등록 2007. 3. 9~3. 24
Final Submission(Stage 1) 2007. 5. 7~5. 14	1단계 작품접수 2007. 5. 7~5. 14
Announcement of Finalists 2007. 5. 29	1단계 심사결과 발표 2007. 5. 29
Announcement of Design Guideline for Stage 2 2007. 6. 8	2단계 설계지침 제공 2007. 6. 8
Entry Submission(Stage 2) 2007. 8. 13~8. 16	2단계 작품접수 2007. 8. 13~8. 16
Announcement of Winners 2007. 8. 28	2단계 최종당선작 발표 2007. 8. 28
See the results on-line	**공식홈페이지**
www.macopenspace.or.kr	www.macopenspace.or.kr

한국토지공사와 행정중심복합도시건설청이 공동으로 주최한 행정중심복합도시 중앙녹지공간 국제설계공모의 수상작이 발표되었다. 2단계 국제 설계 공모 방식으로 진행된 이 공모전은 총 92개팀의 작품이 출품된 가운데 2007년 5월 29일 2단계 설계에 참가할 최종 후보 10개팀(국내 5팀, 국외 5팀)이 선정되었으며, 8월 28일 최종 당선작으로 노선주의 'Ancient Futures' 가, 2등 수상작으로 김종규의 'Long Views', 3등 수상작으로 민현준의 'Carved Park' 가 선정되었다. 당선작은 행정중심복합도시 한 가운데의 광활한 땅이 인간과 자연이 공생하는 지혜로운 공원으로 미래를 위한 생태적 인프라스트럭처로 거듭날 수 있음을 제시하고 있으며, 휴식 및 여가 기능을 제공하는 기존 공원의 기능을 넘어서 생산적인 기능까지 수용하는 새로운 공원 패러다임을 제시하고 있어 21세기 지속가능한 공원상을 잘 제시했다는 평가를 받았다.

1 1st Prize_Ancient Future
2 2nd Prize_Long Views
3 3rd Prize_Carved Park
4 Finalist

Under the join auspices of Korea Land Corporation and the Multifunctional Administrative City Construction, prizewinning works of International Design Competition for Central Open Space in Multi-functional administrative City of Korea had announced. On the final stage designated 10 teams among all participated 92 teams' works on 29th of May at the competition, Noh, Sun Joo's work, 'Ancient Futures', was the prizewinner. And Kim, Jong Kyu's work, 'Long Views' was placed second and Mihn, Hyun Jun's work, 'Carved Park' was the third-place on the competition. The final prized work, Ancient Future, indicates, once given the rebirth of bioregion as the infrasturcture for the bright future, the wise park living together with human and nature at the centre of main land and represents the paradigm of new technical park accomodated with productive capacity beyond the normal functional relief and leisure.

1st Prize

ANCIENT FUTURES

Noh Sun Joo 노선주 · Haiin, Land+Scape Design ㈜조경설계 해인
Partners _ Seo Mi Kyoung 서미경, Han Moo Young 한무영, Min Byoung Wook 민병욱

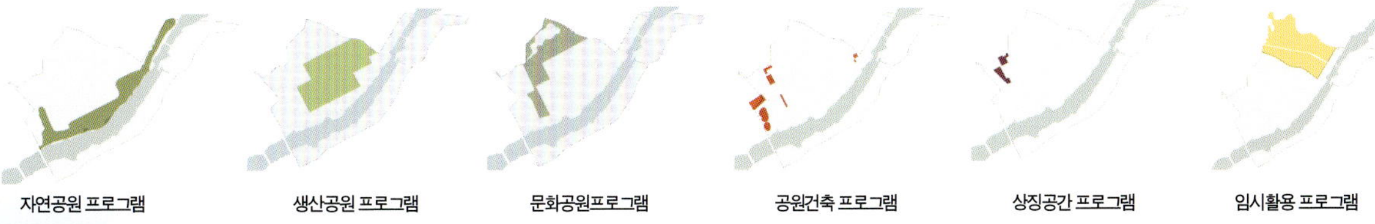

터 만들기

우리는 행정중심복합도시 한 가운데에 위치한 이 땅이, 더불어 함께 살아왔던 오래된 지혜를 지속시키는 대지로서 이어질 것을 제안한다.
미래의 삶, 도시 전체와의 얼개를 갖고 건강한 공존을 구성하는 생태적, 문화적, 자족적 인프라스트럭처로 거듭날 것을 제안한다.

대지와 물 :
공존의 기반 만들기

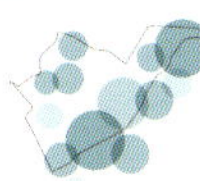

터의 분화 :
다양성의 기반 만들기

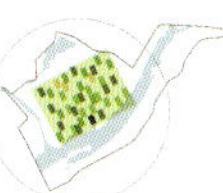

농업가치의 회복 :
자족의 기반 만들기

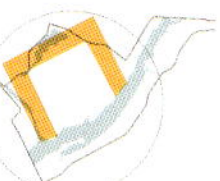

공동체의 회복 :
공동체 기반만들기

1. 공존의 영역 - 경계없는 경계

제방의 위치를 이동시키고 그 형태를 변화시켜서 경사가 완만한 자연적인 하천 호안을 조성한다. 이는 오래된 자연의 지혜로서, 가장 적극적인 자연과 도시와의 공생 방식이다.
금강으로의 단절된 접근을 회복하는 역할을 할 뿐만 아니라 가장 자연적인 방식으로 금강하류 도시지역의 홍수 부담을 나누어 준다.
장래 세계도시문화유산을 꿈꾸는 행정중심복합도시의 자연과의 공존을 대표하는 강력한 상징이 될 것이다.

2. 마주 봄과 마주 섬의 경관

제방에 의해 시각적으로 단절되었던 금강으로의 조망을 중앙녹지공간 전 지역에서, 특히 많은 공원 활동 프로그램이 배치될 공원 경계부에서 경험할 수 있게 된다.
공원 전 지역에서 도시, 강, 숲, 하늘로의 열린 경관을 주고 환상형의 도시지역에서 공원으로 열린경관을 준다.
공원과 공원 내부의 문화시설로부터 주변의 전월산 및 원수봉 등의 수려한 산을 모두 조망할 수 있는 시각적 개방감을 최대한 보장한다.
오랜 시간에 걸쳐 자연과 사람이 공존하여 만들어 온 한국적이고 전원적인 고향의 품과 같은 전통의 농업경관을 계승한다.

3. 도시공원 문화밴드

도시상징정원, 방문자 / 가족 / 장애우정원, 커뮤니티 센터, 아트센터, 공공미술 및 예술로서의 장소, 숲속 여가 및 헬스센터, 희귀식물원 및 들풀식물원, 야외카페촌, 인공호수 및 인공 백사장, 세가지 기능 및 경관으로 분화된 친수지역과 워터프론트, 장소의 다리, 에너지정원, 역사공원, 민속공원, 복합공연장 등이 집적된 문화 밴드를 도시와 직접적으로 연결되는 중앙녹지공간 외곽에 배치함으로써 종합적인 도시공원의 기능을 담고, 도시와 오픈스페이스의 소통을 원활하게 한다.
대안 보행교, 보행 및 자전거 동선을 체계화하여 도시와의 연결을 편리하게 하며, 중앙녹지공간 내부에 무공해 차량을 이용한 순환 교통 체계를 구축하여 도시관광 프로그램과 병행한다.
공원 유지 관리에 소요되는 비용에 도움을 주는 다종의 공공상업시설 및 수익모델을 제시한다.

4. 생산적 경관

생산을 도시 및 지역공동체를 활성화 시키는 강력한 도구이자 공원프로그램으로서 제안한다.
현재의 농지를 보전하되 다채로운 경관적 특징을 지니는 공원내 생산녹지로 분화화고 변모시킴으로써 생태적인 생물 다양성이 있는 생산적 경관을 형성토록 제안한다.
소극적인 휴식 및 레크리에이션만을 제공하는 전통적인 공원을 넘어서는 생산적 공원을 조성하여 , 새로운 형식으로 분화된 공원 패러다임으로 제안한다.

5. 자족적 인프라스트럭처

자연과 더불어 살아 온 삶의 터전, 대지에 축적된 오래된 삶의 지혜를 통해 도시의 자족성과 공원의 자족성을 향상시키는 인프라 스트럭처를 제안한다. 공원의 유지 관리 비용을 절감하며, 공원의 생산기능, 도시의 바이오 매스를 활용하는 자연적이고 무공해적인 에너지를 도입한다. 물 소비를 유지시키며 자원의 순환을 통해 도시내의 자족적 에너지 순환과 하나의 시스템으로 작동하는 계통을 형성한다. 도시와 금강의 광역적인 수계의 일부로써 귀속되는 총량 개념의 물순환 시스템을 제안한다. 인접 도시지역의 우수 및 중수시스템과 연결된 인공호수의 유지 용수 및 공원 관리, 레크리에이션 활용수를 제안한다.

Brief history of the Site

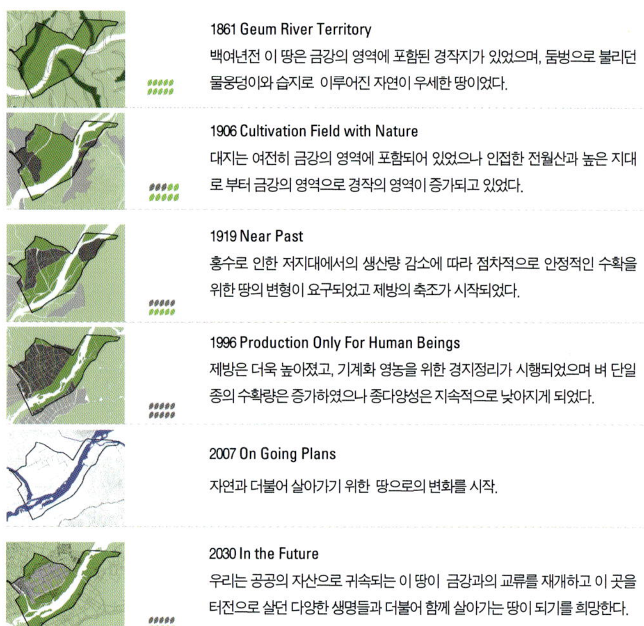

1861 Geum River Territory
백여년전 이 땅은 금강의 영역에 포함된 경작지가 있었으며, 둠벙으로 불리던 물웅덩이와 습지로 이루어진 자연이 우세한 땅이었다.

1906 Cultivation Field with Nature
대지는 여전히 금강의 영역에 포함되어 있었으나 인접한 전월산과 높은 지대로 부터 금강의 영역으로 경작의 영역이 증가되고 있었다.

1919 Near Past
홍수로 인한 저지대에서의 생산량 감소에 따라 점차적으로 안정적인 수확을 위한 땅의 변형이 요구되었고 제방의 축조가 시작되었다.

1996 Production Only For Human Beings
제방은 더욱 높아졌고, 기계화 영농을 위한 경지정리가 시행되었으며 벼 단일 종의 수확량은 증가하였으나 종다양성은 지속적으로 낮아지게 되었다.

2007 On Going Plans
자연과 더불어 살아가기 위한 땅으로의 변화를 시작.

2030 In the Future
우리는 공공의 자산으로 귀속되는 이 땅이 금강과의 교류를 재개하고 이 곳을 터전으로 살던 다양한 생명들과 더불어 함께 살아가는 땅이 되기를 희망한다.

Flowing Pattern Changes of Miho Stream and Geum River

대상지 상류 미호천과 금강이 만나는 지점은 '금강' 과는 별도로 '삼기강' 이라는 별칭으로 불리워졌으며, 이는 광역적인 세지방의 물줄기가 연결되는 지점으로 문화적, 생태적으로 중요한 위치를 표현하던 명칭이기도 하다. 댐의 건설과 영농 등을 위한 제방정비와 하도정비에 따라 물의 흐름이 점차 단순화되었고 종다양성 또한 단순화되는 방향으로 진행되어 왔다. (1920, 1960, 1970, 1980, 1990)

This land has been shaped as a place of coexistence through the relationship with Geum River since a former age of civilization. The ground has provided the great environment of settlement for swelling and production with its abundant resources and biodiversity, and gradually changed according to expending settlement and cultivation. However, its overexploitation of the land for the maximum output from homogeneous crop cultivation has led to major changes of the topography disconnection between human beings and Geum River by building dikes, and eventually the peak of human use of the land.

Major issues

우리는 이 땅을 전체의 일부인 '부분의 대지' 이자 '과정의 대지' 로 인식한다.
사회적인 연계와 더불어 자연적, 공간적, 물리적인 모든 연계와 대지와 땅의 변형으로 인한 인과관계에 대하여 생각한다.

녹지 _ 단절되고 섬화 될 것이다
Green _ It will be disconnected and isolated.

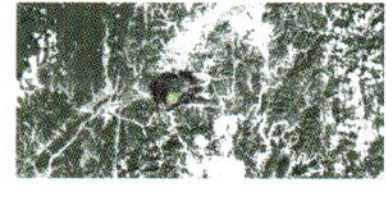

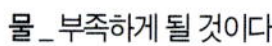

물 _ 부족하게 될 것이다
Water _ The site will lack water supply.

생산 _ 최대의 수확보다 건강과 가치의 나눔이 요구 될 것이다
Crop Fields _ Distribution problums will occur.

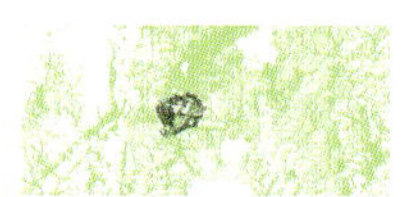

도시 _ 소비가 증가 될 것이다
Urban _ Consumption will be increased.

Earthwork & Stabilizing

대상지의 현재 지형은 단일종 경작인 쌀의 생산성 확보와 홍수위에 대한 대책에서 비롯되어 제방이 조성되게 되었다. 방형의 경지는 기계영농을 도입하며 운영관리 및 생산성의 극대화를 위해 구획되고, 금강물을 끌어들여 콘크리트 수로를 통해 관수하는 현재의 체계를 갖추게 되었다. 공존의 기반을 조성하기 위한 대지의 변화는 제방허물기와 콘크리트 수로의 변화로부터 시작되며, 대상지의 대부분을 이루는 직선 구획의 논은, 시간의 경과와 더불어 안정화된 상태를 유지하므로 재구성의 과정을 거치지 않으며, 생산을 위한 대지에서 자족성을 향상시키면서 자연과의 공존을 위한 다양한 계통으로 분화되도록 한다. 도시공원 프로그램이 구성되는 지역에 대해서는 200년 주기(500년 주기 극한 강우) 홍수에 대한 안정성을 확보하면서 대규모 토목공사를 감소시키는 방향으로 계획한다.

- Planning for the landform is to have a good fit with Korean traditional way.
- Functionally meaningless land manipulation is to be avoided. We believe that it is the best way to respect and take over the traditional landscape.

- 중앙녹지공간의 경계와 외형적 틀, 홍수로부터의 안전성을 확보하기 위한 대지조성계획이 한국적이고 전원적인 전통의 농업경관에 대해 낯설지 않은 것이 되도록 계획한다.
- 특별한 기능적 의미를 담지 않는 한 대지의 낯선 조형적 변형은 계획되지 않아야 하며 이것은 오랜 세월 자연과 인간이 공존하며 이곳에서 만들어 온 전통적인 경관을 유지 계승하는 것이다.

Flooding Simulation for 200 Year Frequency Flood Event

Landscape and Prospect programs

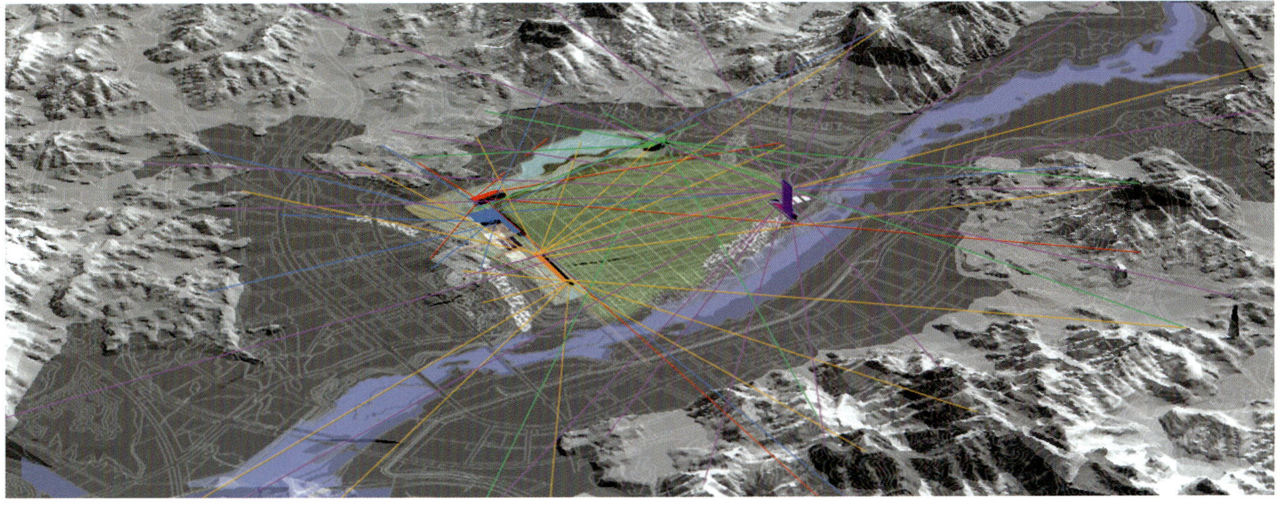

우리는 낯설지 않은 변화가 되기를 희망한다.
옛것과 새것이 함께 하며, 너른 들과 너른 하늘 그대로의 모습이 존중되기를 희망한다.
광대하고 트인 경관으로 막히고 닫힌 곳을 열어 교류하게 하며, 중앙녹지공간에서
도시를 보며, 도시에서 중앙녹지공간으로 균등히 마주보게 되기를 희망한다.

광대하고 트인 경관을 보다

도시와 농촌의 결합을 보다

풍경과 자연변화를 보다

교류의 재개와 공존의 변화를 보다

이웃과 어울림의 즐거움을 보다

옛것과 새것이 함께 함을 보다

LAKE SIDE CAFE (COMMON COMMERCIAL UTILITY)
URBAN EDGE / RECREATIONAL WATER FRONT / MARKET PLACE
COMMUNITY CENTER(1) : MAIN, ROOF & GROUND LEVEL FAMILY & VISITOR' S GARDEN / ENABLING
MARKET PLACE for ALL CITIZEN, BIG CANOPY
PROMENADE & HYDRO, ELECTRIC, BIO DIESEL SHUTTLE BUS, BUMBUNG(FIELD) STATION
SOLAR GARDEN for ENERGY SUFFICIENT & STUDENT EDUCATION
PLACE as ART : PUBLIC ART CENTER, LANDSCAPE & STREET ARCHITECTURE, CROSS CIRCUIT DECK
RIVERSIDE WATER FRONT / FOG, SUNSET, RAIN, WIND & MOON LIGHT
NURSERY HILL FOREST : VARIETY of RESREATION AND EXPLORING ACTIVITIES
COMMUNITY CENTER + AGRICULTURE MUSEUM + CITY TOUR VISITOR' S HOTEL + CULTIVATING MANAGEMENT
GOLDEN FROG LAGOON : VARIETY HABITATS
FLOOD EVENT, HYBRID GROWING FIELD, RIVERSIDE TRAIL
CHECK-DAM(CANAL/CASCADE)
+27.0
ENTRANCE PROMENADE
WATER FLOWER
WATERFRONT CAFE DECK
PARKING(SUNKEN)
ENERGY GARDEN(ROOF)
ESPLANADE
COMMUNITY CENTER (1)
/ BIG GROUND AMPHITHEATER
+27.0
WATER FRONT / MARKET PLACE
COMMUNITY CENTER(1) : MAIN
MARKET PLACE
+22.0
RECREATIONAL WATER FRONT
+20.0
+21.0
FAMILY, VISITORS, ENABLES GARDEN
MARKET PLACE BIG CANOPY
SOLAR (ENERGY GARDEN)
RECREATIONAL WATER FRONT
+27.0
+27.0
ESPLANADE
FLOOD EVENTS (200/500 YEARS SAFETY)
PLACE as ART
CROSS CIRCUIT DECK
MARSH(2.5~3KM)
PUBLIC ART CENTER, LANDSCAPE & STREET ARCHITECTURE
+27.0

Urbanpark Programs

Urban park Program-banding

Sharing the edge that has sequential accesses from the city, the program band of recreational activities is arranged in the park. Having cultural facilities, forest, lake, and hills, it is located on the level which is ever secured from the flood event. People can keep the broad field of vision on the band toward the city, the landscpe of Jangnam Plain, and Geum River. The program band as a urban park provides various activities through lake, esplanade, and waterfront via continuous paths. Insividualized paths meet at the nodes of communityfield and get scattered again. It also can be perceived as landscape hill where people experience the seasonal changes of the crop field.

Public Art Opportunity

The axis of paths passing through the art center, waterfront, and lakeside consist of various public art activities. It gives an opportunity that people communicate within the environment of nature, culture, and history, as well as takes art as the main theme of the park so that it improves people's experience of landscape and a sense of community. The sequential landscape, river field city forest lake, provides people with more pracically experiential landscape.

공원프로그램의 연속적 구성

도시지역과 맞닿아 다발적인 접근과 경계를 공유하며, 문화 및 여가 활동을 담는 프로그램 밴드를 구성한다. 안정성을 확보한 레벨상에 문화시설과 숲, 호수와 언덕을 포함하여 강과 들, 도시를 조망하는 언덕으로, 장남평야의 농업경관과 금강으로의 폭넓은 경관조망이 가능하도록 한다. 도시공원으로서 프로그램 밴드는 호수와 esplanade, 워터프론트 등 길의 연속과 함께 다양한 프로그램을 제공한다. 개별화된 체험의 경로는 각 결절에서 공동체의 장으로 묶여지고, 다시 흐트러지기를 반복한다. 프로그램 밴드는 또한 경작지의 계절변화와 함께하는 도시민의 휴식을 위한 경관언덕으로 정의된다.

공공 예술의 가능성

아트센터와 수변 그리고 호수 주변으로 이어지는 '길의 장소' 경로에 다양한 공공예술 활동의 가능성을 제공한다. 공공예술 활동을 담을 수 있는 기반을 통해 도시민과 소통을 시도하고 행정중심복합도시의 자연, 문화, 역사 등을 환기시킬 수 있는 기회를 제공하며, 예술을 도시공원 활동의 주요 개념으로 설정하여 경관가치를 향상시키고 공동체에 생기를 불어넣는다. 시민으로 하여금 창조적 체험을 영위할 수 있도록 한다. 강-들-도시-숲-호수로 이어지는 평온한 대자연이 공공예술의 일부로서 보여지는 높은 수준의 경관을 형성한다.

Big Ground Amphitheater 도시 규모의 축제, 스포츠, 공연 및 일상여가를 수용

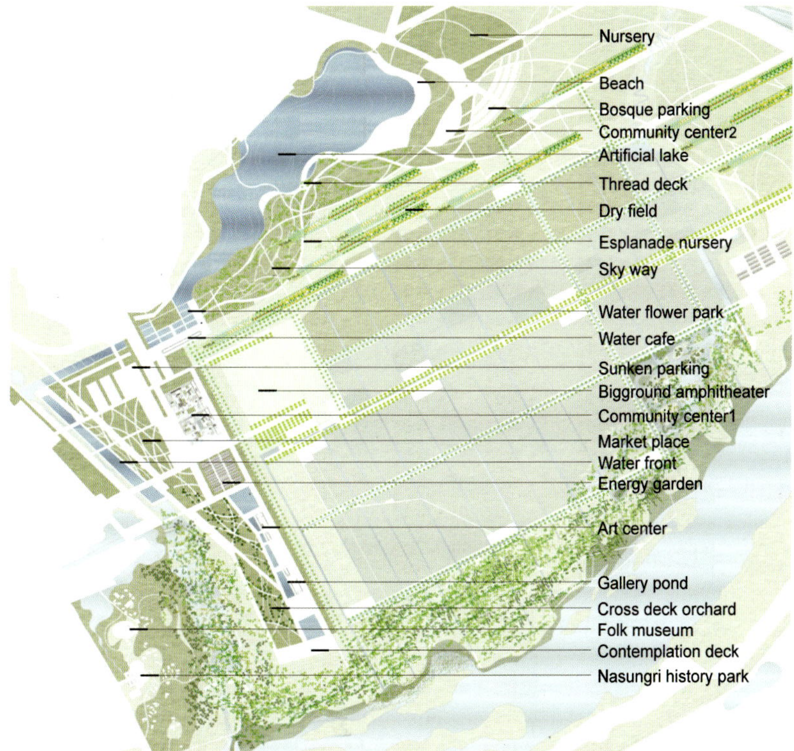

Productive land Programs

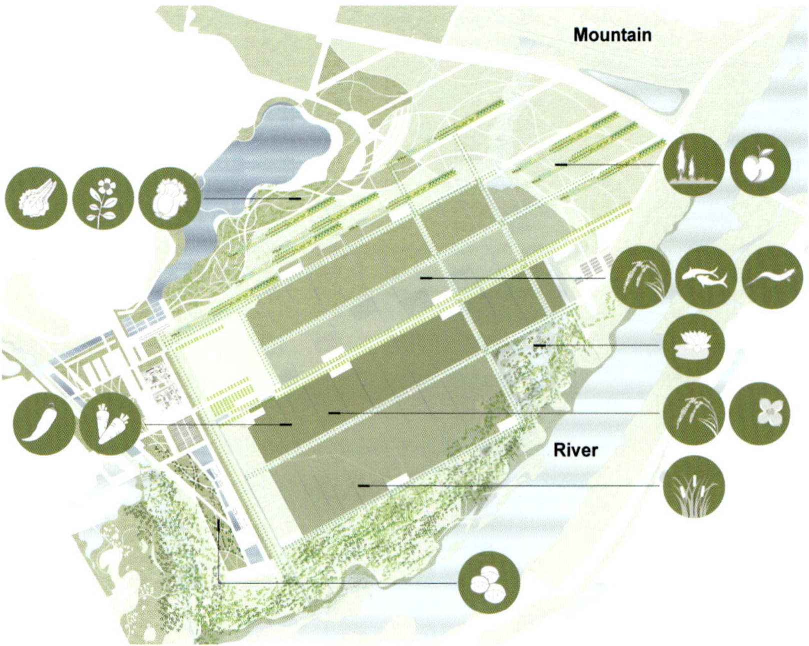

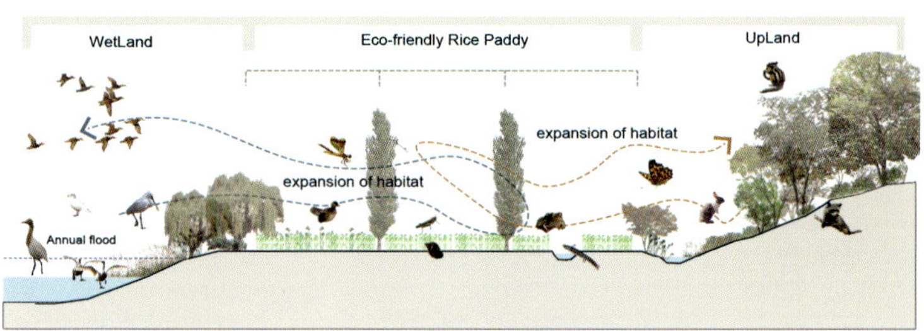

논은 생산녹지의 가장 큰 면적으로 계승되며, '논' 스스로의 생산을 위한 자원 소비에 대하여 자족률을 향상시키며 공존의 바탕이 되도록, 경작이 가능한 방식을 채택하면서 각각 고유의 기능과 특성을 부여하고 분화되게 한다. 외곽부분으로 작물원과 도시원예를 비롯한 다양한 생산지를 형성하고, 1년단위로 시민에게 임대되는 채원을 제공한다.

기존에 행해지던 관행 농법을 대체하는 친환경 농법으로의 전환과, 논의 생산성을 유지하기 위한 가장 큰 소요 자원인 물에 대한 자족률 향상 시스템을 구성하여, 금강의 물을 덜 쓰고 건강하게 돌려주도록 한다. 생산녹지 전체 및 도시지역에서 발생되는 바이오 매스를 수집, 자원화하는 플랜트를 농업박물관과 병행하여 설치하고 영농운영에 관한 기능을 담당하도록 한다.

논의 분화 - 둠 벙논 : 둠벙(물웅덩이)에 물을 저장하고 영농에 사용하는 방식

- 과거 둠벙뜰로 불리울 정도로 많은 자연 습지와 둠벙을 가지고 있던 장남평야의 땅 이야기와 연결된 농업용수 공급과 붕어, 잉어 등 담수 어류의 생산, 논 레크리에이션 등의 기능을 부여한다.
- 둠벙 환경을 서식처로 삼는 동식물을 허용하는 공존과 다양성의 바탕을 제공한다.

논의 분화 - 유기농법 논

- 관행농법으로부터의 생산방식 전환과 생산량을 줄이더라도 더 건강하고, 더 지속가능할 수 있도록 한다.
- 현재 개발되어 있는 수많은 유기농법 중 장남평야의 특성을 가장 잘 알고 있는 농부, 지역 영농 전문가 그룹의 협력을 통해 공존과 다양성의 바탕을 유지하면서 생산을 공유하는 친환경 농법 유기논으로 운영토록 한다.

논의 분화 - 물을 덜쓰는 논(건답)

- 논의 유지 및 생산에 소요되는 물 사용량을 현저히 낮추는 농법의 경작방식으로 운영한다. 둠벙논, 유기농법 담수 논, 물을 덜 쓰는 농법인 건답농법으로 생산과 공존하는 다양한 생물 서식 및 먹이 수집공간으로 동시에 제공하여 다양성 및 건강성에 기여토록 한다.

Seasonal Program in Rice field

'24절기' 는 태양의 변화에 의한 계절, 기후 및 기상환경을 전통적인 농경문화와 일상생활에 접목시킨 오랜 지혜로, 현재도 파종, 작물 및 식물의 성장, 결실 등의 시기를 가늠하는 지표로 널리 사용되고 있다.
절기는 농경문화의 뿌리를 공유하고 있는 한국인들에게 농업활동 여부에 관계없이 일상생활의 지표로서 자주 인용되는 척도이다. 농업경관 및 경작운영을 매개로 동체 프로그램을 도입하며, 전통적 농업문화와 영농 운영, 공동체 축제개념을 연계하여 절기에 따라 한국 특유의 놀이 문화인 '마당' 과 '일판(놀판)' 과 결합된 공동체 절기 프로그램(필드 스테이션)을 제안한다.

Using 24 solarterms has been old wisdom that considers seasonal and climate changes accorsing to the solar movement in traditional agriculture and everyday life. It is used even nowadays in order to decide the seedtime, growing crops, and harvesting. It ahs been a major indicator for Korean people, who have the long agricultural tradition, in everyday life whether it is related to agricultural activities or not. Community program is introsuced through agricultural landscape and cultivating administration. Regarding the traditional agriculturral culture, agricultural management, and public event, special program od community which includes the concept of traditionalKorean recreation culture, 'Madang' and 'Ilpan(Nolpan) is proposed.

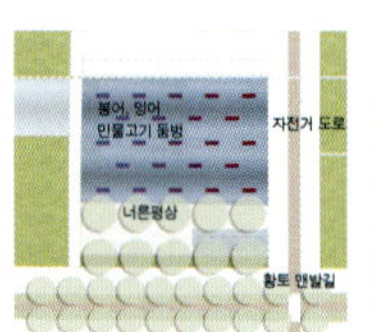
Dumbung station

Monitoring field station

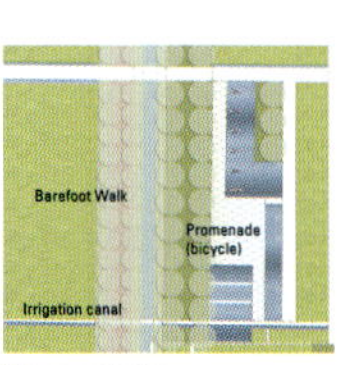

Doore station

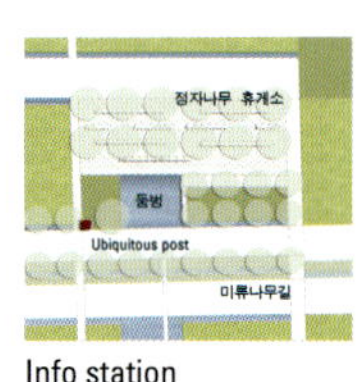

Info station

May 21th

June 6, 7th

September October 23rd

December 7th

System

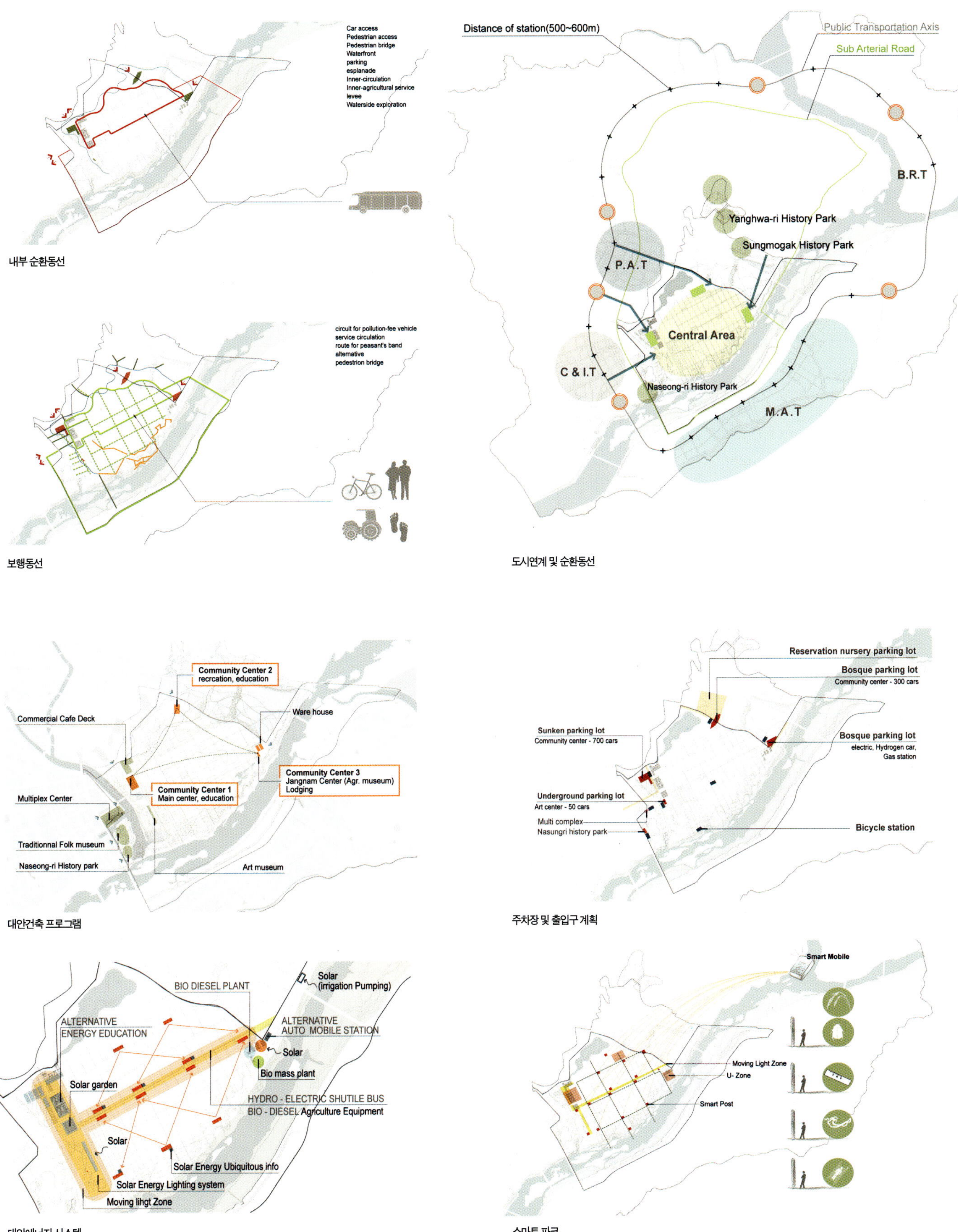

Car access
Pedestrian access
Pedestrian bridge
Waterfront
parking
esplanade
Inner-circulation
Inner-agricultural service
levee
Waterside exploration
내부 순환동선
circuit for pollution-fee vehicle
service circulation
route for peasant's band
alternative
pedestrion bridge
보행동선
Distance of station(500~600m)
Public Transportation Axis
Sub Arterial Road
B.R.T
Yanghwa-ri History Park
Sungmogak History Park
P.A.T
Central Area
C & I.T
Naseong-ri History Park
M.A.T
도시연계 및 순환동선
Community Center 2
recrcation, education
Commercial Cafe Deck
Ware house
Community Center 3
Jangnam Center (Agr. museum)
Lodging
Community Center 1
Main center, education
Multiplex Center
Traditionnal Folk museum
Naseong-ri History park
Art museum
대안건축 프로그램
Reservation nursery parking lot
Bosque parking lot
Community center - 300 cars
Sunken parking lot
Community center - 700 cars
Bosque parking lot
electric, Hydrogen car,
Gas station
Underground parking lot
Art center - 50 cars
Multi complex
Nasungri history park
Bicycle station
주차장 및 출입구 계획
BIO DIESEL PLANT
Solar
(irrigation Pumping)
ALTERNATIVE
ENERGY EDUCATION
ALTERNATIVE
AUTO MOBILE STATION
Solar
Bio mass plant
Solar garden
HYDRO - ELECTRIC SHUTILE BUS
BIO - DIESEL Agriculture Equipment
Solar
Solar Energy Ubiquitous info
Solar Energy Lighting system
Moving lihgt Zone
대안에너지 시스템
Smart Mobile
Moving Light Zone
U- Zone
Smart Post
스마트 파크

생산과 소비가 주요한 가치로 독려되고 권장되던 시기를 지나 더 먼 미래를 향해 천천히 함께 가길 원하는 시기를 맞이합니다.
우리는 오랜 시간 이 땅에 살며 식량을 자급하고, 자녀를 공부시키고, 이웃들과 교류하며,
태양과 달의 변화에 맞추어 농부로서 살아왔던 오래된 생활방식에 대한 생각을 합니다.
이 땅에 맞추어진 채로 수천년의 오랜 시간동안 세대에서 세대로 이어져 서서히 진화해 온 오래된 지혜를 구합니다.
우리는 그 지혜를 빌어 함께 나누어 공존하며 선조들이 그래 왔던 것처럼, 오래된 지혜에 지혜를 더하여 세대로 이어주는 방법에 대해 생각합니다.
할아버지가 손자의 손을 잡고 미소로 들려 주는 옛이야기처럼 세대에서 세대로 이어지도록
과거로부터 함께 하며 이어져, 지금 함께 하듯.

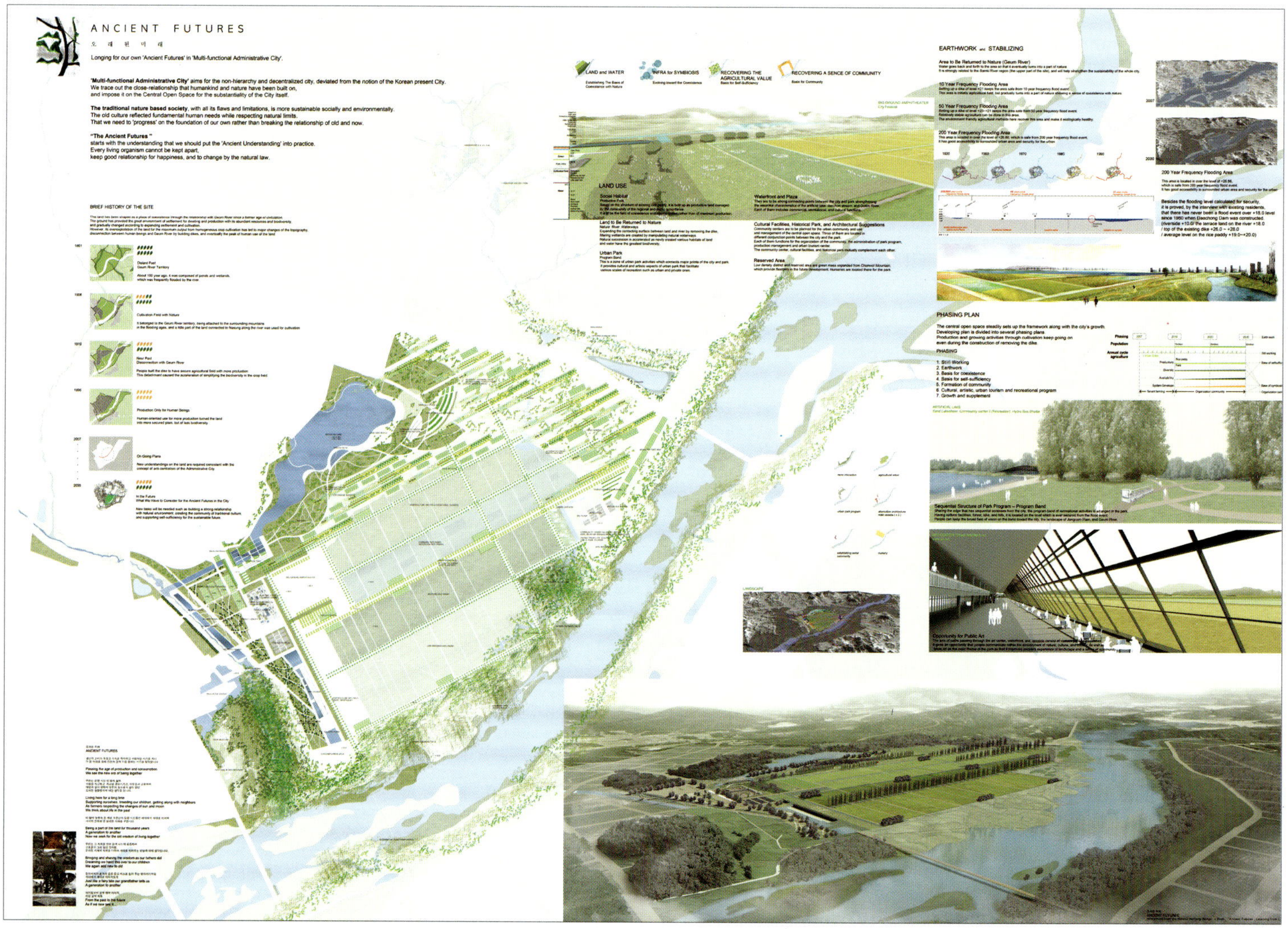

'Ancient Futures' , Referenced from the Helena Norberg-Hodge's book, 'Ancient Futures : Learning from Ladakh (Rider, 1992)

Passing the age of production and consumption, We see the new era of being together.
Living here for a long time Supporting ourselves, breeding our children, getting along with neighbors
As farmers respecting the changes of sun and moon, We think about life in the past.
Being a part of the land for thousand years a generation to another.
Now we seek for the old wisdom of living together.
Bringing and sharing the wisdom as our fathers did, Dreaming we hand this over to our children, We again add new to old
Just like a fairy tale our grandfather tells us, A generation to another.
From the past to the future As if we now see it···

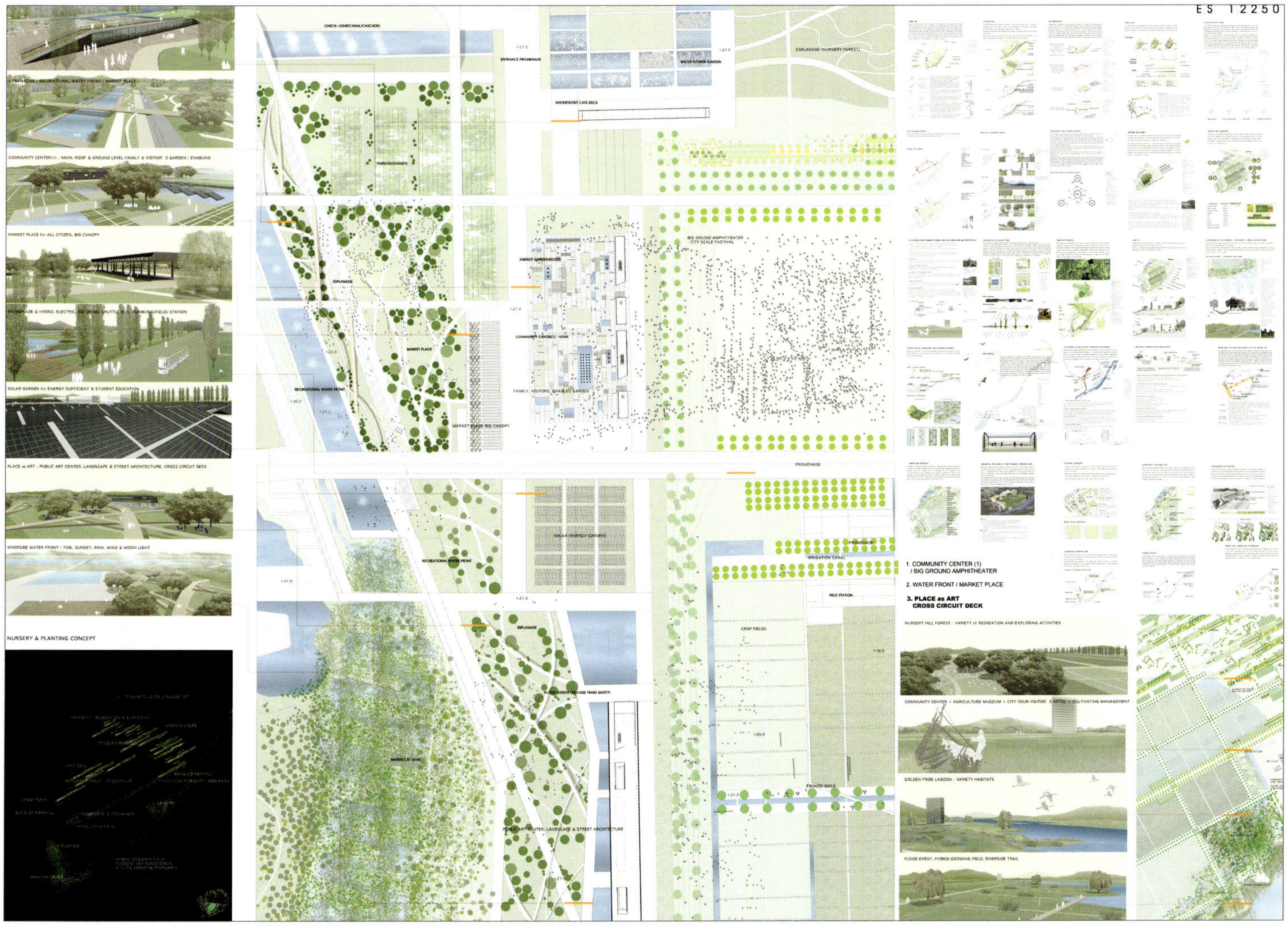

Ancient Futures 오래된 미래, 2단계 현상설계 제출 도판(A0× 8)

2nd Prize

LONG VIEWS

Kim Jong Kyu 김종규 · M.A.R.U / Korea National University of Art, Korea
Partners _ Ahn Jong Hwan 안종환, Min Jun Kee 민준기, Jang Byul 장별, Kim Jae Yoon 김재윤

Long views

In the large central open space of the annular ring city, a major impression is a grand 'openness' of Jangnam Plain and Geum-gang backgrouned by Jeonwol-san and Wonsu-bong. The Jangnam plain is protected from the river by an embankment which runs along the south and the west side of the plain. In addition to the topographical attractions, the unfinished Road #96 branched away from the south embankment and stopped in the middle of the plain. These topographical elements generated by both the nature and human provide a strong character of 'flatness'.
On the embankment and the Road #96, you are immediately able to see the entire site without any obstacle. This overwhelming long view smoothly introduces all the geological and ecological components of the existing landscape by both the flatness and the level difference. They are caused of the water and the agriculture.

환상형 도시의 중앙녹지공간에서 가장 주요한 인상은 전월산과 원수봉을 배경으로 하고있는 장남평야와 금강의 '개방성' 이다. 장남평야는 평야의 남쪽과 서쪽을 따라 흐르고 있는 제방에 의해 강으로부터 보호되고 있다. 아직 완성되지 않은 #96도로는 남쪽 제방으로부터 뻗어나와 평야의 가운데 지점에서 끝난다. 이러한 지형적 요소들은 자연과 인간에 의해 생겨났으며, 편평함이라는 큰 특징을 보여준다.

#96도로와 제방에 올라서면 곧바로 그 어떤 장애물도 없는 전체 부지를 바라볼 수 있다. 편평함과 레벨차이에 의한 이 압도적인 전망은 기존 경관의 모든 지질적, 생태적 구성요소를 설명해주고 있다.

The level of Geum-gang contains the wetland ecology and the flood field used for tree nurseries and dry-field farming. The level of Jangnam plain enclosed by the embankment contains agricultural field, paths and water supplying canals. The level of the embankment and access road contain the figures of artificiality and also it will be the new ground level for the new city.

금강레벨은 습지생태와 나무종묘장과 밭농사로 이용되는 범람원을 포함하고 있다. 장남평야의 레벨은 농경지, 통로 및 물을 제공하는 수로에 의해 둘러싸여 있다. 제방과 접근로의 레벨은 인공적인 형태를 담고 있으며, 이는 향후 새롭게 거듭날 도시의 지평이 될 것이다.

Landscape levels in long views

The existing level difference resulted from the agricultural background separates the site. And it interrupts a continuity of the existing ecology and human accessibility. The steep level difference needs to be loosened down for the continuity of nature and human accessibility. Namely, the urban level(embankment and road level: ~+27) will be sloped down to the natural level(agricultural field level: + 27 ~ +18). And the wet level(Guem-gang and flood field level: +18 ~ +15) will be connected by breaking the west side of embankment.

The ecology of the Guem-gang wet land will flow into the agricultural field through the water. It also help to naturalize the field as a natural field. The slope-downed Road #96 helps human to access to the natural field from the urban level in a same way of nature coming into.

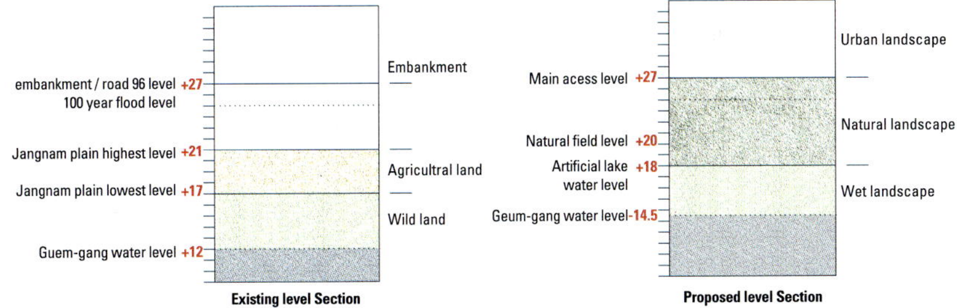

현재의 레벨차이는 농업적 배경이 부지를 갈라 놓은 것으로부터 생겨난 것으로 이는 현재의 생태와 인간의 접근 및 현재 생태의 연속성을 저해하고 있다. 가파른 레벨차는 자연 및 인간 접근성의 연속을 위해 완만하게 낮출 필요가 있다. 즉, 도시레벨(제방과 도로레벨 : ~ +27)은 자연레벨(농경지 레벨 : +27 ~ +18)로 경사지게 해야하며 습지레벨(금강과 범람원 레벨 : +18 ~ +15)은 서쪽 제방을 허무는 것으로 연계되어야 한다.

금강의 습지 생태는 물을 통해 농경지로 흘러들어가게 될 것이다. 그것은 또한 농경지를 자연상태의 벌판처럼 자연화시키는데 도움을 줄 것이다. 경사가 낮춰진 #96도로는 인간이 도시 레벨에서 자연의 벌판으로 접근하는 것을 도와준다.

Land articulation strategy

The strategy of land articulation starts with delivering the wildness landscape to agricultural field in order to help naturalizing the open space. Turning back the agricultural field to the nature is also a way of making the planed artificial lake, at the same time. The soil gaining from breaking the west embankment and the excavation for the lake will be used to make three artificial hills on the field and to loosen the slops of both sides of Road #96 and the south embankment to the wildness landscape for increasing the accessibility between the levels.
These three hills are on a continuity of the urban hills and mountains under the urban structure. In addition, they symbolize the open space by a man-made plantation with keeping viewing quality to surroundings. The loosened slops are not only an access, also a huge venue for public events.

토지구획 전략은 오픈스페이스의 자연화를 돕기 위해 야생경관을 농경지로 전달하는것으로 시작한다. 또한 농경지를 자연상태로 되돌리는 것은 동시에 인공호수를 조성하는 한 방법이다. 서쪽 제방을 허물고 호수를 조성하기 위해 파낸 흙은 논 위에 3개의 인공언덕을 만드는 것, #96도로의 양옆 경사지를 완만하게 하는 것, 레벨 사이의 접근성을 높이기위해 남쪽 제방을 자연경관으로 만드는 것에 사용될 것이다.

이 3개의 인공언덕은 도시구조 아래에 있는 언덕과 산의 연속으로써 주변 조망의 보존과 함께 인공 조림지에 의해 오픈스페이스를 상징한다. 완만해진 경사는 접근성 뿐만 아니라 공공 이벤트를 위한 거대한 장이기도 하다.

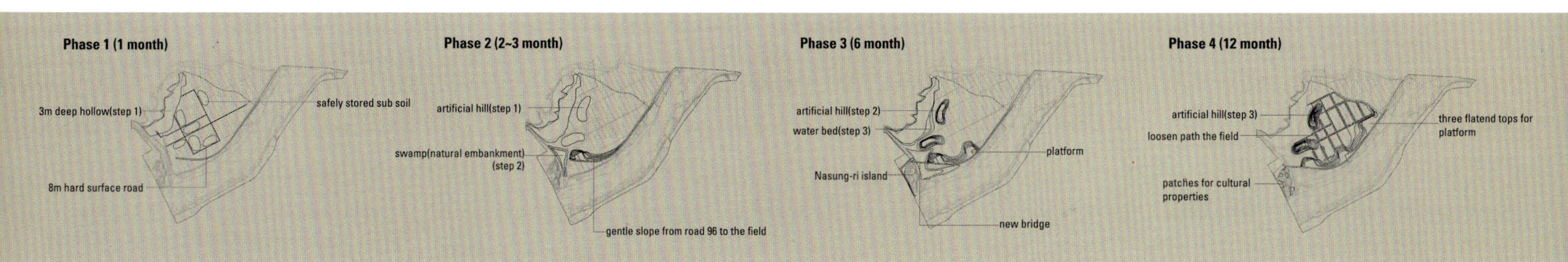

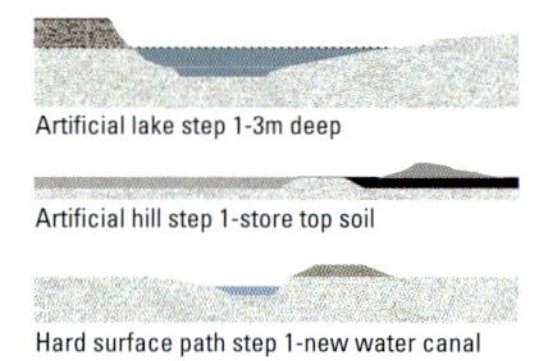

Artificial lake step 1-3m deep

Artificial hill step 1-store top soil

Hard surface path step 1-new water canal

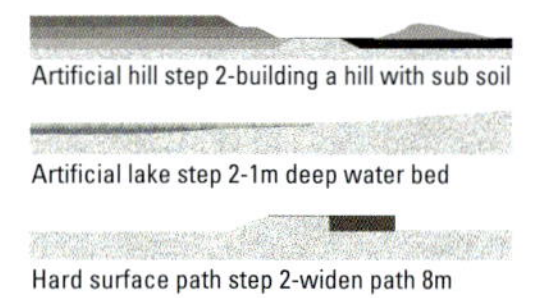

Artificial hill step 2-building a hill with sub soil

Artificial lake step 2-1m deep water bed

Hard surface path step 2-widen path 8m

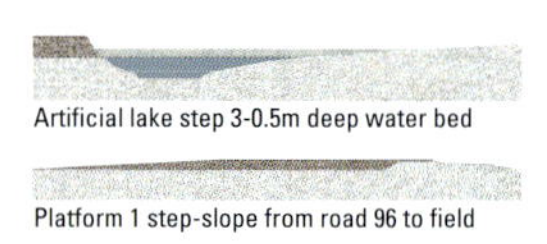

Artificial lake step 3-0.5m deep water bed

Platform 1 step-slope from road 96 to field

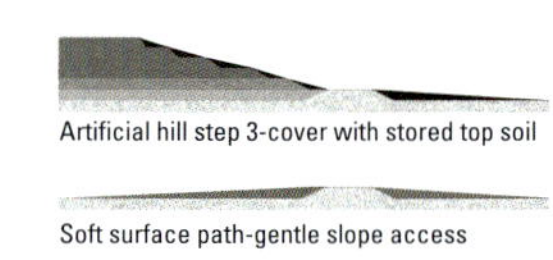

Artificial hill step 3-cover with stored top soil

Soft surface path-gentle slope access

Land naturalising

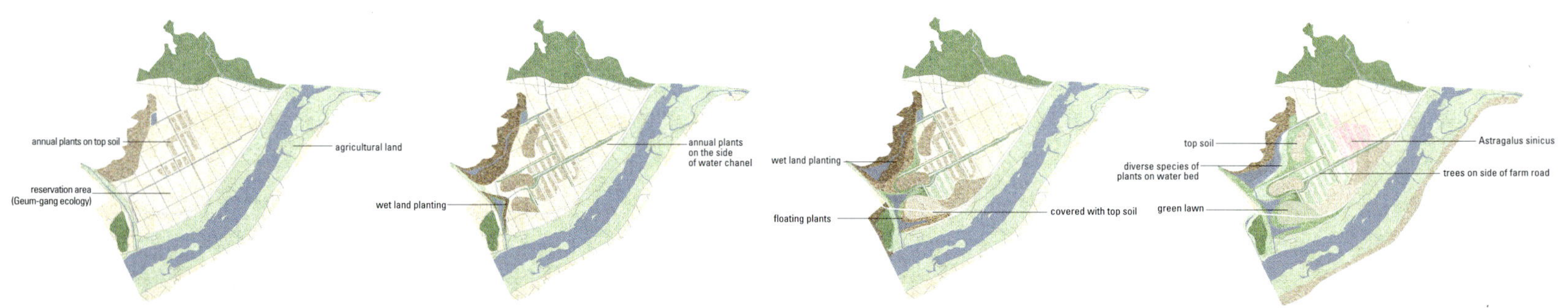

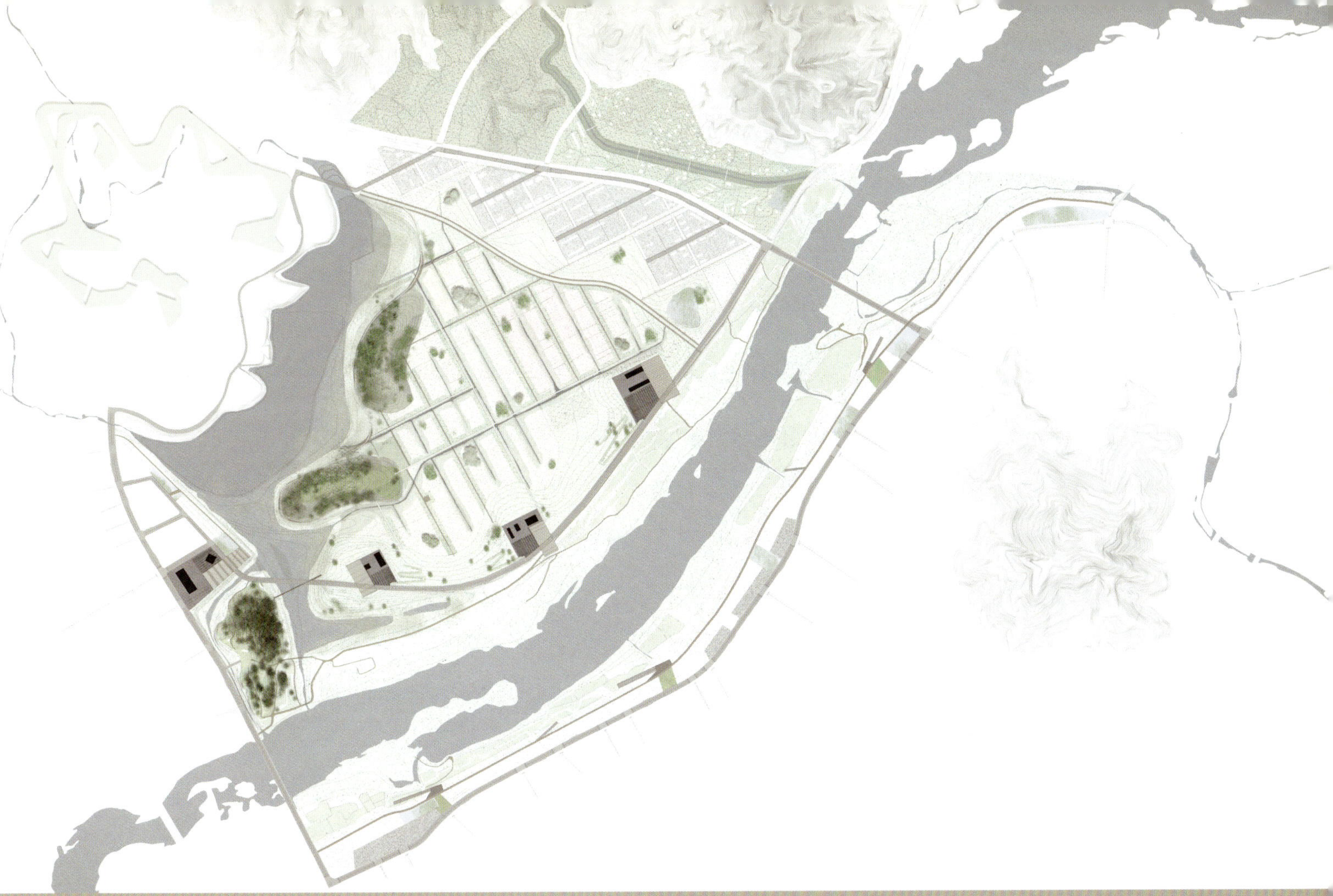

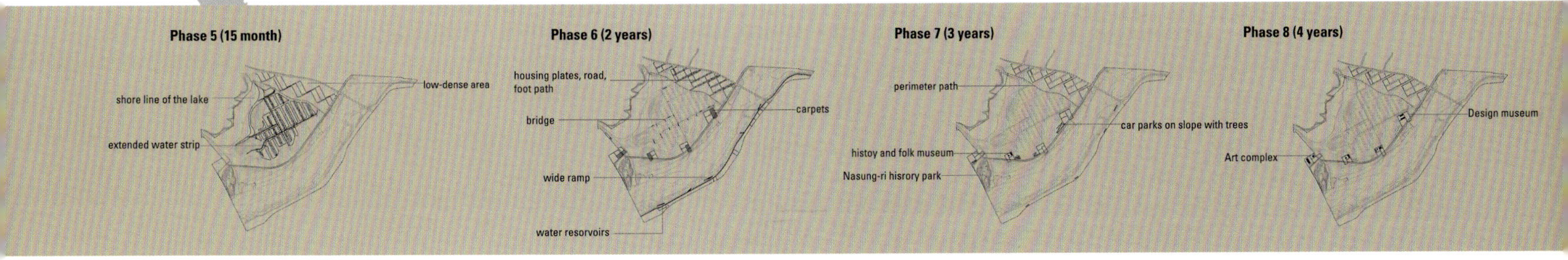

Phase 5 (15 month)
shore line of the lake
low-dense area
extended water strip
Phase 6 (2 years)
housing plates, road, foot path
bridge
carpets
wide ramp
water resorvoirs
Phase 7 (3 years)
perimeter path
car parks on slope with trees
histoy and folk museum
Nasung-ri hisrory park
Phase 8 (4 years)
Design museum
Art complex

Water canal-widen existing irragation canal
Hard surface path step 3-gentle slope access
Low-dense area 1 step-rising land
Low-dense area 2 step-housing plate
Hard surface path step 4-bridge

wild animals on river side
trees
green lawn
more green lawn
green lawn
green water front
river side park
green corridor
trees between car park road

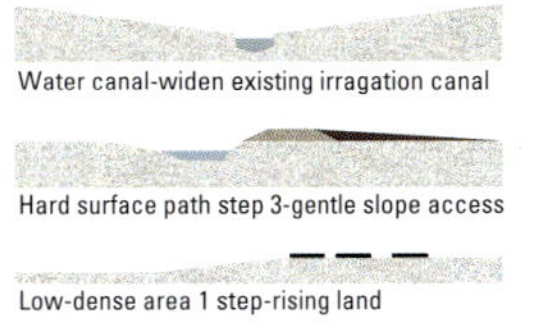

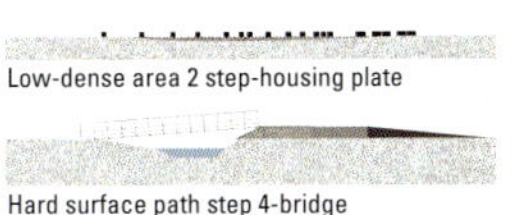

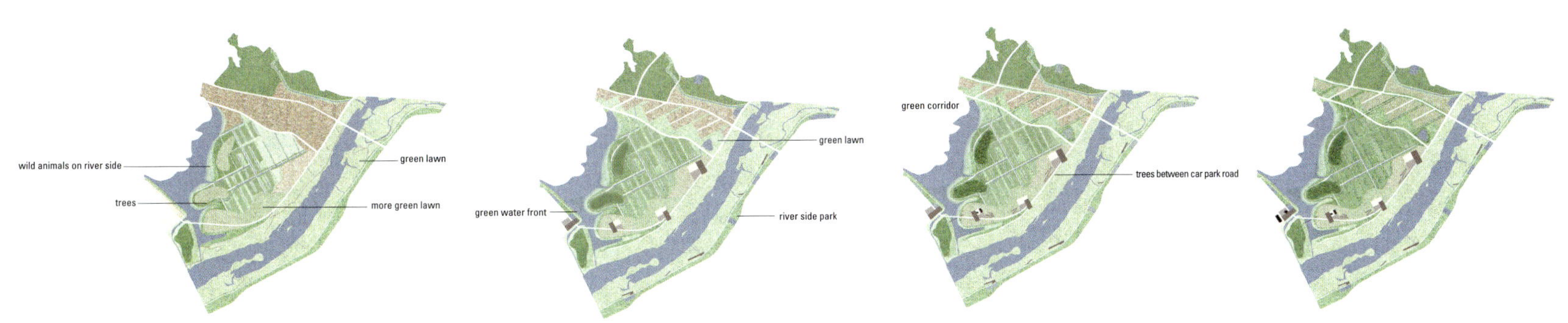

Long views in urban open space

A urban landscape is formed by contents of the society and the nature visually and spiritually. It is also a spatial experience of the reflection of the culture and an interpretation of landscape contents.
The contemporary urban landscape is no longer defined either predicted, because of it's cultural complexity. Therefore it is often characterized by the phenomenon appearing in the park by unexpected uses of the people. This certainly changes the use and the character of what we traditionally consider as urban public space, which is not necessarily the space where everybody meets everybody any longer. In a particular case of the MAC, the urban park needs to play a role to intervene between the rural and the urban sphere.

도시경관은 사회적 컨텐츠와 눈에 보이는 자연과 감수성에 의해 형성된다. 이는 또한 경관이 담고 있는 요소들에 대한 해석이자 문화 반영의 특별한 경험이다.

동시대의 도시 경관은 복합적인 문화에 의해 더 이상 제한되거나 예언되지 않으므로 늘 공원에서 사람들의 예상치 못한 사용에 의해 나타나는 현상에 의해 특징지어진다. 이것은 더 이상 공원이 사람과 사람이 만나는 장소가 아니듯 우리가 전통적으로 생각하는 도시 공공공간의 이용과 성격을 확실하게 변화시킨다. 행복도시와 같이 특별한 경우의 도시공원은 도시와 전원 사이의 역할을 필요로 한다.

Instead of drawing out the programs and zones, we try to minimize the determined space and to let the nature to grow itself in a same way that the culture discovers uses of the undetermined urban void.
The text plan shows only ostensible atmospheres over the urban park, which is not necessarily related to certain activities. These atmospheres could be read through the long views.

그러므로 프로그램과 구역을 나누어 제안하는 대신, 확정된 공간을 최소화하고 확정하지 않은 도시공간에서 발견된 문화같이 자연이 스스로 성장할 수 있도록 하고자 했다. 확정된 계획은 도시공원에 대한 표피적인 분위기만을 보여주게 된다. 실제 분위기는 장기적인 전망long view을 통해서 읽혀지게 된다.

water network as a naturalizing catalyst

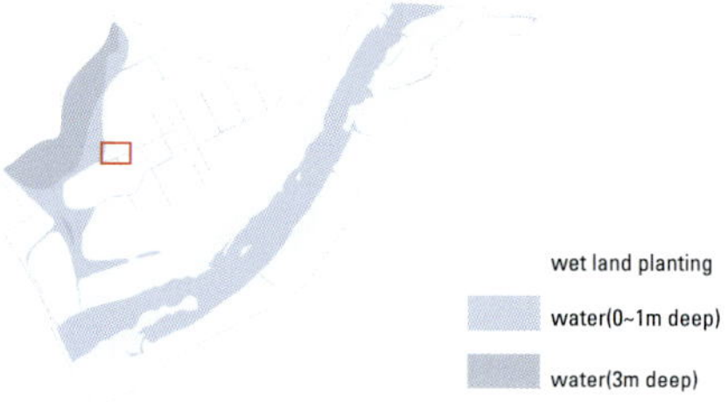

wet land planting
water(0~1m deep)
water(3m deep)

wet land of the artificial lake edge

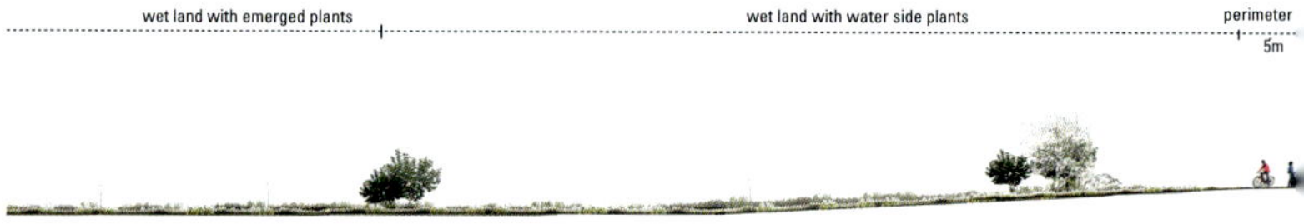

wet land with emerged plants
wet land with water side plants
perimeter
5m

soil seperation for planted/native vegetation

new topsoil(slope)
new topsoil(plain)
existing topsoil

naturalised field as a urban park

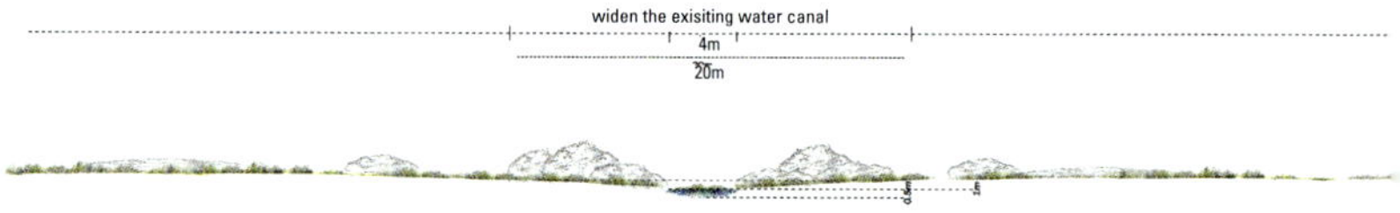

widen the exisiting water canal
4m
20m

inhabitation on activity slope and road 96

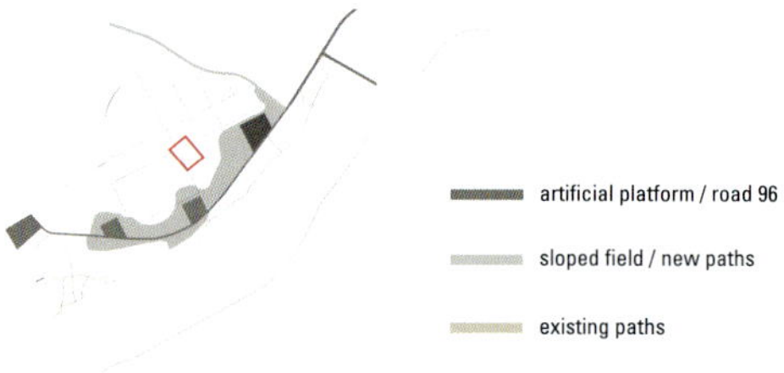

artificial platform / road 96
sloped field / new paths
existing paths

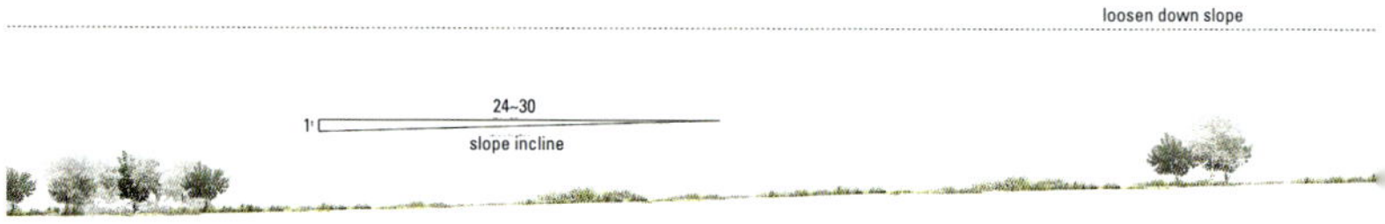

loosen down slope
24~30
1:
slope incline

section a - a'

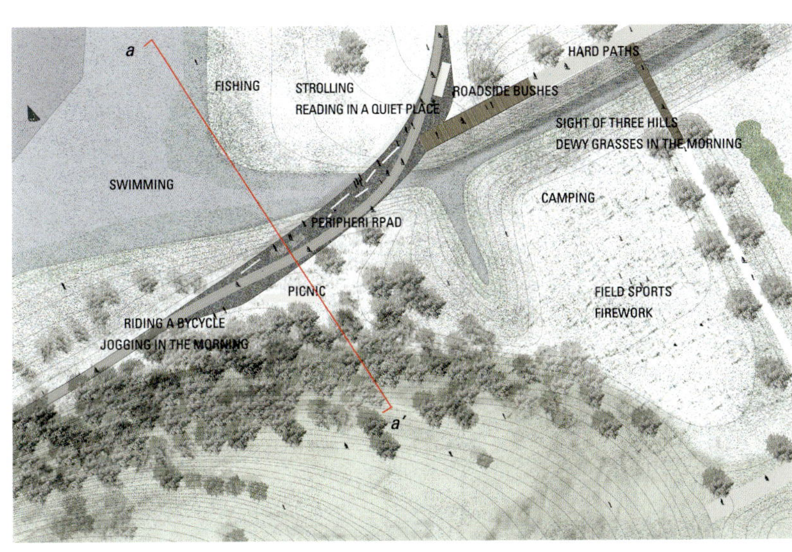
a
a'
FISHING
STROLLING
READING IN A QUIET PLACE
ROADSIDE BUSHES
HARD PATHS
SIGHT OF THREE HILLS
DEWY GRASSES IN THE MORNING
SWIMMING
CAMPING
PERIPHERI RPAD
PICNIC
FIELD SPORTS
FIREWORK
RIDING A BYCYCLE
JOGGING IN THE MORNING

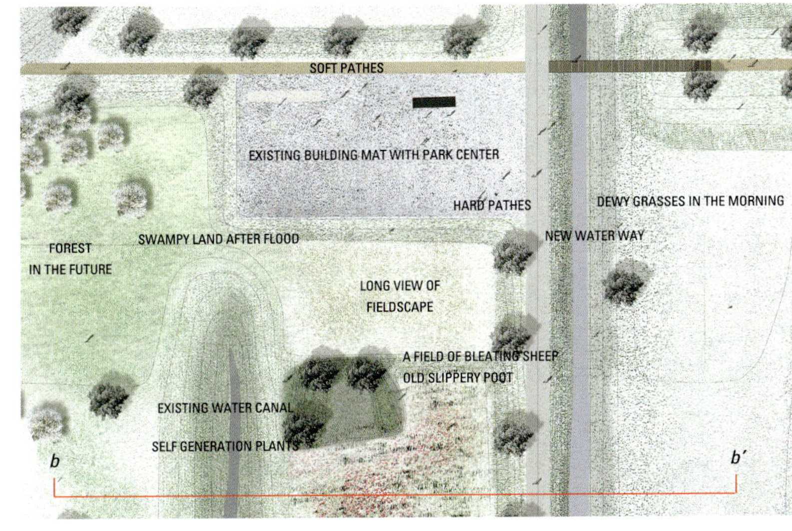
SOFT PATHES
EXISTING BUILDING MAT WITH PARK CENTER
HARD PATHES
DEWY GRASSES IN THE MORNING
FOREST
IN THE FUTURE
SWAMPY LAND AFTER FLOOD
NEW WATER WAY
LONG VIEW OF
FIELDSCAPE
A FIELD OF BLEATING SHEEP
OLD SLIPPERY POOT
EXISTING WATER CANAL
SELF GENERATION PLANTS
b
b'

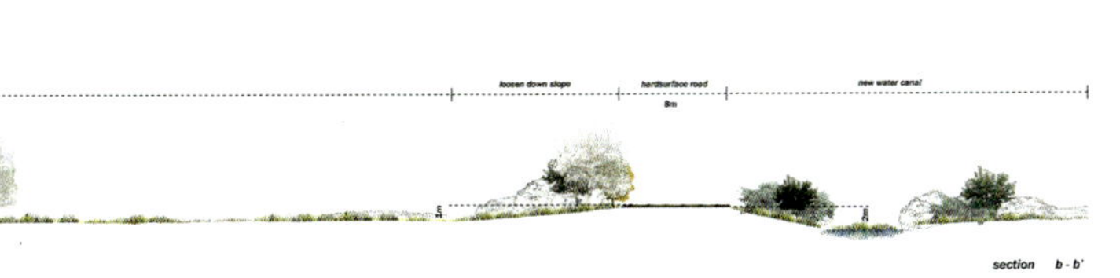
section b - b'

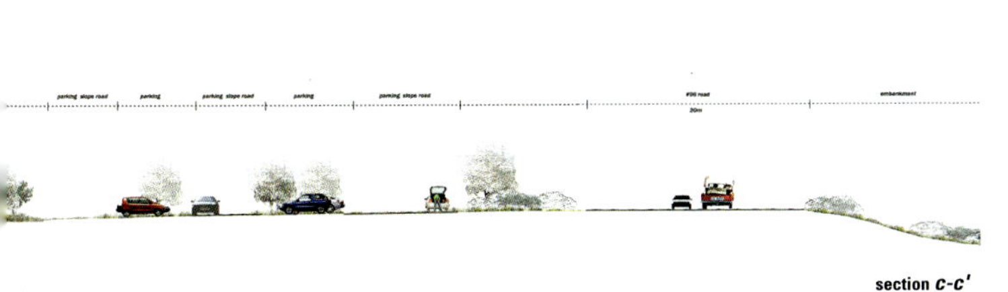
section c-c'

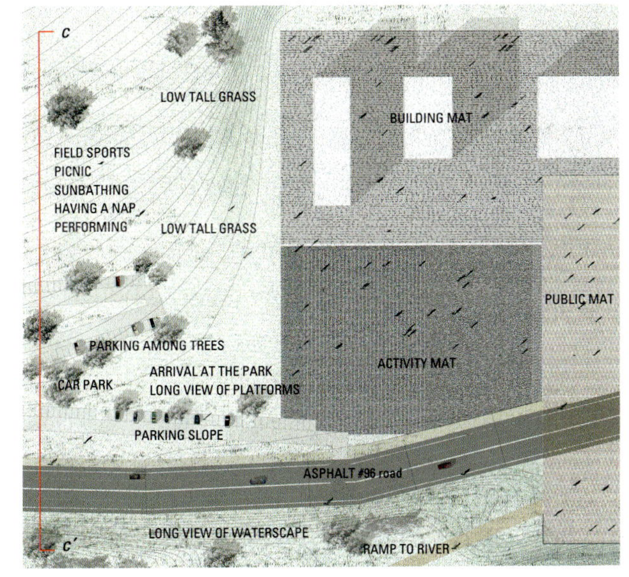
c
LOW TALL GRASS
BUILDING MAT
FIELD SPORTS
PICNIC
SUNBATHING
HAVING A NAP
PERFORMING
LOW TALL GRASS
PUBLIC MAT
PARKING AMONG TREES
CAR PARK
ARRIVAL AT THE PARK
LONG VIEW OF PLATFORMS
ACTIVITY MAT
PARKING SLOPE
ASPHALT #96 road
c'
LONG VIEW OF WATERSCAPE
RAMP TO RIVER

3rd Prize

CARVED PARK

Mihn Hyun Jun 민현준 · Hongik University / Lab. Kwang Poong, Korea
Partners _ Cho Sung Ik 조성익, Cho Seung Man 조승만, Kwon Young Sang 권영상

Carved Park is

1. flexible, easily adapting to the urban plan and evolving with the city.
2. economical, utilizing the existing landscape.
3. community-oriented, shaped by citizens' active participation.

1. 도시계획에 유연하게 순응하며, 도시와 함께 발전하는 공원
2. 기존 경관을 이용하는 경제적인 공원
3. 커뮤니티를 지향하고, 시민들의 참여로 만들어 가는 공원

The main idea of the Carved Park is consistent with the original master plan of the surrounding looped city-preserving the rice field in the center of the city. The Carved Park is not imposed on but carved into the rice field, preserving the wide horizon of the landscape. Rather than turning the entire surface of the rice field into a park, the carved lines and spots are incorporated into the minimal landscape manipulation.

'농경지에 음각된 공원' 이라는 이 작품의 주요 특징은 환상형 도시의 중앙부의 농경지 경관을 보호하는 것이다. 농지 전체를 공원으로 조성하는 것보다 부분적으로 형태를 만들고 장소를 조성하는 방법을 통해서 기존의 수평적인 경관의 특징을 보존하고자 한다.

mud

mud

Agri-scape

Program-scape

Carved Park in the Rice Field

Hidden Park

The project goal is preserving the horizon of the peaceful, extensive rice field (Agri-scape). The Carved Park is not imposed on the field, but planted and hidden in the field (Program-scape). Park facilities and promenades are carved out and will not disturb the horizon, minimizing the impact on the ecological system. The park is visually and functionally subordinated to the rice field.

이 프로젝트의 목표는 광활한 평야의 평화롭고 수평적인 경관을 보존하는 것으로, 음각된 공원은 농지 위로 드러나지 않지만, 농지 안에 세워지고 감춰진 공원Program-scape이다. 공원시설과 산책로는 음각되지만 수평성을 깨지 않으며, 생태계의 충격을 최소화 한다. 공원은 시각적으로나 기능적으로 논 아래에 있게 된다.

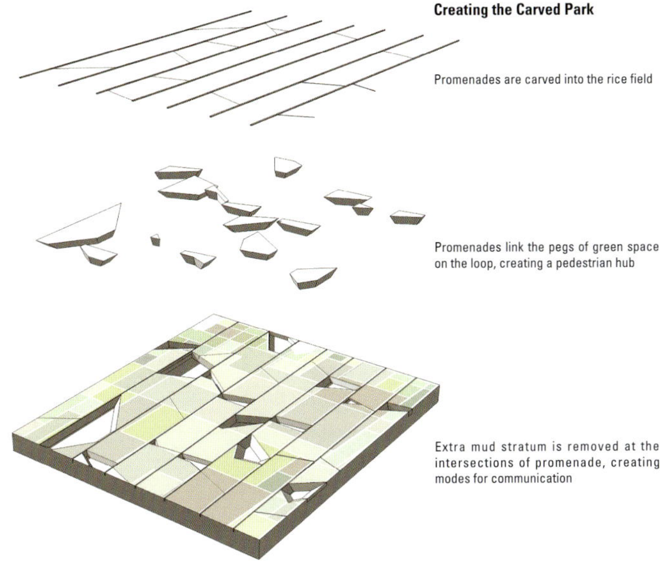

Creating the Carved Park

Promenades are carved into the rice field

Promenades link the pegs of green space on the loop, creating a pedestrian hub

Extra mud stratum is removed at the intersections of promenade, creating modes for communication

Pedestrian Hub

The ring-shaped road system that principal identity of the city can cause major traffic congestion and prolong travel time between spots on the loop. The Carved Park connects the pegs of green spaces on the loop, and creates a network through which pedestrians and bikers can effectively move in the city. The park as a pedestrian hub will be incorporated into the daily lives of citizens, and make the city a pedestrian paradise.

행복도시의 정체성이라 할 수 있는 환상형 도로 시스템은 심각한 교통체증을 유발하며, 지역간 이동시간을 늘어나게 할 가능성이 있다. 음각된 공원은 환상형 도로상의 쐐기모양의 녹지공간들을 연결하고, 이들 간의 네트워크를 구축함으로써 보행자와 자전거 이용자들이 보다 쉽게 도시로 이동할 수 있도록 한다. 보행자 축으로서의 공원은 시민들의 일상생활과 밀접하게 되고, 보행자들의 천국인 도시가 될 것이다.

Green Pegs for Pedestrian

Shapeless Park

Unlike a conventional park that is confined to certain boundaries, the Carved Park is shapeless. The carved spots of the park are scattered on the field, and connected to each other like a chain by promenades, creating functionally distinctive groups.
There is neither a startingpoint nor an ending point. New spots and promenades can proliferate, while existing spots and promenades can be easily covered and turned back to the rice field. The entire park can be easily expanded or shrunk depending on the execution of the city plan. The Carved Park, hence, is an ideal plan that can easily adapt to the evolution of the surrounding city in the long run.

경계가 확실한 전형적인 공원과 달리 이 공원은 형태가 없다. 공원의 음각된 지점은 농지 위에 흩어져 있으며, 사슬처럼 연결된 산책로에 의해 연결되고, 기능적으로 독특한 그룹을 만든다. 여기에는 시작점과 끝점이 없다. 새로운 지점과 산책로들이 증가할 수 있는 한편 기존 지점과 산책로들이 쉽게 조각되기도 하고 다시 논으로 환원될 수 있다. 전체 공원은 도시계획의 실행에 따라 쉽게 확장되거나 축소될 수 있다. 따라서 이 계획안은 오랜 기간동안 도시 주변부의 발달과정에 따라 쉽게 대응할 수 있는 개념적인 계획이다.

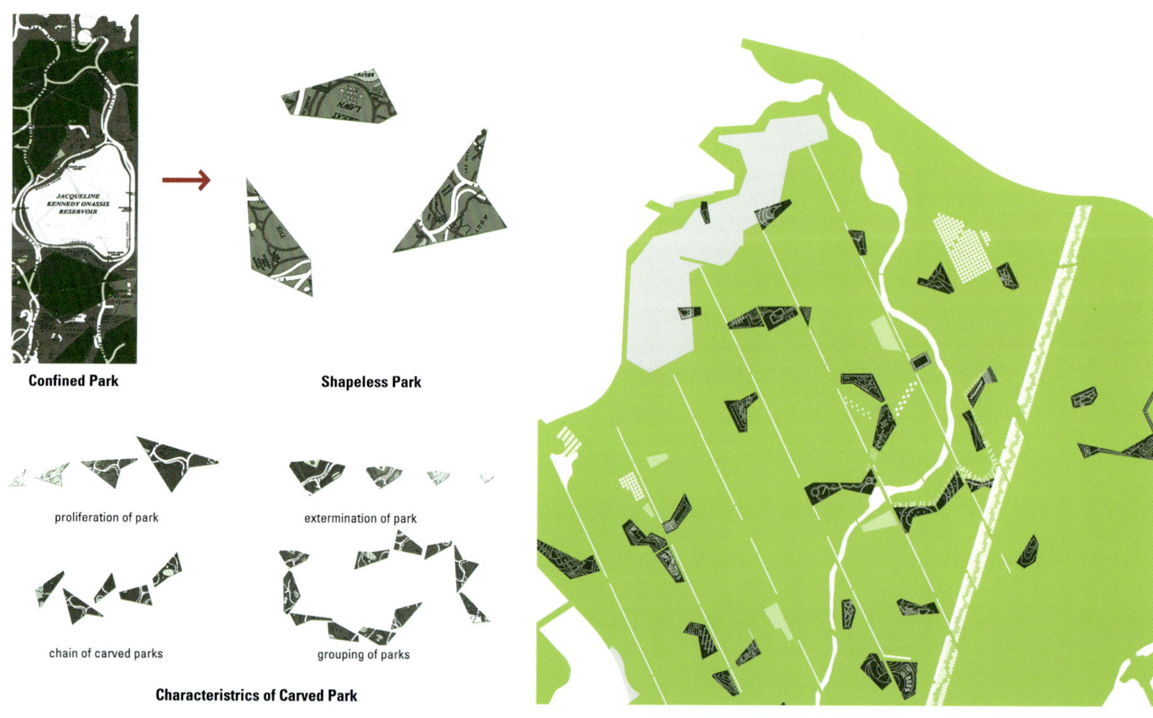

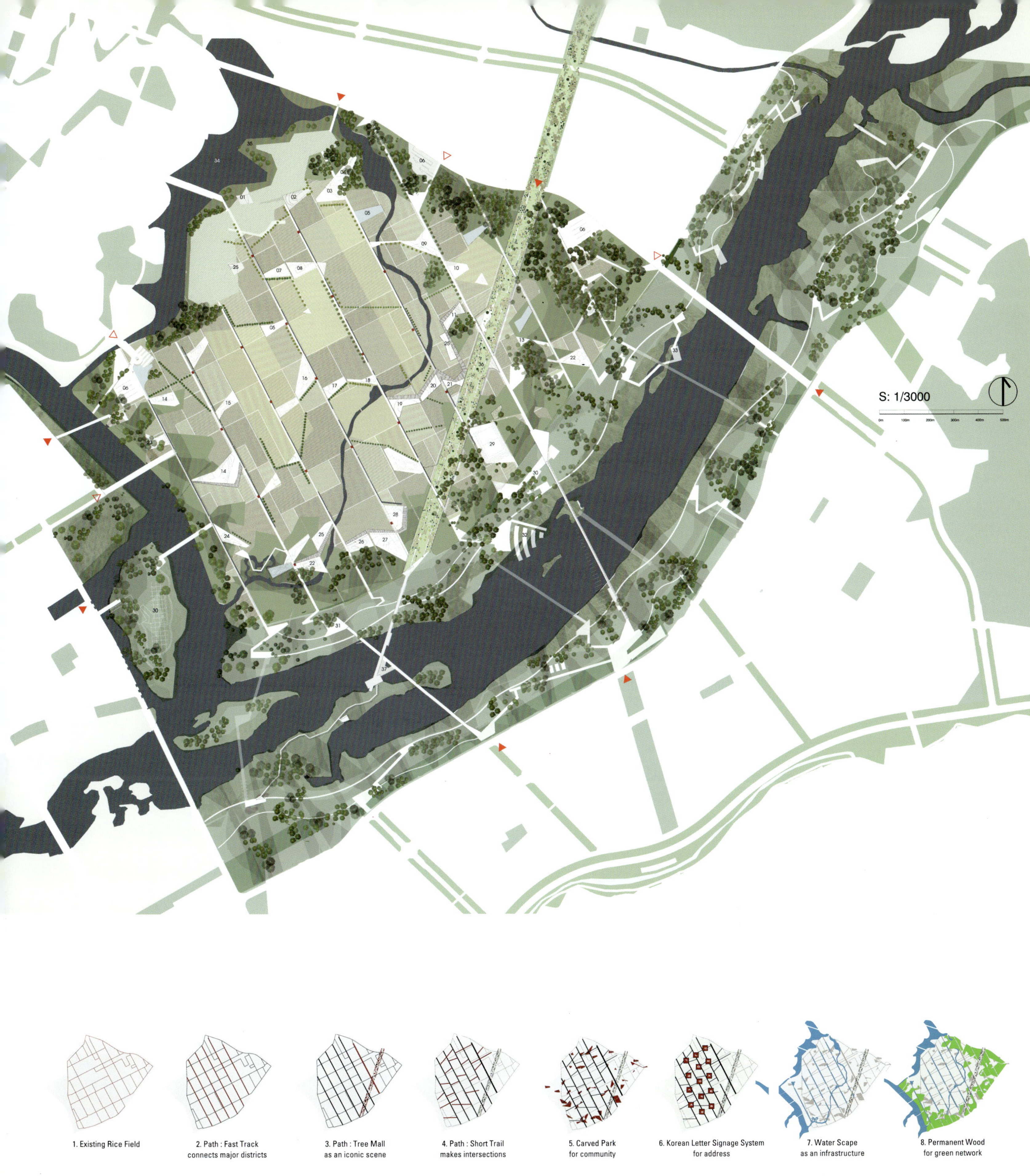
S: 1/3000
1. Existing Rice Field
2. Path : Fast Track connects major districts
3. Path : Tree Mall as an iconic scene
4. Path : Short Trail makes intersections
5. Carved Park for community
6. Korean Letter Signage System for address
7. Water Scape as an infrastructure
8. Permanent Wood for green network

Promenades

On the existing paths on the field, tree-lined promenades are created. These promenades connect the pegs of green spaces on the loop, and function as a pedestrian network that effectively connects different parts of the looped city. City residents can commute through these avenues but also stroll, exercise, and enjoy the nature on them.

기존 통로에는 수목이 심겨진 산책로가 조성된다. 이 산책로들은 환상형 도로에 놓인 쐐기모양의 녹지축들을 연결하며, 환상형 도시의 다른 지역을 효과적으로 연결하는 보행자 네크워크 기능을 가진다. 도시 거주자들은 이 거리를 통해 출퇴근 및 통학 할 수 있으며, 자연속에서 산책, 운동을 즐길 수 있다.

Plan Cultural Chain : Library and Cultural Center

Nodes for Communication

Where these promenades meet, nodes for communication are created. The promenades will be widened or deepened to create a functional space depending on the needs. Such a node is a flexible space that can host varieties of performances and activities, and function as a place for communication where visitors can meet with others and participate in diverse events. This makes the Carved Park a place for the community.

산책로가 만나는 곳은 커뮤니케이션을 위한 결절점들이 만들어진다. 산책로는 필요에 따라 넓어지거나 깊어질 수 있는 기능적 공간으로 조성된다. 이 결절점은 다양한 퍼포먼스와 활동을 위한 유연한 공간이자 방문객들이 서로 만날 수 있고 다양한 이벤트에 참여할 수 있는 장소이다. 이것이 커뮤니티를 위한 공간인 음각된 공원이다.

Architecture

Architectural spaces in the Carved Park are not in the form of conventional masses on the land. They are embedded in the rice field, utilizing different floor levels and walls created by carving. For example, a space for an art museum can be generated from carved promenades. Art pieces can be exhibited on the outdoor promenades while keeping only small amount of art pieces indoor. Only a sign post will stand on the field level, allowing visitors to easily find the art museum from the distance.

이 공원에서 건축공간은 땅 위의 전형적인 건축물 형태가 아니라 음각에 의해 형성된 벽과 다른 지형레벨들을 이용해 논 아래 묻혀있는 형태로 조성된다. 예를 들어 박물관은 음각된 산책로에 세워진다. 예술조각들은 산책로 외부에 전시되고, 소량의 예술품만 실내에 보관된다. 오로지 표지판만 논의 수평면상에 세워지며, 방문객들은 먼거리에서도 쉽게 박물관을 찾을 수 있다.

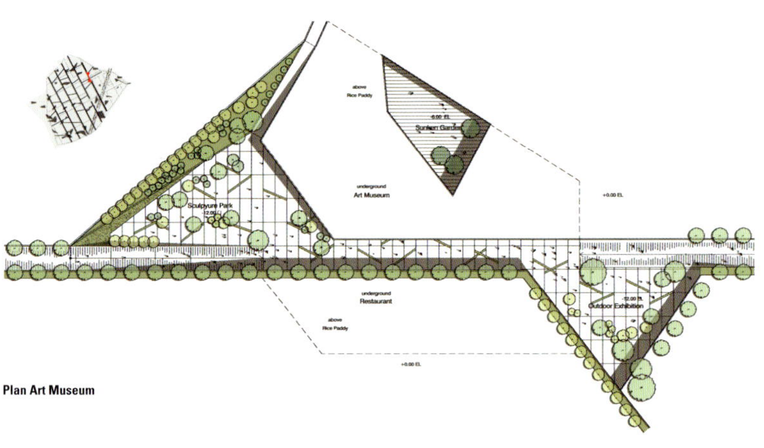

Plan Art Museum

Tree Mall

The tree mall is a metaphor of incorporating the commercial element of the city into the park. The widest carved promenade from the Wonsu-bong is utilized as a tree mall, filled with diverse kinds of trees inhabit all over Korea. Citizens can be educated about the ecology of trees and purchase and plant their own trees.

트리몰은 도시의 상업적 요소를 공원안으로 비유적으로 도입한 것이다. 원수봉으로부터 오는 가장 넓게 음각된 산책로는 한국에 자생하는 다양한 수목들이 채워진 트리몰로 활용된다. 시민들은 나무 생태학 교육을 받을 수 있으며, 자신만의 나무를 구입할 수 있다.

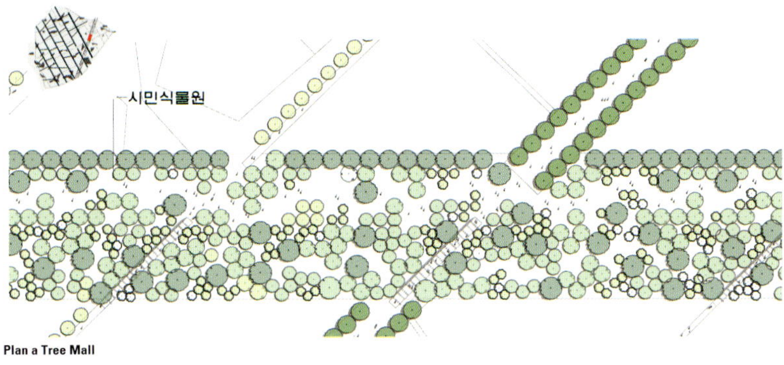

Plan a Tree Mall

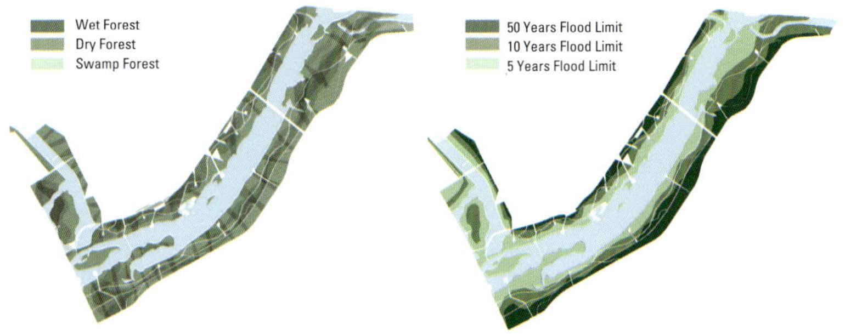

Multi-Level Waterfront Ecology Development Strategy

The riverside areas are divided into three flood zones. In the lowest flood zone, flooded about every five years, eco-friendly swamps with limited human access will be created. The species endangered by the creation of new city can return to their natural habitats in this zone. In the second lowest flood zone, flooded about every ten years, spaces where human being and nature can coexist, such as bike paths and trail, will be created. In the final flood zone, which is flooded about every one hundred years, architectural structures such as cafes will be built to let people observe the wildlife. The Flood Control Structure on the water front can also be used as a wildlife education center.

수변지역은 3개의 범람원으로 나뉜다. 가장 낮은 범람원은 5년 주기 홍수에 범람하며, 인간의 접근이 제한된 생태친화적인 습지로 조성되어 신도시 조성에 따라 멸종위기에 놓인 종들의 새로운 서식지가 되는 곳이다. 두 번째로 낮은 범람원은 10년 주기 홍수에 범람하며, 자연과 인간이 공존하는 공간으로 자전거도로와 오솔길이 만들어진다. 마지막 범람원은 100년 주기 홍수에 범람하는 지역으로 사람들이 야생을 관찰할 수 있는 카페같은 건축물이 들어서는 곳이다. 이러한 수변의 홍수를 조절하는 구조물은 야생교육센터로 이용할 수 있다.

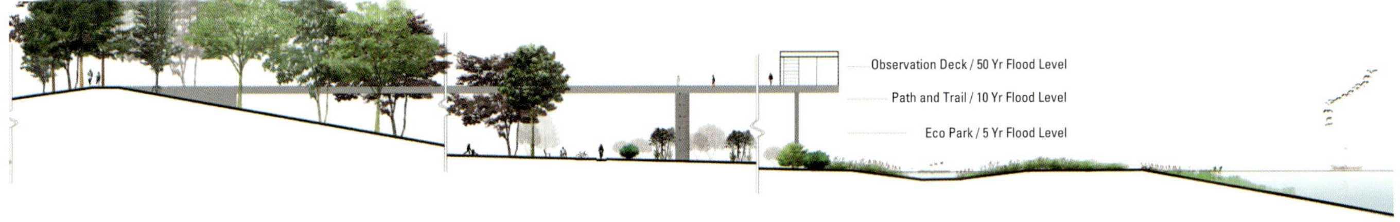

Finalist

1-1=1+1

Yu Eui Yoal 유의열

Partners _ Kim Do Kyong 김도경, Choi Won Man 최원만, Choi Jae Hyuk 최재혁, Hong Keun Pyo 홍근표

앞으로의 공원은 자연의 땅을 밟을 수 있는 오픈스페이스로서 자연스럽고 다소 비어보이는 계획을 통해 복잡한 도시 조직의 해방구를 제공해야 할 의무가 있다. 도시 오픈스페이스가 제공하는 어메니티는 필수이며, 녹지와 시원한 그늘이 일상이 된 현실에서, 새로운 도시 중앙부에 광활한 들녘은 도시민에게 이상적이며 일탈의 해방구를 제공하기에 충분하다. 중앙부 100만 평의 평야를 보전하는 과감한 계획을 제안한다.

결선작은 2단계 결선에 진출한 10개의 작품 중 수상작 3점을 제외한 7점의 작품을 무작위 등록번호PIN 순으로 소개한다.

Finalist

THE LAST PIECE

Kim Kwang Bae 김광배

Partners _ Choi Hyung Jin 최형진, Han Jea hee 한재희, Park Myung Kweon 박명권, Shin Ji Hoon 신지훈, Song, Young Tak 송영탁, Choe Yi Kyu 최이규

MAC는 환상형 도시개념 하의 6개의 개발계획을 가지고 있으며, 이 계획들은 구역에 따른 도시 전반을 나누고, 그 지역에 상응하는 프로그램들을 채워나가는 방법이 된다. 만약 이 계획을 퍼즐에 의해 인지하게 된다면 도시 전반은 한 개의 독립적인 그림으로 볼 수 있고 계획된 프로그램들은 이 그림을 완성하기 위한 조각으로 볼 수 있다. 각각의 조각은 이들 단위, MAC가 완성되면 독특한 정체성을 지니게 된다. 이러한 퍼즐 조각들 가운데, 도시의 중심부에 비워진 공간이 중앙오픈스페이스 조각이 된다. 이 조각은 정체성을 가지며 한 개의 조각으로써 이 도시에 살고 있는 사람들의 휴식공간 역할을 한다. 프로그램들은 중앙오픈스페이스 조각 안에서 여러개의 조각들로 나누어진다. 그 내부의 조각들은 각각의 프로그램을 지니며, 모든 조각들이 모아져 MAC이라는 그림이 완성된다.

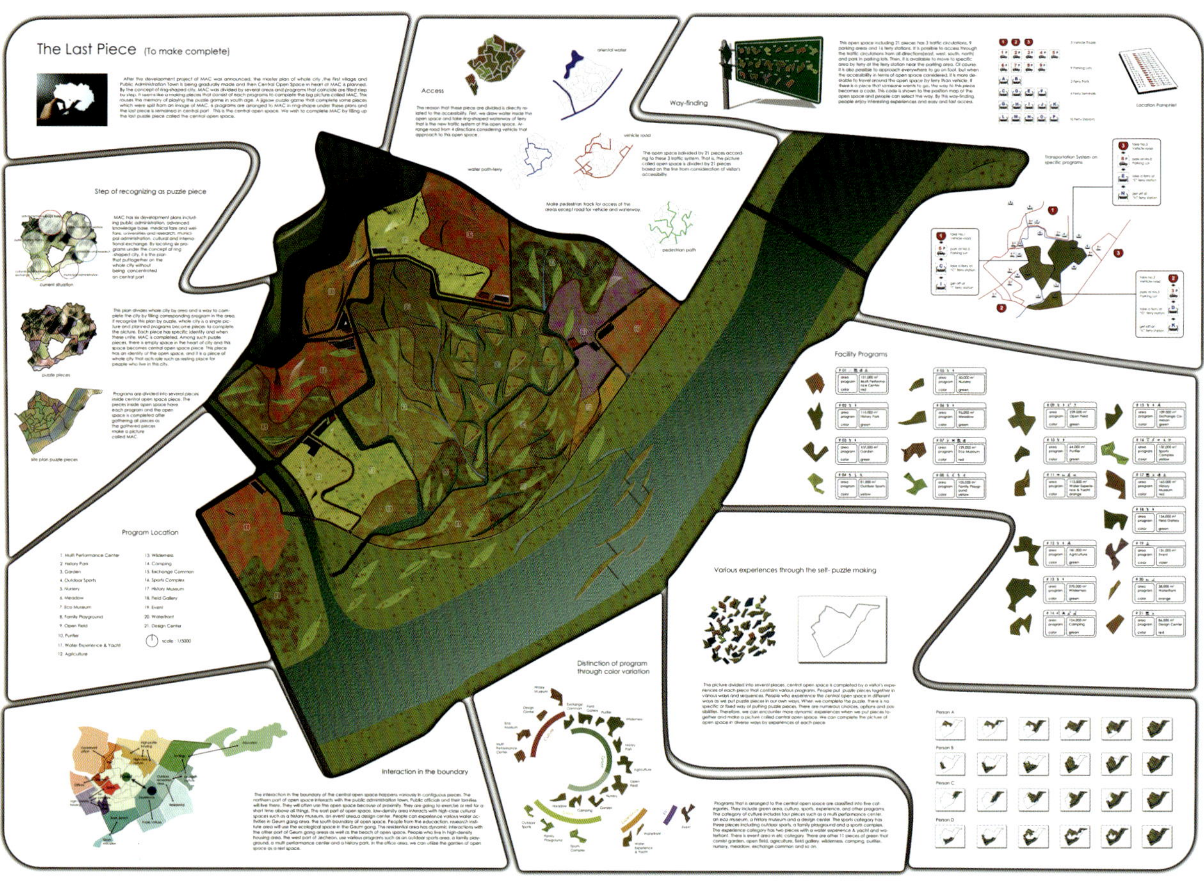

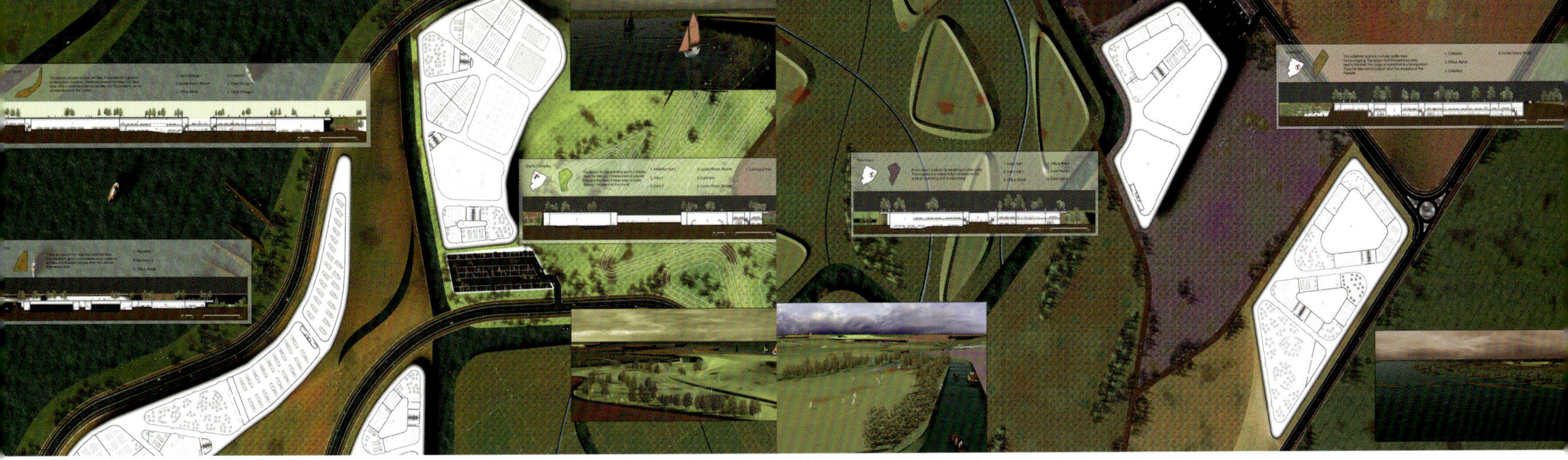

Water
Forest
Program
Path
Service

The child draws a picture.
The picture is the one open space.
The open space is divided by several puzzle pieces.
The child wants to complete this picture by putting each piece together.....

Finalist

URBAN DACHA

Yoon Woong Won 윤웅원
Partners _ Kim Jeong Joo 김정주, Park Joo Hyun 박주현, Jung Sang Hyun 정상현, Kim Na Young 김나영, Saur Jean-luc, Ahn Chul Hwan 안철환

과거 공공의 공간이란 자발적 참여가 결여된 공간이었으며 어느 누구의 것도 아니었다. 새로운 공공의 공간은 자발적인 참여로 채워지고 개인의 목적과 공공의 매너가 공존하는 모두의 공간으로 재탄생한다. 이에 시민의 공원Citizen Garden 개념을 도입하여 공간을 수많은 조각으로 나누고 있는 개인의 Kitchen Garden과 공공의 Park의 밀접한 관계를 유도하며 개인과 공공의 교감을 통해 만들어지는 새로운 도시공원을 제시하고자 한다.

MPPAT
Plan of Area Lower than Dike

llow land
Commune center
Farmers' market

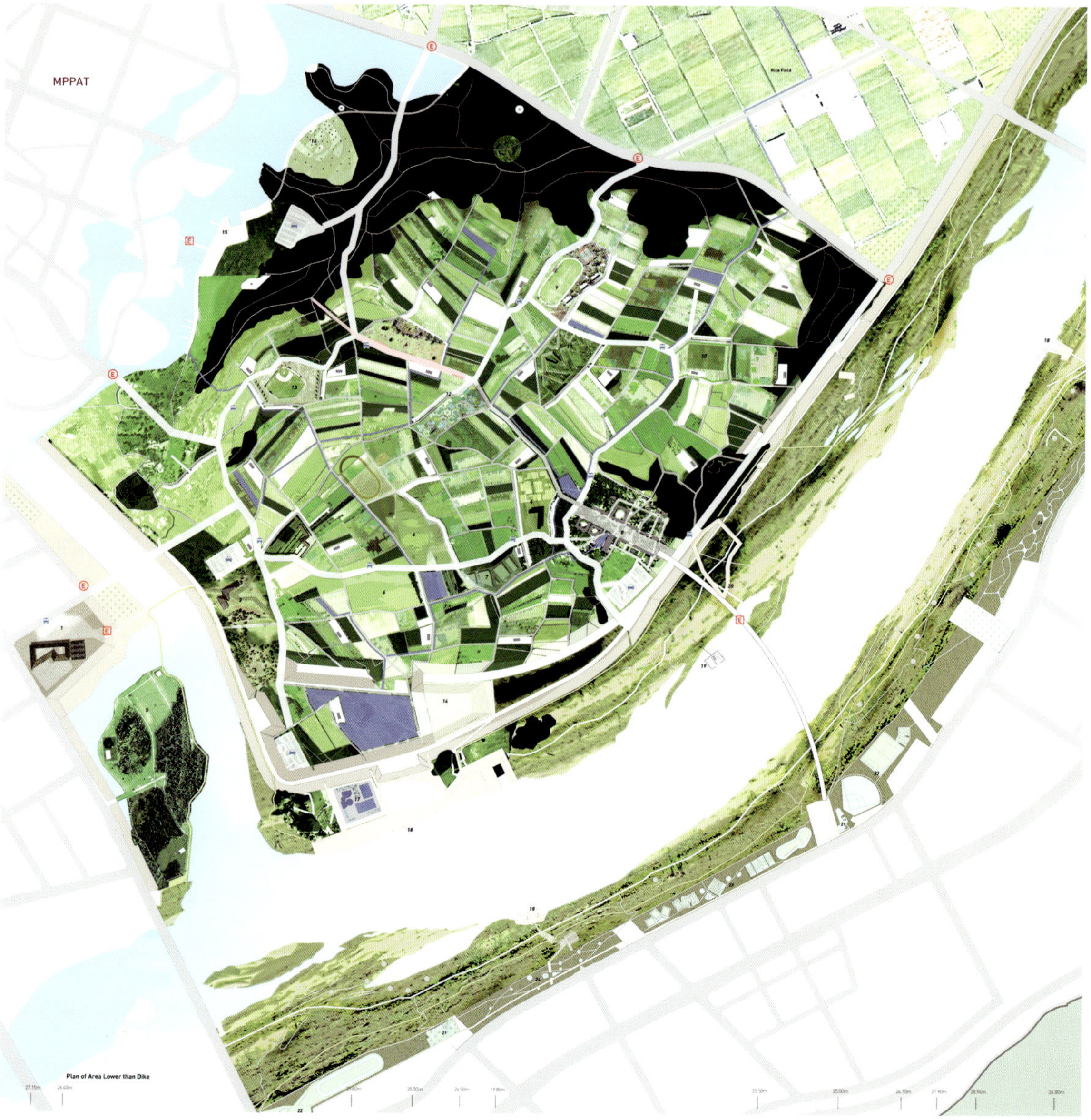

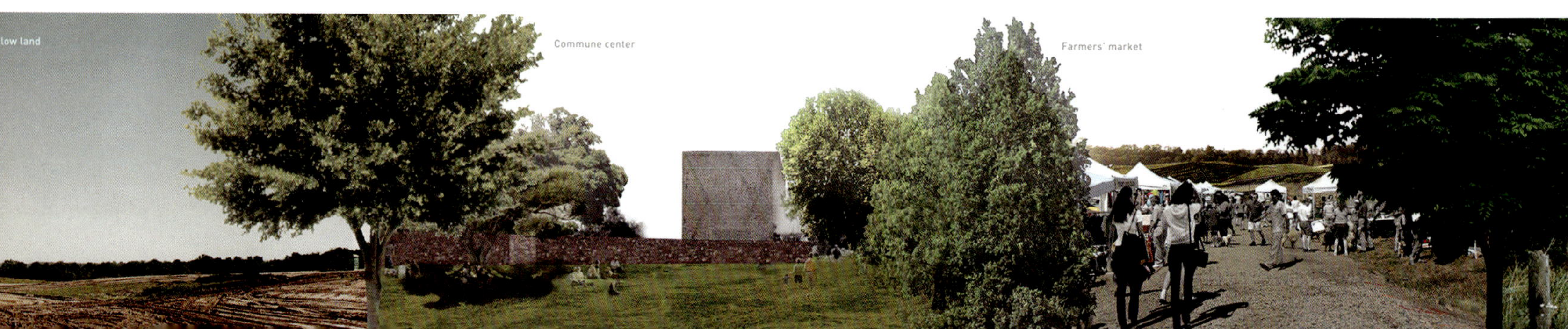

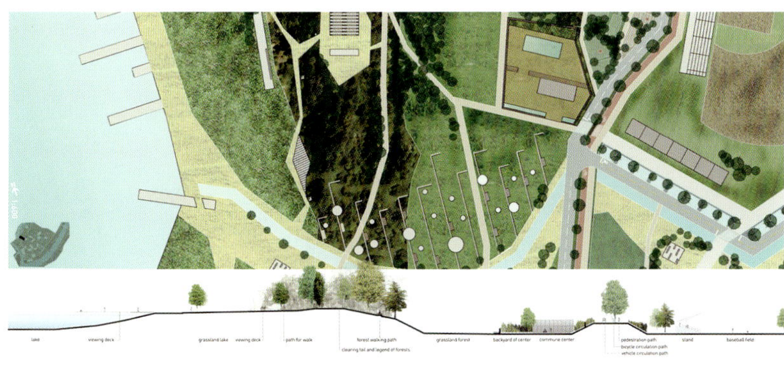

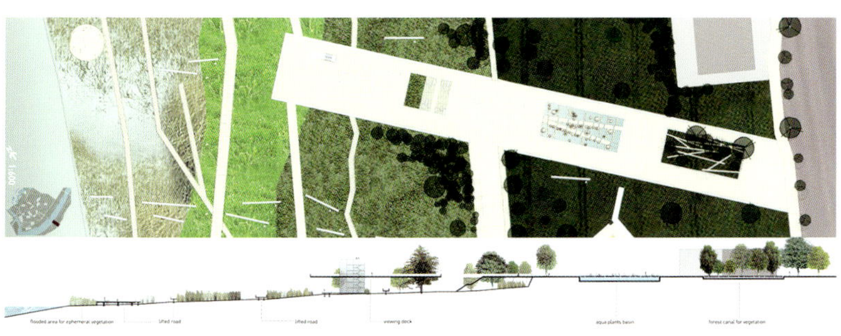

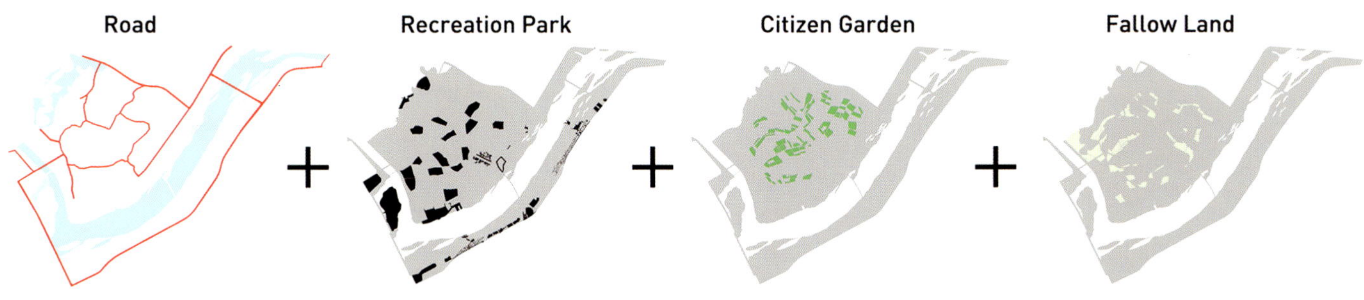

The half of the park is a planned area, and the other half is left without any design. This is a strategy for cost efficiency. At the same time, this is a challenge to beauty in landscape design. Limiting the boundary of FALLOW LAND and CITIZEN GARDEN as a path is the one and only design. Time and farmers' choices will influence this boundary area.

Finalist

NATURE MOSAIC

James Corner | **Partners** _ Cha Tae Wook, Flynn Michael, Yoon Hee Yeon, Jung Jay Yun, Harada Yoshiki, Buro Happold

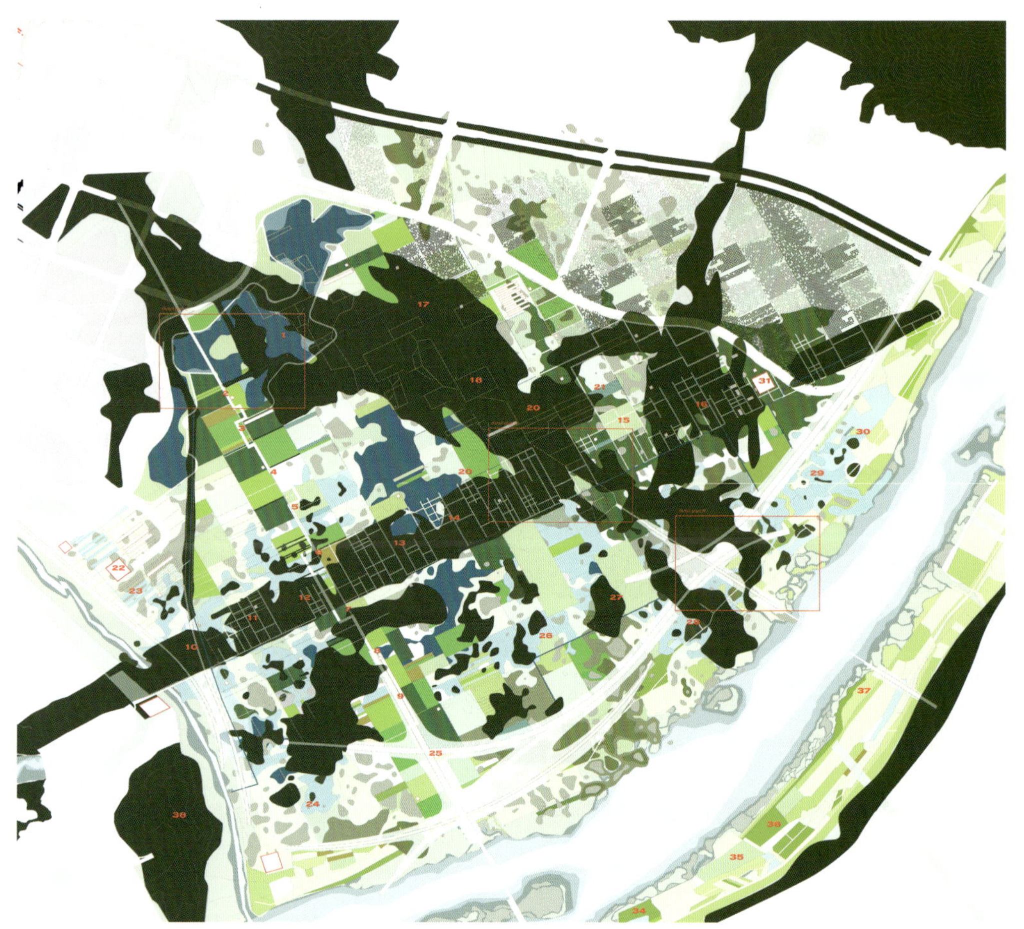

대상지가 가지고 있는 매력있는 잠재성인 '자유에너지Free Energy' 를 통해 다이나믹하고 지속가능한 생태계를 만들어내고, 자연이 중심이 된 21세기 어바니즘의 새로운 모델을 창조하고자 한다. 이것은 새로운 생태계와 문화적 장소들의 총체로서 산과 숲을 강으로, 자연의 맥박을 도시로 연결하는 강력한 상징적 경관이 된다.

Finalist

LANDSCAPE CARPET

Alessandro-Carmine Console | **Partners** _ Gina Oliva, Claudia Streuli

랜드스케이프 카펫은 추상적인 정사각형에 의해 만들어지는데, 각각의 정사각형은 산, 나무, 벌판, 호수와-개울을 상징한다. 카펫의 구성은 이 4개의 항목을 다중적으로 결합시킨 결과이다. 이렇게 구성된 모듈을 대상지에 중첩시켜 실제의 형태로 나타낸다. 이것은 새롭고 규칙적인 구조이며, 대규모 공원의 특성을 나타내는 기본구조이며, 새롭게 정착된 심장의 고동이다.

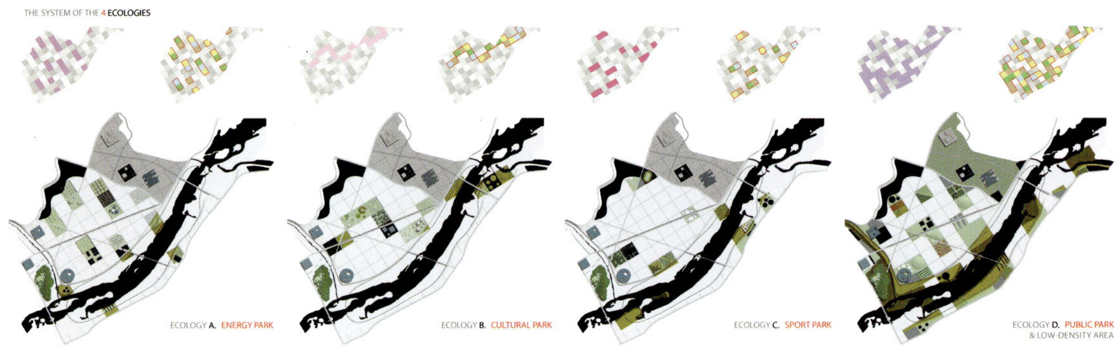

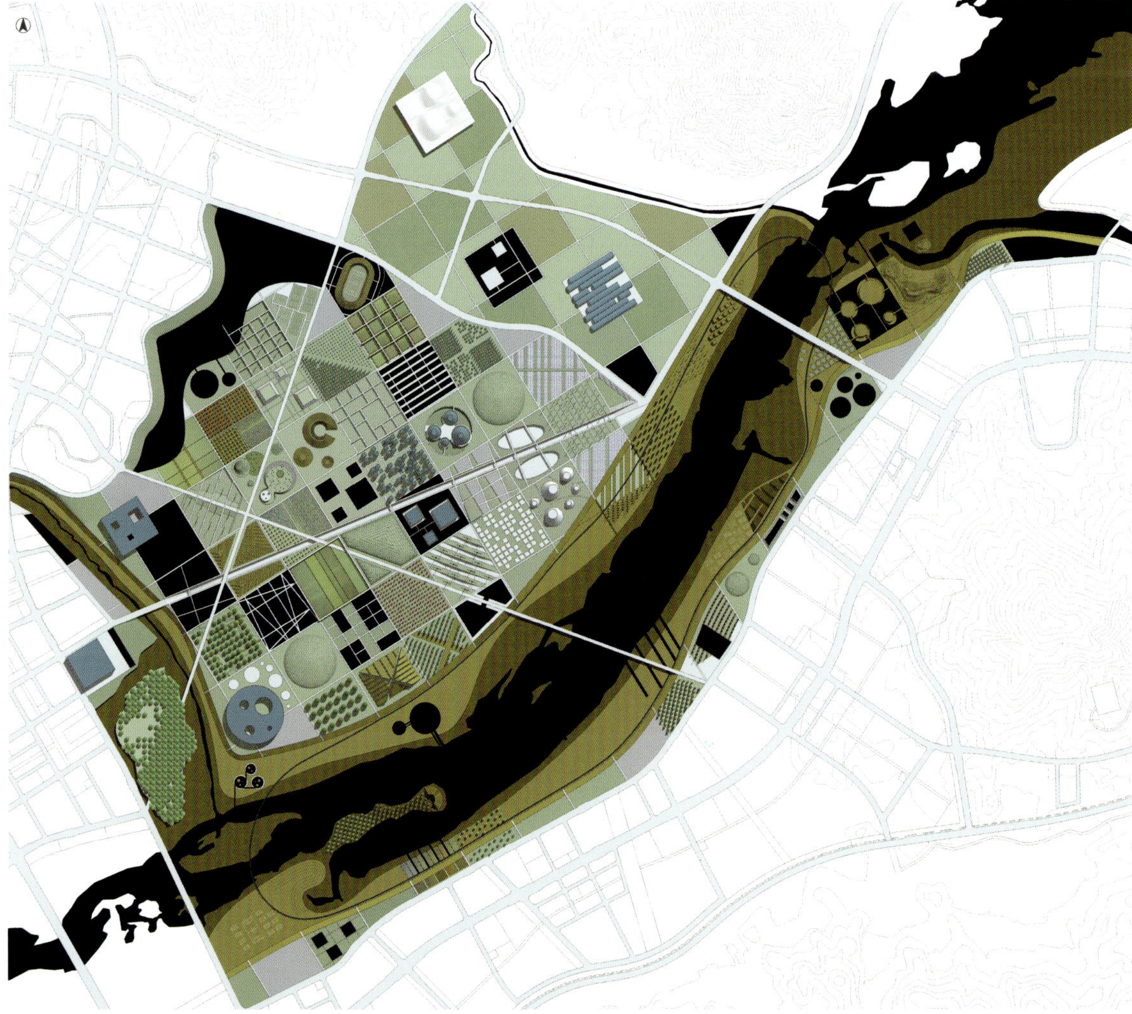

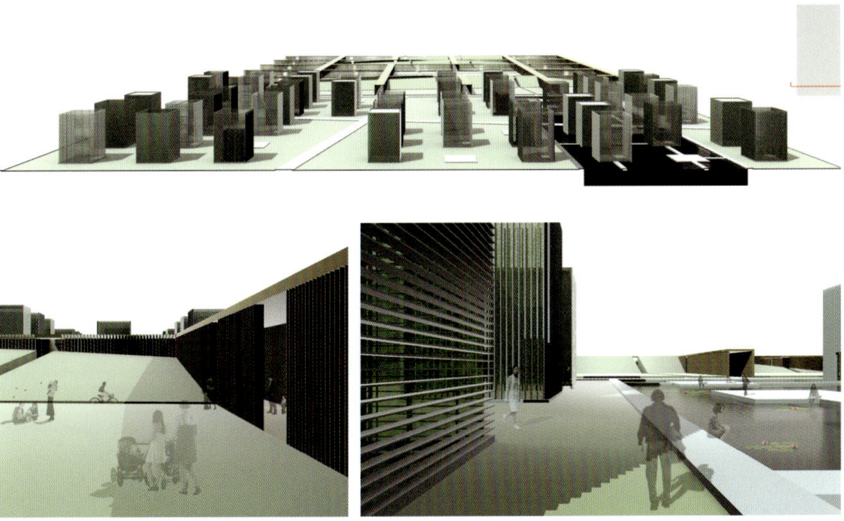

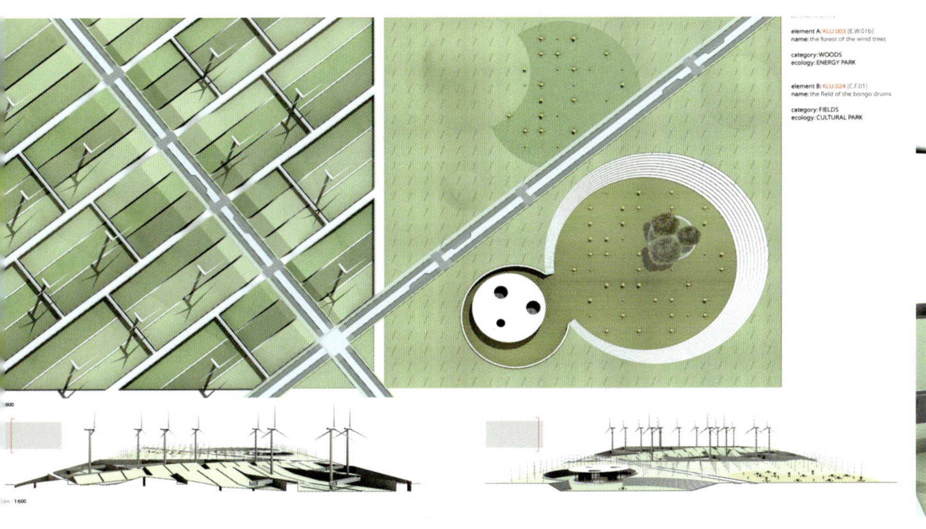

element A: KLU.003 [E.W.01b]
name: the forest of the wind trees
category: WOODS
ecology: ENERGY PARK
element B: KLU.024 [C.F.01]
name: the field of the bongo drums
category: FIELDS
ecology: CULTURAL PARK

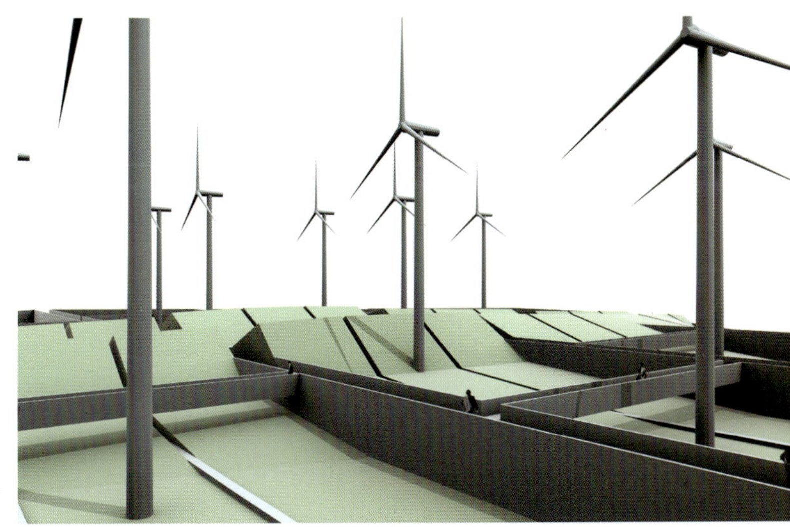

Finalist

ARABLE KITCHEN-GARDEN PARK

CJ Lim | **Partners** _ Studio 8, Matthew Wells, Fulcrum Consulting

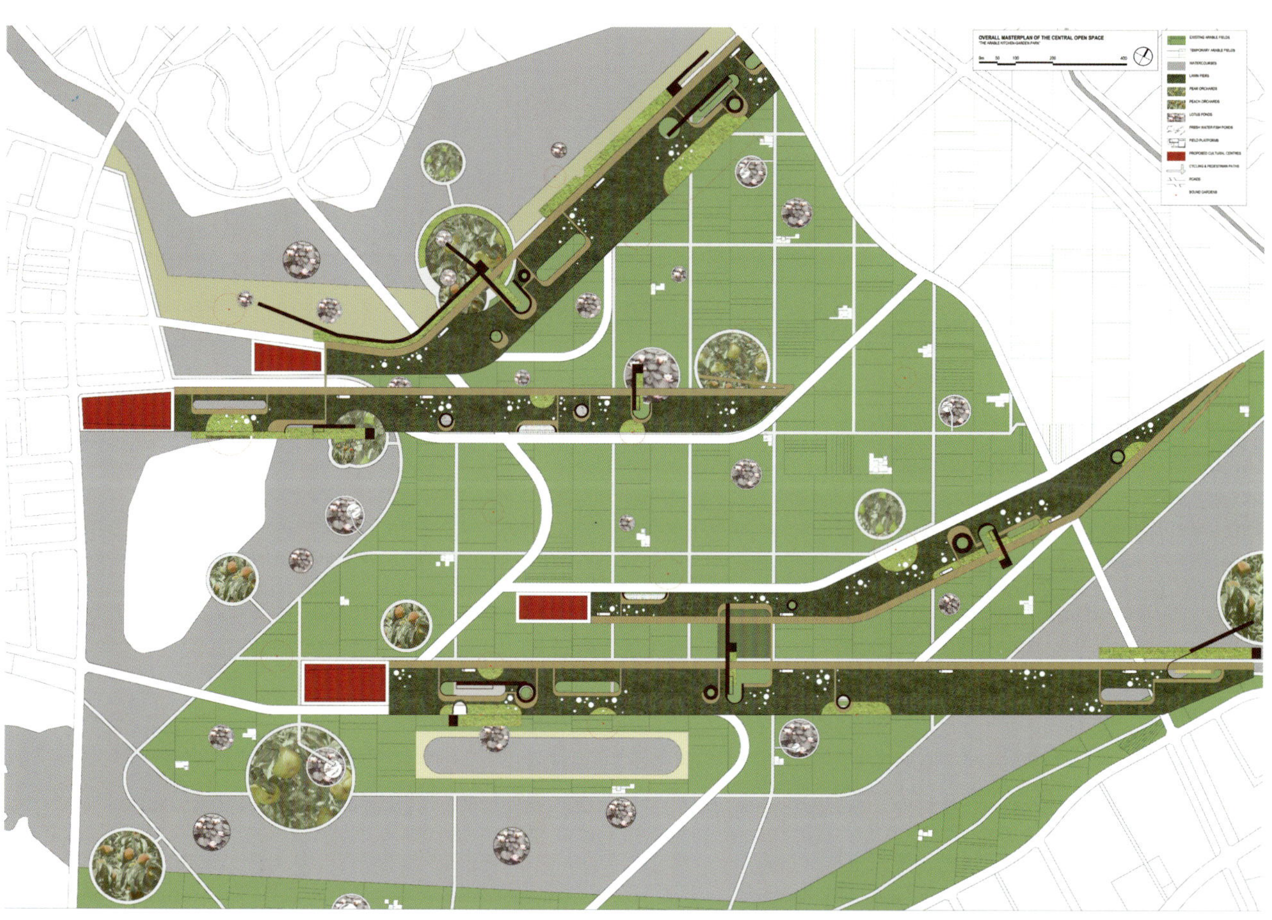

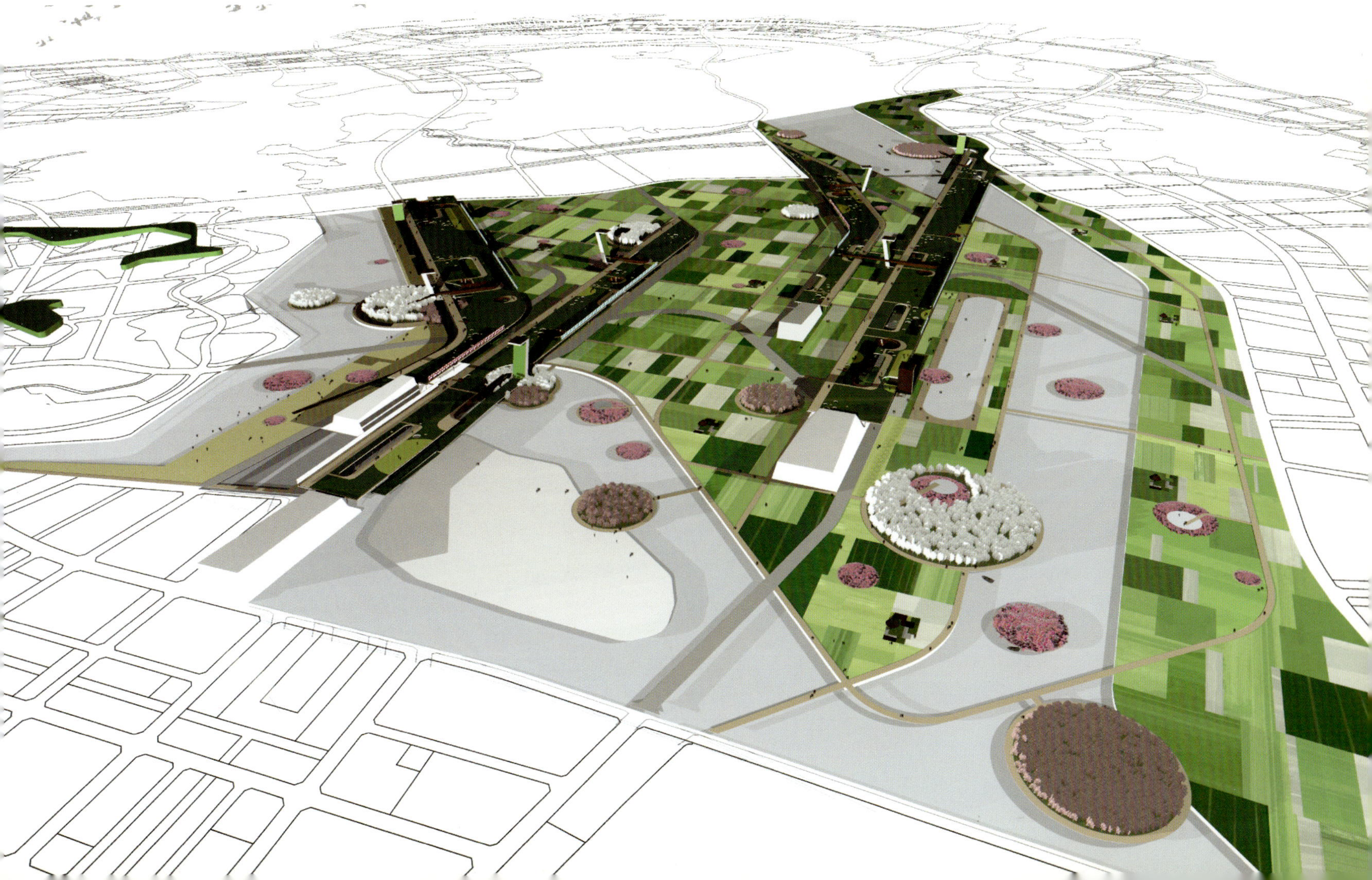

Finalist

10LI PARK

Diana Balmori | **Partners _** Mark Thomann, Kim Sang Mok, ARUP, MAD OFFICE, Joel Sanders, Balmori associates, inc

10리의 원형은 거대하고 수평적이며, 웅대하고 강력한 경관이다. 또한 이는 가치있는 경관이 되며 비워져 있고 열려있는 경관이기도 하다. 한국 고유의 측량 단위인 리의 표현과 추상적 개념으로써, 이 원형은 과거와 현재, 자연과 인위에 대한 뚜렷한 해석이다. 이 원형 안의 논은 경험적인 기술을 바탕으로 한 생산적인 활동으로서 새로운 생명을 지니고 있다. 논은 각 계절마다 드라마틱하게 바뀌는 초현실적이고 아름다운 경관을 만들고, 그 원형을 따라서 공공과 개인, 구조물과 경관, 예술과 일상 사이의 거리를 메우는 세 개의 박물관과 여러 활동을 수용하는 구조물이 생성된다.

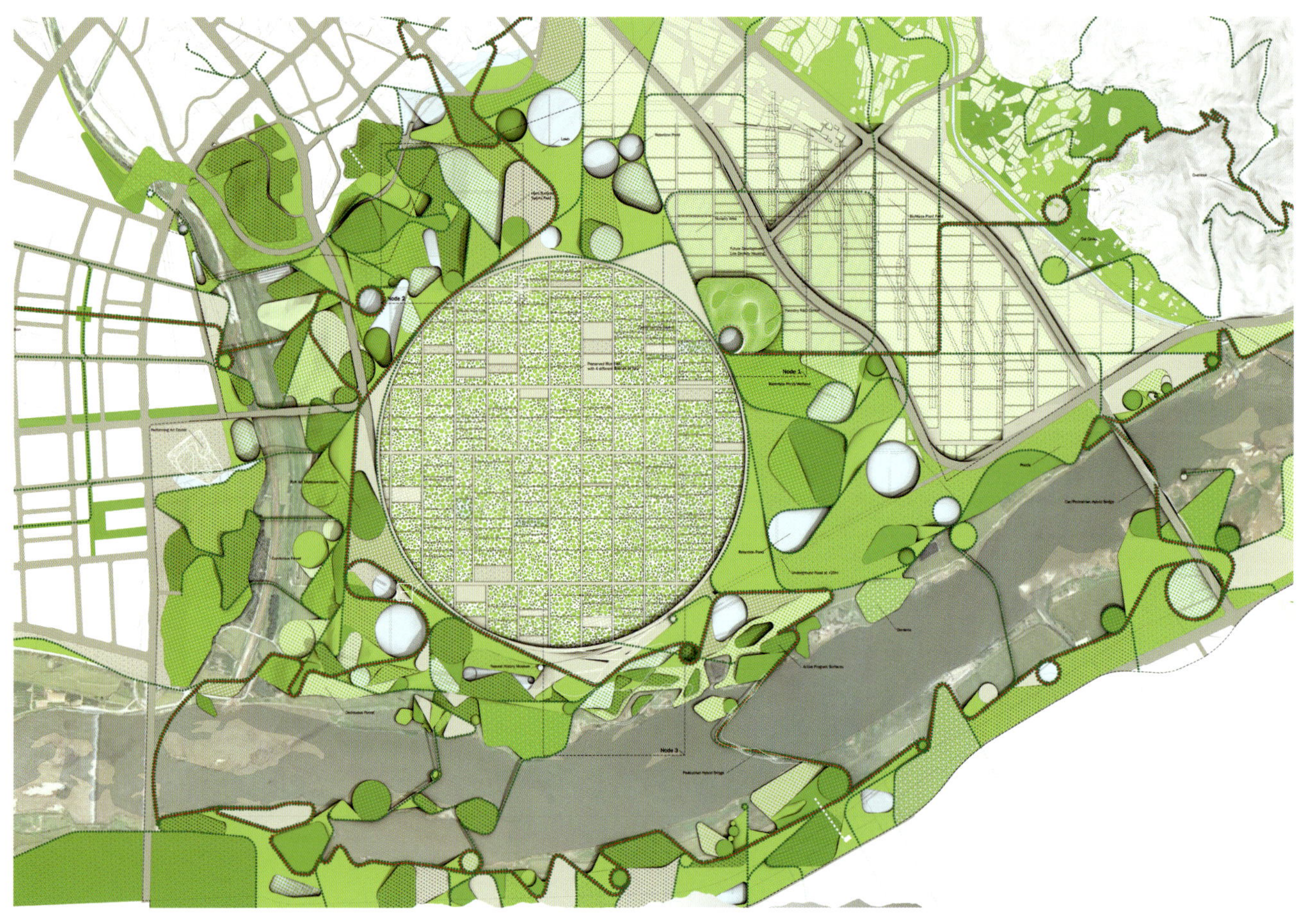

FT 10204
8
C
7
Section C

B
6
Section B

WORKS OF LANDSCAPE ARCHITECTURE

Pentagon Memorial
Location _ Arlington, Virginia, U.S.A.
Size _ 2 acres
Landscape Design _ Lee+Papa Associates
Completion _ September 11, 2008
Consult _ ela no.254

Seattle Olympic Sculpture Park
Location _ Seattle, U.S.A.
Size _ 9 acres
Landscape Design _ Weiss/Manfredi Architects with Charles Anderson Landscape Architecture
Completion _ January 20, 2007
Awards _ The 10 Best Architectural Marvels
Consult _ ela no.247

Sydney and Walda Besthoff Sculpture Garden
Location _ New orleans, Louisiana, U.S.A.
Size _ 4.76 acres
Landscape Design _ Sawyer/Berson Architecture & Landscape Architecture, LLP
Consult _ ela no.238

Santa Monica Airport Park
Location _ Culver city, California, U.S.A.
Size _ 8.3 acres
Landscape Design _ Ah′ be Landscape Architects
Completion _ April 29, 2007
Consult _ ela no.236

Jinji Lake Park
Location _ Suzhou, China.
Size _ 513 ha
Landscape Design _ EDAW
Consult _ ela no.236

Balsley Park
Location _ New York, NY, U.S.A.
Landscape Design _ Thomas Balsley Associates
Consult _ ela no.251

Capital Plaza
Location _ New York, NY, U.S.A.
Landscape Design _ Thomas Balsley Associates
Consult _ ela no.250

Tide Point
Location _ Baltimore, Maryland, U.S.A.
Size _ 13 acres
Landscape Design _ W Architecture & Landscape Architecture, LLC
Consult _ ela no.252

West Harlem Piers Park
Location _ New York, NY, U.S.A.
Size _ $9,800m^2$
Landscape Design _ W Architecture & Landscape Architecture, LLC
Awards _ 2004 ASLA Professional Awards(Masterplan)
Consult _ ela no.253

Copenhagen Harbour Bath
Location _ Copenhagen, Denmark
Size _ 2,500m^2
Landscape Design _ PLOT=BIG(Bjarke Ingels Group)+JDS(Julien De Smedt Architects)
Consult _ ela no.253

Museo del Acero Horno^3 Monterrey
Location _ Monterrey, Mexico
Size _ 275 acres
Landscape Design _ Surfacedesign, Inc. + Harari Arquitectos
Awards _ ASLA 2009 professional awards(Honor Award)
Consult _ ela no.246

Sasakwa Hill Lodge
Location _ Serengeti, Tanzania, Africa
Landscape Design _ WATG + The Really Useful Landscape Company
Consult _ ela no.238

Lost Dog Wash Trailhead
Location _ Scottsdale, Arizona, U.S.A
Size _ 275 acres
Landscape Design _ Floor Associates, Inc.
Awards _ ASLA 2008 professional awards (General Design Honor Award)
Consult _ ela no.244

Nicolai School Centre
Location _ Kolding, Denmark
Size _ 3,800m^2
Design _ Arkitekt Kristine Jensens Tegnestue
Completion _ October, 2007
Consult _ ela no.251

Gudbrandsjuvet Viewing Platforms & Bridges
Location _ Gudbrandsjuvet, Norway
Size _ 350m^2
Landscape Design _ JSA(Jensen & Skodvin Architects)
Consult _ ela no.253

Erman Residence Garden
Location _ San Francisco, California, U.S.A
Landscape Design _ Surfacedesign, Inc.
Awards _ ASLA 2007 professional awards (Residential Design Honor Award)
Consult _ ela no.246

Baroda Garden
Location _ San Francisco, California, U.S.A
Size _ 5 acres
Landscape Design _ Rios Clementi Hale Studios
Consult _ ela no.251

Manhattan Townhouse Garden
Location _ New York, U.S.A
Landscape Design _ Sawyer/Berson Architecture & Landscape Architecture, LLP
Consult _ ela no.241

INTERNATIONAL DESIGN COMPETITION IN KOREA

Gwanggyo Lakeside Park, Suwon 光教湖公园
Location _ Suwon-si, Gyounggi-do, Korea
Size _ 2,043,858m²
Reference Homepage _ www.lakesidepark.kr
Consult _ ela no.249

In December, 8, 2008, International Design Competition for Gwanggyo Lakeside Park, Suwon, Korea lead by Governments of Gyeonggi-do, Suwon-si, Yongin-si, and Gyeonggi Urban Innovation Corporation has announced the result of awards. To make up a world level Lakeside Park as a focal point of natural and cultural of Gwanggyo new town, 8 companies located at local and foreign which submitted a the request for qualification(RFQ) for participation got a designated invitation.

The prizewinner, 'Urban Reservoir', was awarded by Synwha Consulting, '8Land-scapes' as 2nd by Field Operations and Group Han, and 'Ribbon 55 & Community 76' as 3rd prized winner awarded to Seo-Ahn Total Landscape.

Magok Waterfront, Seoul 麻谷滨水区
Location _ Magok-dong, Gangseo-gu, Seoul, Korea
Size _ 1,170,780m²
Consult _ ela no.244

The point of Magok prize competition was to utilize the purpose of specialized space formation that enables cultural dynamic power and energy of the existing Magok district, maximize the sources of Han River and the Lake Park and harmonize all the factors for the project progress. The 1st prize winning 'Heart of Magok is Nature of Living Water', which is based on continuously flowing water suggests not just one perfectly completely completed park but a park with the possibility of further development and expansion under the goal of making park where nature and human beings show active life together in harmony. This forms an urban center to support water, nature, culture and community and enables Magok district to work as the gateway of Han River.

Gangbuk Large Park, Seoul 江北大型公园

Location _ Gangbuk-gu, Seoul, Korea

Size _ 900,000㎡

Awards _ 2009 AIA New York Design Awards(project)

Consult _ ela no.242

Seoul city hosts 'International Master Plan Contest for Gangbuk Park' for Gangbuk area which has insufficient green zones. This site has caused the place to become a slum, losing the role of a Park. In this competition, the 1st prize winning 'Open Field', The proposal was so nice to show the Korean cultural sense of rural landscape which was quite compatible with face water system of ponds and water ways with soft landscape of existing forests and grasses.

The proposal of cultural facility represented by the art center, because of its urban sense of attractiveness, was evaluated as of the symbolic existence of this park in the future which attracts many people to visit here.

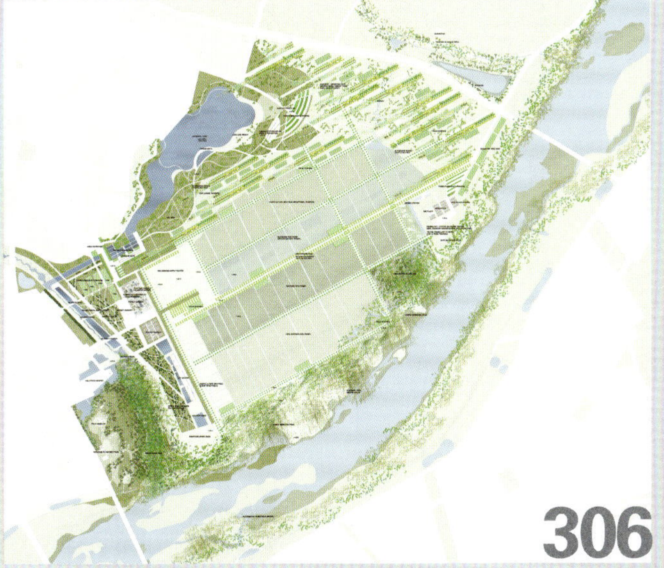

Central Open Space in Multifunctional Administrative City

行政中心复合城市　中央绿地

Location _ Yeongi-gun, Chungcheongnam-do, Korea

Size _ 6.982㎢

Consult _ ela no.234

Under the join auspices of Korea Land Corporation and the Multifunctional Administrative City Construction, prizewinning works of International Design Competition for Central Open Space in Multifunctional administrative City of Korea had announced. On the final stage designated 10 teams among all participated 92 teams' works on 29th of May at the competition, Noh, Sun Joo's work, 'Ancient Futures', was the prizewinner. And Kim, Jong Kyu's work, 'Long Views' was placed second and Mihn, Hyun Jun's work, 'Carved Park' was the third-place on the competition.

Fantastic 水 Waterscape

Creator of Culture ! Plus Fountain

Human, Nature, Music, Fountain and Creation of Culture!!

MoNA is a bridge linking peoples and cultures.
Sometimes, this communication represents meaning beyond speech.
As time goes by, MoNA is giving deeper impressions
in space and bringing comfort to your life.

Solar Sound Internet USB Plant

New innovations for space ...

Propose new experiences among cultures and peoples in nature together.

IT communication

MoNAdesign Tel +82 2 338 2807 Fax +82 2 338 7490